Digital Maturity in Hospitals

Armin Scheuer • Jörg Studzinski
Editors

Digital Maturity in Hospitals

Strategies, Frameworks, and Global Case Studies to Shape Future Healthcare

Springer

Editors
Armin Scheuer
Lemonmint HealthTech Communications GmbH
Berlin, Berlin, Germany

Jörg Studzinski
University Hospital Leipzig
Leipzig, Sachsen, Germany

ISBN 978-3-031-80703-9 ISBN 978-3-031-80704-6 (eBook)
https://doi.org/10.1007/978-3-031-80704-6

This Springer imprint is published by the registered company Springer Nature Switzerland AG
The registered company address is: Gewerbestrasse 11, 6330 Cham, Switzerland

Digital Maturity: Rethink Care, Disrupt Now!

Digital maturity is a critical factor in improving the quality of care for people with chronic diseases. It not only changes how doctors and patients communicate but also influences the behavior and health choices of individuals daily.

From a medical perspective, the fact that digitalization has reached the "pocket" of the patient presents a tremendous opportunity. Today, healthcare providers are theoretically closer to their patients than ever before. However, if we view therapy through traditional lenses and use digital tools merely as additional access points, we will miss the full potential of this revolution. Digital maturity allows us to rethink our interventions and therapeutic strategies, adapting them closely to the real-time needs and behaviors of our patients.

Building the Foundation: Infrastructure and Processes

Before we can explore the advanced opportunities that digital maturity offers, it is essential to establish a robust infrastructure and processes that support this evolution. Digital maturity is not achieved overnight. It requires a foundational commitment to integrating digital tools and technologies at all levels of healthcare. This includes ensuring that the necessary digital infrastructure is in place, from secure data storage systems to reliable and user-friendly workforce and patient interfaces.

Moreover, we must develop processes that allow for the seamless integration of these tools into existing workflows. This means training healthcare professionals to utilize digital tools effectively and ensuring that patients have the digital literacy needed to engage with these technologies.

Unleashing the Digital Potential for Patients

Once the foundational infrastructure and processes are established, the true potential of digital maturity can be unleashed. Consider this: each of us makes approximately 1,800 decisions every day, and 450 of those decisions are related to our lifestyle. If a digital therapeutic tool could influence just 10 of these lifestyle decisions each day—providing the right prompts at the right time, with the right motivational strategies—it could lead to significant health outcomes, such as a 10 kg weight reduction over a year. This simple example illustrates how digitalization, when applied with a high level of maturity, can be aligned with patient needs to achieve meaningful therapeutic effects.

Achieving such outcomes requires digital maturity on multiple levels. First, patient health literacy is crucial. Reaching patients who are not inherently interested in health is a challenge. We must now explore digital ways to engage these patients by addressing their specific needs and utilizing anchor motivations that can lead to positive health and therapeutic outcomes. Additionally, digital literacy among patients is essential. Modern, intelligent digital tools must be designed to intuitively meet patients where they are, making digital therapy not an additional burden but an enhancement of their health outcomes.

Disrupting Traditional Approaches to Medicine

As we consider digital maturity as a continuous process, it is essential to rethink and sometimes disrupt traditional approaches. Chronic diseases are typically diagnosed based on biochemical biomarkers. However, many of these conditions are lifestyle-related, and lifestyle can be effectively monitored through digital trackers. Emerging studies suggest that behavioral data, such as online purchasing habits, can predict the onset of conditions such as hyperglycemia and depression months before traditional symptoms appear. Yet, in our conventional medical system, it can take years before a diagnosis is made. This raises the question: could digital biomarkers become a disruptive force in our diagnostic toolkit and the ultimate expression of digital maturity in modern medicine?

The answer is a resounding yes. We have a unique opportunity to diagnose lifestyle-related chronic diseases based on everyday behaviors, and we should be open to leveraging digital biomarkers to detect risks early. By doing so, we can intervene with digital therapeutics long before a patient becomes aware of their condition. Of course, this approach also raises ethical questions about whether such early interventions constitute medical treatment or preventive care and to what extent patients need to be informed about these strategies. Medical guidance in these scenarios remains crucial and can itself be delivered through digital means.

The Future has Started Already

While these ideas may seem futuristic, they are not far off. In countries like Rwanda, where digital infrastructure is robust, and the healthcare system is struggling to manage a surge in chronic diseases, digital diagnostic and therapeutic interventions offer almost limitless possibilities. Here, the ability to deploy intuitive digital solutions could transform healthcare delivery, filling gaps that traditional systems cannot.

In my view, digital therapeutics represents a revolution in medicine akin to the introduction of antibiotics. We now can be closer to our patients than ever before, and we must embrace this potential with creativity, openness, and a relentless focus on the needs of our patients. Digital maturity in the treatment of chronic diseases means moving beyond merely extending the doctor's reach. It requires developing autonomous, intelligent solutions that engage patients where they are—tailored to their intuitive needs and integrated seamlessly into their daily lives.

As the President of the International Diabetes Federation and a doctor who has dedicated my career to battling chronic diseases, I want to thank every CEO, CIO, and CMIO, every researcher and IT professional, and every doctor and nurse for their unwavering commitment to exploring and using digital technologies to improve the lives of our patients. I also want to thank the governments for investing in the digital transformation of their health systems and for encouraging their population to join the digital revolution in healthcare. And, I want to express my gratitude to the editors and authors of this book for advancing this critical issue on a global scale.

Together, we can make a difference—building healthcare systems that prioritize prevention over treatment, deliver the best possible care to our patients, and ensure no one is left behind.

Let's unite and make this happen now.

International Diabetes Federation (IDF) Peter Schwarz
Dresden, Germany

The original version of this book has been revised. The correction to this book can be found at https://doi.org/10.1007/978-3-031-80704-6_20

Acknowledgments

We extend our deepest gratitude to the more than 50 authors from across the globe who have generously contributed their expertise, time, and insights to this work. Your diverse perspectives and dedication to advancing digital maturity in healthcare have been instrumental in shaping this book. #LeadershipMatters

The creation of this book would not have been possible without the input of this network of highly knowledgeable and ambitious digital enthusiasts. Interdisciplinary networking and ambition drive humanity forward and enable transformation. This also applies to the digital transformation in the healthcare sector, where collaboration is essential. #CollaborationMatters

We also express our heartfelt appreciation to all the healthcare professionals, administrators, IT specialists, and policymakers who work tirelessly every day to raise digital maturity within their organizations. Your commitment to keeping the needs of patients and the healthcare workforce at the forefront of digital transformation is truly commendable. #DedicationMatters

A special thank you goes to Springer Nature for publishing this book. We are especially grateful to our Adviser and Project Manager, Dr. Alina Ceprega, whose guidance and support have been invaluable throughout this journey. We would also like to thank our Assistant Editor, Hazel D'Coutu, for her meticulous work and dedication to ensuring the success of this project. #ProffesionalismMatters

We believe that knowledge is global, and healthcare provision is local. In sharing this work, our hope is to inspire and empower healthcare leaders around the world to adapt and implement digital innovations that meet the unique needs of their communities. Together, we can improve patient outcomes and operational efficiency, one step at a time. #ImpactMatters

Contents

Part I

Digital Maturity Strategies and Measurements

Navigating the Path to Digital Maturity in Healthcare

Armin Scheuer and Jörg Studzinski

Abstract

This chapter provides an introduction to the book *Digital Maturity in Hospitals*, which explores the critical role of digital maturity in healthcare. It discusses the concept of digital maturity, its benefits, and limitations, as well as its impact on healthcare delivery and organizational efficiency. The chapter further examines key models such as the WHO Global Digital Health Monitor, CHIME's Digital Health Most Wired Survey, and HIMSS EMRAM, along with other specialized models. Additionally, it delves into national initiatives like Germany's DigitalRadar, Saudi Arabia's SeMA, and NHS England's Digital Maturity Assessment. The book also addresses strategic aspects of digital maturity, including the importance of selecting appropriate models, overcoming implementation challenges, and leveraging these models to drive continuous improvement in healthcare systems.

Keywords

Digital maturity · Digital transformation · Digital maturity models · National health assessments · Health IT benchmarking

A. Scheuer (✉)
Lemonmint HealthTech Communications, Berlin, Germany
e-mail: armin.scheuer@lemonmint.org

J. Studzinski
University Hospital Leipzig, Leipzig, Germany
e-mail: joerg.studzinski@medizin.uni-leipzig.de

A. Scheuer, J. Studzinski (eds.), *Digital Maturity in Hospitals*,
https://doi.org/10.1007/978-3-031-80704-6_1

Abbreviations

AMAM	Adoption Model for Analytics Maturity
C-COMM	Community Care Outcomes Maturity Model
CCMM	Continuity of Care Maturity Model
CHIME	College of Healthcare Information Management Executives
DIAM	Digital Imaging Adoption Model
DMA	Digital Maturity Assessment
EMRAM	Electronic Medical Record Adoption Model
EPR	Electronic Patient Record
HIMSS	Healthcare Information and Management Systems Society
INFRAM	Infrastructure Adoption Model
IT	Information Technology
SeMA	Saudi e-Health Maturity Assessment
UAE	United Arab Emirates
WGLL	What Good Looks Like
WHO	World Health Organization

1 Introduction

In this book, more than 50 digital health experts present a collection of international best practices, proven strategies, and lessons learned on the path to digital maturity. Their shared mission is to spread global knowledge and enable hospitals worldwide to drive digital health transformation for the benefit of healthcare workers and patients. Their innovative ideas and global experiences will help revolutionize the provision of health and care, preparing our societies for the challenges and opportunities of the future.

The global journey toward digital maturity has been anything but straightforward. At times, these efforts have been celebrated as the cornerstone of digital transformation; at others, they've been criticized as unnecessary or burdensome. Yet, in recent years, the tide has turned. There is now widespread recognition that digital maturity is a key indicator of healthcare excellence. This shift in perspective has been driven by widespread adoption in countries like the United States, Sweden, Denmark, or Israel and bolstered by national initiatives in countries like the UK, Turkey, and the UAE, where governments have actively promoted digital maturity roadmaps and measurements to transform their health systems.

A tipping point in global digital health maturity measurement was Germany's DigitalRadar assessment, which evaluated 1624 hospitals within just 3 months. We, the editors of this book, played a leading role in this project, transforming it into the world's largest and fastest digital maturity assessment of hospitals.

When the German government initially announced plans for a nationwide hospital assessment, it sparked intense debate among healthcare stakeholders. Questions arose: Was such an assessment truly necessary? Would it place an undue burden on hospitals? Which approach should be used, and what were the real intentions behind it? In a healthcare system as complex as Germany's, where diverse stakeholders hold varying interests, navigating these dynamics became our primary focus—well before we could even begin to tackle the technical aspects. Fortunately, we weren't alone in this effort. We drew upon a global network of digital maturity experts—individuals who have dedicated their careers to creating a better healthcare system that is fit for the challenges of the future.

1.1 Knowledge Is Global. Healthcare Systems Are Local

Our work at HIMSS and our experience with the DigitalRadar project further strengthened our conviction that, while knowledge may be global, healthcare systems are inherently local. Every healthcare system operates within a unique context, influenced by regional and national regulations, cultural norms, and specific needs. In Germany, the discussions around adopting a nationwide digital maturity tool often centered on whether it could be adapted to fit an established healthcare system—one that is already among the best-performing in the world, with its own legacy systems, established processes, and a complex, decentralized political landscape. There was a recognition that while no existing model was a perfect fit for Germany's needs, simply reinventing the wheel would risk losing valuable global insights. The solution was to develop a tailored system that incorporated the best elements of various established maturity models—a "best of breed" approach that was key to achieving success and gaining stakeholder buy-in.

Like every digital maturity expert contributing to this book, we undertook this work with a purpose: to support the healthcare workforce and improve patient care. For healthcare professionals, our goal is to foster a digital transformation that truly works for them—one that makes their lives easier, reduces administrative burdens, and equips them with the tools they need to provide the best possible care.

For patients, we recognize the urgent need to modernize healthcare systems to better meet the evolving needs of our societies. Digital maturity is about more than just adopting new technologies—it's about creating healthcare systems that are more responsive, personalized, and efficient, ultimately leading to better patient outcomes and experiences.

As you delve into the chapters ahead, we invite you to explore how digital maturity can be a powerful catalyst for transforming healthcare, improving patient outcomes, and creating more resilient health systems. Whether you are a healthcare leader, policymaker, or practitioner, we hope this book provides you with the insights and practical guidance you need to navigate your own digital maturity journey toward healthcare excellence.

A key objective of this book is to provide a comprehensive overview of the various dimensions relevant to hospital digitalization. To determine which dimensions are most pertinent, it is essential first to define the characteristics of a digital hospital, understand how its digital maturity can be measured, and identify the tools available for this purpose.

1.2 Characteristics of a Digital Hospital

Finding a clear and widely accepted definition of a "digital hospital" in the literature is challenging. Instead, various paraphrases attempt to approximate the concept using keywords. These keywords emphasize the use of digital technologies—particularly electronic medical records—and a focus on optimizing care, work efficiency, connectivity, personalization, and the mobilization of information (Canfell 2024). Recent publications also reference terms such as "smart hospital" and the use of artificial intelligence when describing digitally mature hospitals (Omid 2024). Often, the focus on the patient—or more broadly, the citizen or care consumer—plays a critical role in these descriptions. However, it's worth questioning whether this focus on the patient is truly new in the context of a digital hospital or if it is simply a continuation of a principle as old as healthcare itself.

The downside of relying on keywords is that terminology evolves over time, especially in response to technological and social changes. Therefore, it may be more effective to adopt a two-part approach: (a) using a simple and concise definition, and (b) employing a more detailed, sophisticated, and proven framework that explains and measures the various dimensions of digital maturity. In the context of this book, we propose the following short and simple definition of a digital hospital:

> A digital hospital leverages technology and organizational culture to optimize patient care, staff working conditions and research capabilities in a holistic way that can be objectively and externally validated through a digital maturity model.

This book offers best practice examples from individual hospitals and governmental initiatives that have utilized digital maturity models to achieve externally validated recognition for their digital transformation successes. It also highlights exemplary practices across various dimensions of digital maturity frameworks, including IT capability, strategy and management, and patient-centered care. Additionally, the book explores strategic aspects that will shape the future of digital maturity, such as green IT, ethical considerations, social determinants of health, and workforce development.

2 Digital Maturity in Hospitals

Closely related to the term "digital hospital" is the concept of "digital maturity." When applied to hospitals, digital maturity refers to the extent to which healthcare institutions have adopted and integrated digital technologies *"as an enabler of high-quality care through supporting improvements to service delivery and patient experience. Digital maturity is multi-faceted, and encompasses not only technology, resources and capability but also the digital literacy, ability and motivation of staff and patients to use new technologies"* (Duncan 2022).

Duncan et al. conducted a systematic literature review, analyzing various digital maturity models to distill the most common dimensions that form the cornerstones of digital maturity. These were further consolidated into a final maturity model framework comprising the following seven key dimensions:

1. Strategy
2. Information technology capability
3. Interoperability
4. Governance and management
5. Patient-centered care
6. People, skills, and behavior
7. Data analytics

While the naming convention for these key dimensions may vary from model to model, organization to organization, or country to country, we believe it is helpful to keep these keywords in mind when managing, supporting, or assessing the digital capabilities of individual hospitals or larger health systems. It is not possible to achieve higher levels of digital maturity if some of these key dimensions are neglected and lag behind. A holistic approach that breaks down the digital transformation process into relevant sub-aspects has proven successful in practice for many healthcare organizations.

2.1 "You Can't Manage What You Don't Measure"

"You can't manage what you can't measure." This statement, often attributed to management consultant Peter Drucker, is particularly relevant to the digital transformation process in healthcare organizations. Governments and healthcare administrators require objective data on the state and impact of digitization to make informed decisions, ensuring that their strategies are effective and aligned with the goals of improving patient care and operational efficiency. From a strategic point of view, it is worth considering at an early stage how digitally supported systems, processes, and services can be monitored and managed with the help of key performance indicators and purpose-specific frameworks. For this purpose, digital maturity models are generally helpful.

Digital maturity models serve as essential tools for identifying an organization's digital strengths and weaknesses, helping decision-makers understand their position on the digital transformation journey. These assessments can be conducted at any stage of the journey. When used at the beginning of a major project—such as an EMR replacement within a hospital, the introduction of a regional health data exchange platform, or the implementation of a national funding program for digital health services—they provide all stakeholders with an objective, easily understood, and shared understanding of the starting point.

Reassessments can measure progress at key milestones; some organizations choose to evaluate their success at the end of a significant digital transformation initiative. Regardless of the approach, the most effective frameworks offer digital transformation leaders strategic tools to measure their current status and provide insights into their potential destinations, including guidance on prioritizing actions to reach their goals.

Governments and funding bodies can leverage maturity assessments to allocate resources efficiently and support digital health initiatives effectively. Additionally, these assessments often offer valuable benchmarking opportunities, allowing comparisons across organizations of varying sizes—from single institutions to entire regional or national health systems. When used correctly, digital maturity models create a unique opportunity for healthcare organizations to connect, share insights, and learn from one another, paving the way for the development of learning health systems.

While digital maturity models are valuable tools, they come with inherent limitations:

1. *Reliance on Self-Assessments*: Many models depend on self-assessments to minimize costs and effort, which can introduce bias and affect the accuracy of the results.
2. *Limited Measurement of Outcomes*: These models often focus on the presence and use of digital technologies rather than measuring the outcome quality or usability of the digital solutions and services implemented.
3. *Lack of Financial Insights*: Digital maturity assessments typically do not provide insights into the (direct) return on investment (ROI), making it challenging to assess the financial impact of digital transformation efforts.
4. *Restricted External Impact Analysis*: These models generally do not explain how the internal digital maturity of an organization influences external stakeholders or the broader healthcare system, limiting their utility in evaluating system-wide effects.

However, these limitations should not prevent decision-makers from utilizing digital maturity models. The benefits they offer far outweigh their shortcomings. There are additional tools available to correlate results from digital maturity assessments with, for example, studies on quality metrics or financial performance.

This book provides insights into how digital maturity assessments have been implemented across various organizations and countries, highlighting both their original purpose and the ultimate benefits achieved through these initiatives.

2.2 Choosing the Right Assessment Framework

A wide range of digital maturity models are available on the market, making it crucial to choose one that aligns well with one's strategic goals. To ensure the selected model meets requirements and expectations, decision-makers should apply several filters before making a choice. The following filter criteria may help narrow down the search for a suitable model:

1. Source of Development and Support

- *Industry*: Solution providers and consultancies often create and apply digital maturity models. While some of these models are practical and sophisticated, they may lack balance and serve vested interests.
- *Academia*: Academic organizations develop more objective models, but these can be overly complex and may have a limited support and maintenance period, often ending with the funding period of the project.
- *Governmental Bodies*: Models developed by or on behalf of governmental bodies, such as ministries of health or digital health agencies, often serve specific purposes, like monitoring special funding programs. They may be mandatory in some cases or flexible in others.
- *Professional Societies*: While not free from potential conflicts of interest, some models developed by professional societies have proven successful globally and over time.

2. Geographical Scope

- *Local and Regional Models*: Some models are designed specifically for organizations operating in a particular location or region, such as a province or a group of countries with shared characteristics.
- *National Models*: These are designed for application within a single country and are often tailored to the specific needs and regulations of that nation.
- *International Models*: A few models are applicable on an international scale. While these models offer broader comparisons and higher reputational value, they may lack the granularity and specificity found in localized models.

3. Target Audience

- *Organization-Specific Models*: Digital maturity models are usually tailored to specific types of healthcare organizations, such as hospitals, ambulatory care providers, or health insurance organizations.
- *Cross-Organization Models*: Some models are designed to be used across different types of organizations, like ambulatory and stationary healthcare providers within a community. However, these models may yield less specific and granular results for individual organizations.
- *Broad Applicability*: Models that apply across a wide range of healthcare system stakeholders tend to produce results that are less accurate and actionable for specific entities. In this book, we focus on digital maturity strategies, use cases,

and measurements for hospitals and other key stakeholders, such as ministries and digital agencies from various countries.

When selecting a digital maturity model, decision-makers should start by clearly defining their strategic objectives and identifying the key stakeholders who need to be convinced and aligned to achieve these goals. This approach will help in choosing a maturity model that effectively guides the digital transformation process. Regardless of the specific model chosen, its application will provide more valuable insights than operating without one. By using a digital maturity model, you can better manage your digital transformation and gain deeper insights that would otherwise be missed.

3 Selected Digital Maturity Models

While various digital maturity models are available, we have curated a selection of the most prevalent and widely utilized models in healthcare. It is important to acknowledge that other models may be more closely aligned with specific strategic needs. However, this list is intended to serve as a foundational reference for your own comparisons and evaluations.

3.1 WHO Global Digital Health Monitor

The World Health Organization (WHO) is using the Global Digital Health Monitor,[1] a survey of 23 indicators designed to assess and track the digital maturity of national health systems worldwide. The monitor provides a standardized framework for evaluating critical areas in seven dimensions, such as governance, strategy, legislation, interoperability, workforce, infrastructure, and the implementation of digital health solutions. First released in 2016, the framework has since been subject to ongoing refinement. It assesses the performance of health systems at the country and larger regional levels on a scale from phase 1 (low capability) to phase 5 (high capability).

Of the more than 100 countries currently tracked, all have moved beyond the initial phase 1 of digital health maturity. The median global maturity level is phase three, with some countries advancing to phase five (Scheuer et al. 2023).

[1] https://digitalhealthmonitor.org/.

Table 1 Certification level definitions for CHIME's most wired survey[a]

"Most Wired" Level	Description
Level 1–3	Early stages of digital infrastructure development involve transitioning to electronic data formats, basic use of EMRs, and focus on end-user adoption
Level 4–6	Expanded IT infrastructure supports strategic initiatives. Basic protections (e.g., firewalls, encryption) are in place. Challenges in data utilization and exchange remain
Level 7–8	Meets "Most wired" criteria. Advanced deployment of technologies (e.g., analytics, HIEs, patient portals). Achieving meaningful clinical and efficiency outcomes
Level 9–10	Industry leaders in digital innovation. Advanced tech use (e.g., telehealth, cost-analysis tools). Significant outcomes include improved care, patient experience, and cost reduction

[a]Adapted from https://www.dhinsights.org/most-wired, accessed 13.9.2024

3.2 International Digital Maturity Models of Professional Associations

Professional US-based associations like the Healthcare Information and Management Systems Society (HIMSS) and College of Healthcare Information Management Executives (CHIME) were among the first to develop digital maturity models and offer them internationally.

3.2.1 CHIME Digital Health Most Wired Survey

One of the earliest digital maturity assessment frameworks in healthcare launched in 1999 is the *Digital Health Most Wired Survey* by CHIME.[2] It is an assessment tool that measures the overall digital health maturity of healthcare organizations. The CHIME survey evaluates a broad spectrum across eight different digital health dimensions, including infrastructure, security, data management, clinical quality, population health, innovation, supply chain, and patient engagement.[3]

The survey employs a 10-point scale to categorize organizations according to their level of digital maturity. Levels 9 and 10 are considered the most advanced, signifying "industry leaders in digital innovation" and conferring eligibility for an award from the CHIME committee (Table 1).

3.2.2 HIMSS Digital Maturity Models

While CHIME offers a broad overview, the HIMSS Digital Maturity Models[4] are tailored to different facets of healthcare IT, allowing organizations to assess and improve specific areas of digital maturity. Here is an overview of the most widely used HIMSS models:

[2]The College of Healthcare Information Management Executives—www.chimecentral.org.

[3]https://www.dhinsights.org/most-wired, accessed 13.9.2024.

[4]https://www.himss.org/maturity-models/. Accessed 13.9.2024.

- *EMRAM (Electronic Medical Record Adoption Model)*: The EMRAM model, launched in 2005, is the foundational HIMSS framework, designed to assess the adoption and maturity of electronic medical records (EMR) systems in hospitals. It consists of eight stages, from minimal EMR use to advanced digital maturity, and has been widely adopted internationally as a benchmark for digital health transformation. EMRAM has not only been used by individual hospitals but also as a strategic tool for policy-makers at regional health authorities and ministries of health in several countries.
- *INFRAM (Infrastructure Adoption Model)*: INFRAM assesses the maturity of a healthcare organization's IT infrastructure, ensuring it is robust enough to support advanced digital health solutions like telehealth and mobile health.
- *AMAM (Adoption Model for Analytics Maturity)*: AMAM focuses on the maturity of an organization's analytics capabilities, helping healthcare providers move from basic data collection to advanced predictive analytics and real-time decision-making. This model emphasizes the critical role of data in improving clinical and operational outcomes.
- *CCMM (Continuity of Care Maturity Model)*: CCMM supports assessing different types of care provider organizations, such as ambulatory care, acute care, nursing homes, community services, and others. It assesses the maturity of these single providers as well as the overarching maturity of the care chain with a focus on collaboration, information exchange, and integrated care planning. As with other HIMSS models, the results are presented on a scale from 0, indicating low maturity, to 7, indicating high maturity.
- *C-COMM (Community Care Outcomes Maturity Model)*: The C-COMM framework is designed to assess the efficacy of digital systems in non-acute care settings. Attaining Stage 7, the highest level of achievement, signifies a comprehensive health information exchange, facilitating seamless patient information sharing among healthcare providers and enhancing care coordination and continuity. This model, derived from the former O-EMRAM framework, has recently been updated with a new focus and question set.
- *DIAM (Digital Imaging Adoption Model)*: DIAM assesses the implementation and integration of digital imaging technologies in healthcare settings. It aids imaging departments in hospitals and specialized centers in achieving a fully digitized, interoperable environment, enhancing service quality and efficiency. Achieving DIAM Stage 7 indicates full compliance with international standards and prioritizes patient health.

3.3 National Digital Maturity Models

Various countries have developed their own national digital maturity models, sometimes using indicators from other models, to address their specific healthcare needs and contexts—in particular the Kingdom of Saudi Arabia, the United Kingdom, and Germany.

Countries develop their own national digital maturity models to address the unique characteristics and needs of their healthcare systems. The following list provides an overview of the objectives and reasoning expressed by various stakeholders with regard to developing a national digital maturity model:

- *Tailored Assessment Criteria:* National models allow countries to customize the criteria to better reflect local nuances, such as regulatory requirements, cultural factors, and the specific operational context of their healthcare system.
- *Alignment with National Objectives:* Developing a national model ensures that digital health strategies are aligned with broader governmental goals, such as improving public health outcomes, enhancing interoperability, and ensuring data security within the local legal framework.
- *Greater Relevance and Effectiveness:* National models are more relevant to the specific challenges and opportunities faced by the country's healthcare system, making them more effective in driving meaningful digital transformation.
- *Digital and Financial Autonomy:* By developing their own models, countries can achieve greater digital and financial autonomy. This autonomy allows them to control and direct the development, implementation, and funding of digital health initiatives in a way that aligns with their national priorities. It also reduces dependence on external vendors or international frameworks, enabling countries to allocate resources according to their own strategic goals and economic conditions.
- *In-Country Know-How Build-Up:* Creating a national model fosters the development of local expertise in digital health, which is crucial for the sustainable growth of the country's healthcare sector. This build-up of in-country know-how supports long-term innovation and reduces reliance on external consultants or solutions.
- *Enhanced Local Engagement:* A national model can increase buy-in from local healthcare providers, who may be more inclined to engage with a model that directly addresses their specific context and needs.

3.3.1 Germany's DigitalRadar

One of the most notable examples is Germany's DigitalRadar, a national initiative that assessed the digital maturity of 1.624 hospitals within just 3 months, making it one of the largest and fastest digital maturity assessments globally (Amelung et al. 2022). DigitalRadar was developed to support the German Hospital Future Act, which provided significant funding to improve the digital infrastructure of hospitals across the country. Developed within the framework of the German Hospital Future Act, the model is based on best practice indicators from various maturity models, including HIMSS EMRAM, as well as tailored items to monitor the success of the national funding program.

DigitalRadar not only provides a snapshot of the current state of digital maturity across German hospitals but also identifies critical gaps and areas for improvement. The results inform national policy decisions and ongoing digital health investments, helping to ensure that Germany's healthcare system remains resilient and capable of meeting future challenges.

3.3.2 Saudi Arabia's National Hospital Assessments: The SeMA Framework

Saudi Arabia has launched the Saudi e-Health Maturity Assessment (SeMA), a framework designed to assess and enhance the digital maturity of the Kingdom's healthcare sector. This initiative is a component of Saudi Arabia's Vision 2030, which emphasizes the transformative role of technology in advancing healthcare delivery. SeMA aims to establish a baseline for digital maturity across Saudi healthcare organizations, facilitating informed decision-making and the integration of new entities into the Health Information Exchange system (Scheuer et al. 2023).

The SeMA framework targets four critical healthcare segments: medical cities, hospitals, primary care centers, and specialized or standalone centers. It is built on four foundational pillars:

1. *Digital Fitness Assessment Framework:* Evaluates the readiness of organizations to adopt and implement digital technologies.
2. *International Digital Assessment Benchmarking:* Incorporates global models like HIMSS EMRAM, the Global Digital Health Index, and NHS benchmarks to provide a comprehensive perspective.
3. *KSA Current State Assessment:* Analyzes the current state of healthcare providers and regulators within Saudi Arabia.
4. *Emerging Technologies*: Focuses on the adoption of cutting-edge technologies, including AI, blockchain, and digital building management systems.

SeMA's approach covers eight core dimensions: alignment with national strategies, leadership alignment, financial efficiency, governance structures, organizational competencies, partnerships, and the integration of emerging technologies.

The assessment process under SeMA includes a detailed scoring system that emphasizes applications and infrastructure (40%) and organizational and workforce competencies (20%). Other important areas such as strategy and governance, patient experience, privacy and security, interoperability, data analytics, and emerging technologies also play crucial roles in the evaluation. The assessment process involves an orientation, a mock assessment, registration in the SeMA portal, online and onsite validation, and a certification process that is renewed every 3 years.

SeMA is not only a tool for assessment but also a strategic guide, helping healthcare facilities in Saudi Arabia to identify and address pain points, set goals, and advance their digital maturity in alignment with the Kingdom's broader digital transformation objectives.

3.3.3 NHS England's Digital Maturity Assessment

NHS England has developed the Digital Maturity Assessment (DMA)[5] to evaluate the digital capabilities of healthcare organizations across the NHS. This tool provides a standardized approach for assessing how well NHS trusts and healthcare providers are using digital technologies to support patient care and improve operational efficiency. The DMA measures digital maturity across seven dimensions, as defined by the "What Good Looks Like" (WGLL)[6] framework, including leadership, infrastructure, data management, and clinical processes.

First Results NHS England The initial results from the 2023 DMA highlighted both progress and challenges within the NHS. Notably, the assessment revealed that only 10% to 30% of NHS trusts with an electronic patient record (EPR) system were fully utilizing key functionalities, such as integrated prescriptions and record sharing with citizens and other hospitals. This finding underscores the need for continued focus on maximizing the potential of existing digital tools across the healthcare system (Solof 2024a).

The DMA results also showed a variation in digital maturity across different regions and types of trusts. Some trusts scored as high as 3.9 out of 5 on the digital maturity scale, while others lagged significantly behind, with scores as low as 1.6. These disparities indicate a pressing need for targeted interventions to raise the overall digital maturity of the NHS by 2025, as part of its broader digital transformation strategy.

In 2024, NHS England expanded the scope of the DMA to include primary care for the first time. This expansion is designed to provide a more comprehensive picture of digital maturity across the entire healthcare continuum, including general practices (Solof 2024b). By integrating primary care into the DMA, NHS England aims to identify and address gaps in digital capabilities at the grassroots level, ensuring that all aspects of the healthcare system are aligned in their digital maturity efforts. This broader approach is intended to improve planning, investment decisions, and the spread of best practices across the NHS.

4 Conclusion

As you explore the contents of this book, it becomes clear that digital maturity is not just a technical achievement but a strategic imperative for the future of healthcare. The journey toward digital maturity, while complex and varied across different regions and healthcare systems, is essential for creating resilient, efficient, and

[5] https://www.england.nhs.uk/digitaltechnology/connecteddigitalsystems/digital-maturity-assessment/. Accessed 13.9.2024.

[6] https://transform.england.nhs.uk/digitise-connect-transform/what-good-looks-like/what-good-looks-like-publication/. Accessed 13.9.2024.

patient-centered care environments. Whether through the broad application of global frameworks like the WHO Global Digital Health Monitor, the specialized models offered by professional associations like HIMSS and CHIME, or the tailored national initiatives seen in countries like Germany, Saudi Arabia, and the UK, the path forward requires a commitment to continuous improvement and collaboration. This book provides a roadmap for that journey, offering practical insights, best practices, and lessons learned from healthcare leaders around the world. As you embark on your own digital transformation journey, may these insights guide you in making informed decisions that enhance the quality of care and health outcomes for all.

References

Amelung V, Angelkorte M, Augurzky B et al (2022) Ergebnisse der ersten nationalen Reifegradmessung deutscher Krankenhäuser [Zwischenbericht]. https://www.digitalradar-krankenhaus.de/download/220914_Zwischenbericht_DigitalRadar_Krankenhaus.pdf. Accessed 13 Sept 2024

Canfell OJ, Woods L, Meshkat Y, Krivit J, Gunashanhar B, Slade C, Burton-Jones A, Sullivan C (2024) The impact of digital hospitals on patient and clinician experience: systematic review and qualitative evidence synthesis. J Med Internet Res 26:e47715. https://doi.org/10.2196/47715. PMID: 38466978; PMCID: PMC10964148

Duncan R, Eden R, Woods L, Wong I, Sullivan C (2022) Synthesizing dimensions of digital maturity in hospitals: systematic review. J Med Internet Res 24(3):e32994. https://www.jmir.org/2022/3/e32994/

Omid R, Khayami SR, Rezaei MS (2024) Smart hospital definition: academic and industrial perspective. Int J Med Inform 182:105304.,ISSN 1386-5056. https://doi.org/10.1016/j.ijmedinf.2023.105304

Scheuer A, McArthur R, Bell J (2023) Mastering large scale digital maturity in healthcare, perspective 12/2023. https://lemonmint.org/wp-content/uploads/2023/11/2023-11-Perspective-Digital-Maturity-FINAL.pdf. Accessed 13 Sept 2024

Solof J (2024a) DHI snapshot report examines trust digital maturity assessment scores. https://www.digitalhealth.net/2024/04/dhi-snapshot-report-examines-trust-digital-maturity-assesment-scores/. Accessed 13 Sept 2024

Solof J (2024b) Digital maturity assessment expands into primary care for 2024. https://www.digitalhealth.net/2024/05/digital-maturity-assessment-to-expand-into-primary-care/. Accessed 13 Sept 2024

Digital Maturity Models for Continuous Development and Evaluation in Healthcare

Jan-David Liebe, Franziska Jahn, Uwe Buddrus, and Ursula Hübner

Abstract

This chapter provides developers and IT decision-makers in healthcare organizations with recommendations on how digital maturity models (DMM) can be developed, selected, and used for the continuous advancement and evaluation of digital maturity in hospitals. Therefore, nine requirements for the development and use of DMMs are formulated as a checklist for the selection of suitable DMMs. Furthermore, the Digital Quality Framework (DQF) is discussed, which represents a holistic indicator system for DMMs. Example indicators for monitoring maturity are provided for each of the nine strategic fields of action of the DQF.

Keywords

Digital maturity model · Digitalization strategy · Hospital information management · Digital quality framework · Digital maturity

Abbreviations

BYOD	Bring Your Own Device
CIO	Chief Information Officer
CPOE	Computerized Physician Order Entry

J.-D. Liebe (✉) · U. Buddrus · U. Hübner
Osnabrück University of Applied Sciences, Faculty of Business, Economics and Social Sciences, Osnabrück, Germany

F. Jahn
Leipzig University - Institute for Medical Informatics, Statistics and Epidemiology, Leipzig, Germany
e-mail: franziska.jahn@imise.uni-leipzig.de

A. Scheuer, J. Studzinski (eds.), *Digital Maturity in Hospitals*,
https://doi.org/10.1007/978-3-031-80704-6_2

DMM	Digital Maturity Model
DQF	Digital Quality Framework
ETL	Extract, Transform, Load
GDPR	General Data Protection Regulation
GMDS	German Association for Medical Informatics, Biometry, and Epidemiology
IM	Information Management
IT	Information Technology
KIS	Krankenhausinformationssystem (Hospital Information System in German)
OR	Operating Room

1 Use of Digital Maturity Models: Aspiration Versus Reality

Digital maturity indicates the extent to which information technology is used to deliver high-quality healthcare (Flott et al. 2016). To measure digital maturity in healthcare settings, a variety of digital maturity models have been developed in recent years for hospitals, practices, health authorities or national health systems. Digital maturity models enable the assessment of the current state and the definition of the target state of the digital maturity of a healthcare setting. Likewise, they provide a roadmap for closing the gap to current best practice in the field. Digital maturity models specify how digital maturity indicators can be determined. Digital maturity indicators are reported either as composite scores or as stage-based scores.

In the context of strategic IT management in healthcare organizations, digital maturity models (DMM) serve the systematic planning, implementation, and evaluation of digitalization strategies. Therefore, the first step is to define the organization's internal digitalization goals and operationalize them using maturity indicators. In a second step, the degree of target achievement is checked using these indicators, whereby standardized questionnaires are mostly used as a survey method (Becker et al. 2009). Based on a target/actual comparison, optimization potential is identified in a third step and used to review and adjust the IT project portfolio. This step can be supported by benchmarking activities in which best practices are identified and adopted for the sound planning and implementation of the organization's digitalization activities (Jahn et al. 2015; Liebe and Hübner 2013). In terms of strategic monitoring, the measurement of the digital maturity level will be repeated at regular intervals and adjustments can be implemented at the tactical and operational level of information management (Fig. 1).

Although over 100 DMMs have been developed for healthcare organizations over the past two decades (Carvalho et al. 2016; Tarhan et al. 2020), most models are rarely used in the above described way (Kücherer et al. 2016). It can be assumed that the reasons for this lie in various conceptual and methodological limitations, which together make the use of DMMs difficult. One of the reasons is the often-rigid selection of maturity indicators, which hardly supports the review of organization-

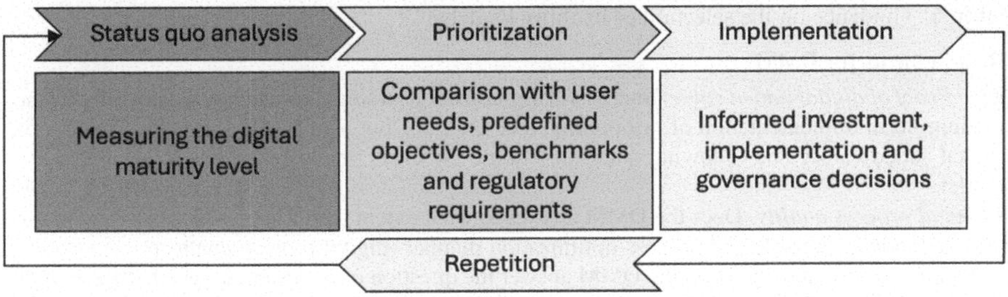

Fig. 1 Continuous maturity measurement to support strategic information management

specific digitalization goals (Liebe and Jahn 2019). Another reason lies in the conceptual equation of digital maturity and the degree of technological implementation. This approach falls short for the evaluation of digitalization strategies, as the availability of IT components only represents a medium-term digitalization goal. Long-term or strategic objectives, on the other hand, relate to the creation of digital added value, i.e. the IT-supported optimization of processes and the facilitation of high-quality and efficient patient care (Winter et al. 2023).

The focus on the technological side of digital transformation and the frequent neglect of socio-cultural aspects, which is evident in many maturity models, also makes it difficult to derive recommendations for action for the successful implementation of digitalization strategies. Thus, digital maturity is often equated with technical availability alone—and not with meaningful use of the technical solutions or their impact on health outcomes. They also don't consider whether physicians, nurses, and patients have the necessary skills for a beneficial use of the technologies or whether the information management of the institution is able to realize local digitalization needs in the context of available resources, regulatory requirements, and strategic priorities remains unanswered (Cresswell et al. 2019). Methodological limitations that presumably prevent the widespread use of maturity models in hospital information management include the lack of transparency and traceability with regard to model development. DMMs are being developed and provided by a number of organizations, including consultancy firms, associations, technology providers, governments, etc. It is therefore often unclear what interests the providers of maturity models are pursuing, which in turn raises the question of the scientific quality and rigor of the survey and analysis methods used.

2 Requirements for Maturity Models to Support Strategic Digitalization Activities

Against the background described above, the working group *Methods and Tools for the Management of Hospital Information Systems* (mwmKIS) of the German Association for Medical Informatics, Biometry and Epidemiology GMDS e. V. (2020) has developed nine requirements for the development and implementation of maturity models (Liebe et al. 2022). Table 1 presents those requirements. Questions are

Table 1 Guidance on the selection of maturity models

Requirements for DMM
#1—Proof of digital added value: Successful digitalization strategies are measured not only by the medium-term implementation of information technologies, but also by the long-term creation of digital added value. The following questions should therefore be examined when selecting a model: • *Digital process quality*: Does the DMM answer to what extent the IT applications already in use are used in a sustainable and process-optimization manner (digital process quality)? • *Digital outcome quality*: Does the DMM answer the question of the extent to which the use of IT applications leads to quality and efficiency benefits for the organization, employees, patients, and other IT stakeholders (digital outcome quality)? • *Undesirable effects*: Does the DMM answer to what extent the availability and use of IT applications also leads to undesirable effects (additional workload, incorrect medication, etc.)?
#2—Derivation of recommendations for action: Simply describing the level of digital maturity is not enough to support strategic digitalization activities. Rather, it should also be possible to derive explanatory approaches, best practices, and recommendations for action for a well-founded and targeted management of digital change. The following questions should therefore be examined when selecting a model: • *Influencing factors*: Does the DMM provide explanatory approaches to the recorded level of digitalization, e.g. by identifying promoting or inhibiting factors of information management and the general organizational framework? • *Best practices*: Does the DMM provide information on best practices for implementing digitalization strategies, e.g. as part of internal or external benchmarks? • *Recommendations for action*: Does the DMM provide recommendations for action, e.g. in the form of weak point analyses or priority lists? • *Configurable information output*: Can the information output be configured so that it is tailored to individual information requirements (e.g., via an addressee-specific dashboard output)?
#3—Transparent presentation of the effort involved: Measuring the digital maturity level requires additional effort from everyone involved. This is particularly true if the process is designed for the long term and addresses several IT stakeholders. Against this background, the acceptance and success of the maturity level measurement depends on a clearly recognizable cost–benefit ratio. The following questions should therefore be examined when selecting a model: • *Additional personnel costs*: Is there a transparent and comprehensive indication of the number of employees or internal person-hours to be planned for each survey cycle (in relation to the size of the facility)? • *Costs and additional services*: Is it transparently and comprehensively shown whether and what costs are incurred for using the DMM and for additional services (consulting, validation and certifications, etc.)? • *Cost–benefit ratio*: Is it possible to derive a realistic cost–benefit estimate for all parties involved based on the available information?
#4—Consideration of local digitalization needs: Healthcare facilities pursue different digitalization strategies against the backdrop of different conditions and framework conditions. Maturity models can only support the planning, implementation, and evaluation of these strategies if they take into account the respective conditions and are not based on a normative gold standard. The following questions should therefore be examined for an optimal model selection: • *Adaptability*: Are the survey contents adapted to organization-specific conditions, e.g. functional areas, service focuses, technological preconditions and digitalization needs (e.g., through conditional query and evaluation logic such as "OR available" vs. "OR not available")? • *Customizability*: Can the survey content be flexibly supplemented with facility-specific metrics (e.g., by selecting individual digitalization goals from a catalog)? • *Relative maturity level calculation*: Is the maturity level calculated relative to the organization-specific conditions and objectives?

(continued)

Table 1 (continued)

#5—Repeatability and visualization of development processes: To support digitalization strategies, the digital maturity level must be measured regularly and in consideration of current developments (Fig. 1). For this reason, the maturity model should be designed for continuous use in terms of content, technology, and organization. The following questions should therefore be examined when selecting a DMM: • *Technical requirements*: Does the conceptual and technological design of the model ensure that it meets the requirements for repeatedly measuring the digital maturity level, e.g. by having an appropriate data model and providing data that has already been collected as a starting point for repeated measurement? • *Up-to-date content*: Is it ensured that the survey content corresponds to the current state of technological development (e.g., through established expert committees and transparent version documentation)? • *Development progression*: Is the development progression of the maturity level parameters presented for the individual organization and at different aggregation levels? Is the comparison of different measurement points supported?
#6—Interdisciplinary view of digital maturity: Digital change is a complex phenomenon, especially in the healthcare sector. A valid measurement of digital maturity therefore takes different perspectives into account. In addition, data arising from system utilization can also be used. The following questions should therefore be examined when selecting a model: • *Multiperspectivity*: Does the data collection consider the perspectives of all relevant stakeholders (i.e., IT decision-makers, internal and external users and patients)? • *Cross-sector applicability*: Can the process be applied across healthcare institutions and sectors? • *Objective data sources*: Does the process consider data from the information system to validate the maturity level (e.g., audit logs)?
#7—Transparent and comprehensible methodology: The broad and continuous use of maturity models depends on the acceptance of the process, which is why the methods used to collect, store, and analyze data should be made transparent and comprehensible to all those involved. The following questions should therefore be examined when selecting a model: • *Data collection*: Are the procedures for data collection and plausibility checks presented in a transparent and comprehensible manner (i.e., identification of the targeted survey participants and data sources, detailed presentation of the questionnaire content and validation procedures, transparent presentation of the survey logistics, etc.)? • *Data storage*: Are the procedures for data storage presented in a transparent and comprehensible manner (e.g., GDPR-compliant data storage, provision of collected data for repeated measurements, etc.)? • *Data preparation*: Are the procedures for calculating the maturity levels and scores presented in a transparent and comprehensible manner (evaluation metrics, weighting, handling of missing data, etc.)?
#8—Proof of scientific quality: Particularly when strategic investment decisions are linked to maturity measurement, a rigorous review of the scientific quality of model development and validation appears appropriate. The following questions should therefore be examined when selecting a model: • *Development domain*: Is the development domain of the maturity model comprehensively represented (e.g., academia, associations, industry, consulting service providers) so that conflicts of interest can be ruled out? • *Development process*: Is the development of the maturity model demonstrably based on recognized reference models (e.g., CMMI, COBIT, ITIL) and scientific process models for the design and implementation of evaluations for information technologies (e.g., STARE-HI) in the healthcare sector? • *Objectivity*: Has the objectivity of the model been scientifically verified, i.e. does the model measure what it is supposed to measure independently of the actors involved?

(continued)

Table 1 (continued)

• *Reliability*: Has the reliability of the model been scientifically verified, i.e. does the model reliably measure what it is supposed to measure, even when used repeatedly and in different settings? • *Validity*: Has the validity of the model been scientifically verified, i.e. does the model measure what it is supposed to measure (e.g., actual IT utilization and not just IT availability)?
#9—Compatibility and connectivity: Maturity assessments have already been carried out in many healthcare facilities. At the same time, it can be important for the purpose of benchmarking to compare with maturity level assessments of other organizations. For this reason, the maturity model should be compatible with existing models and approaches. The following questions should therefore be checked when selecting a model: • *Technical compatibility*: Can the measured maturity level parameters be linked to data from previous surveys (e.g., by supporting ETL procedures)? • *Content compatibility*: Can parameters of the maturity model be compared with those from existing models?

formulated for their review that can support IT decision-makers in selecting a suitable model.

2.1 Strategic Fields of Action and Generic Digitalization Goals: The Digital Quality Framework

The evaluation of digitalization strategies is just as complex as their planning and implementation. For this reason, it makes sense to systematically compare strategic objectives and indicators for monitoring success. In Tables 2, 3, and 4, this systematization is based on three strategic fields of action: The clinical field of application, information management, and the organizational environment. Strategic generic target images were formulated for each field of action based on Donabedian's quality model (Donabedian 1980). The established system follows the Digital Quality Framework (DQF), which was developed as part of the GMDS-GI mwmKIS working group (Liebe et al. 2022; Fig. 2).

3 Conclusion

To date, maturity models have mostly been used to take stock of the available IT components. This chapter shows how maturity level measurement can be used for the continuous (further) development and evaluation of digitalization strategies. The proposed testing scheme is intended to support IT decision-makers in selecting a suitable procedure and provide model developers with information on the design of new approaches. The fields of action and digitalization goals based on the Digital Quality Framework can also be used to systematically develop and monitor the success of IT strategies.

Table 2 Generic digitalization goals and indicators for monitoring success in the clinical application field

Generic digitalization goals	Indicators (examples)
Digital structural quality in the clinical application field: The availability of information technology resources that can be used to support, change, or reorganize medical, nursing, therapeutic, and administrative processes is ensured	• Availability of IT applications to support clinical processes (e.g., patient administration systems, clinical information systems, laboratory information systems, robotic systems, digital dictation and speech recognition systems, CPOE systems) • Availability of IT components to support cross-sector care (e.g., cloud computing, telemedical network structures, patient portals, online-based care data systems for beds) • Availability of IT components for mobile access to patient data and information (e.g., WLAN, end devices, mobile apps) • Degree of integration and interoperability of available IT applications (interfaces, standards)
Digital process quality in the clinical application field: The use of available information technologies leads to the optimization of medical, nursing, therapeutic, and administrative processes.	Processes optimized through IT use: • Admission, transfer, discharge, in particular communication with external providers, service providers and patient • in service requests and results feedback • in the emergency department and emergency care • in medication management and in ensuring AMTS • in surgery (diagnosis, planning, implementation) • in the documentation of care and treatment services (e.g., progress documentation, reports on findings, doctor's letters) • in longitudinal care (e.g., billing, occupancy, room, material and personnel planning)
Digital outcome quality in the clinical application area: A measurable increase in the quality and efficiency of care is achieved thanks to optimized processes. IT users experience the use of information technologies as beneficial against the background of their individual and professional requirements	• General outcome indicators for the development of clinical benefit (e.g., better patient care, improved patient safety) • Process-related outcome indicators of clinical benefit development (e.g., the 6-R rules of AMTS are implemented; the status of orders and prescriptions can be better tracked) • General outcome indicators of user orientation and satisfaction (e.g., less time spent on routine activities; less overtime; more time for patients) • Process-related outcome indicators of user orientation and satisfaction (e.g., higher quality of operating theater documentation; better interdisciplinary cooperation in referral management)

Table 3 Generic digitalization goals and indicators for monitoring success in information management

Generic digitalization goals	Indicators (examples)
Digital structural quality in information management: Information management (IM) is embedded in mature IT governance and has the necessary human, financial, and technological resources to enable the implementation of professional IM activities.	• Sufficient IT budget • Sufficient and qualified IT staff • Availability of information management tools (e.g., project management software, wiki for project documentation, CIO dashboard) • Established structures to ensure IT compliance, in particular information security, data retention, and data protection (e.g., compliance rules, BYOD policy, availability of systems to detect and prevent intrusions) • Established structures to ensure operational security and business continuity (e.g., redundant data centers, emergency power supply, data backup and recovery) • Established structures to ensure IT support (e.g., helpdesk, ticket system)
Digital process quality in information management: Based on existing structures and through the use of digital IM tools, the degree of professionalization of information management for the (further) development and operation of a secure, user-friendly, economical, agile and legally compliant information system is made possible	• IT project management activities (including management of the IT project portfolio, system analysis, specification, selection and introduction, project evaluation) • Innovation management activities • IT monitoring and evaluation activities (e.g., evaluation of system utilization, user surveys) • Information security measures and risk management activities (e.g., maintenance/repair, emergency drills) • IT support activities (e.g., user training, training courses)
Digital outcome quality in information management: Information management ensures the operation of a secure, user-friendly, economical, agile, and legally compliant information system	• Ensuring business continuity (e.g., downtime, data availability in an emergency) • Ensuring IT compliance/information security (e.g., number of attacks averted) • Ensuring IT service performance (e.g., user satisfaction with IT support) • Ensuring the success and cost efficiency of IT projects (e.g., benefit calculations/evidence)

Table 4 Generic digitalization goals and indicators for monitoring success in the organizational environment

Generic digitalization goals	Indicators (examples)
Digital structural quality in the organizational environment: There are institutionalized functions and formal sets of rules that support the cooperative and cross-functional (further) development of the information system and enable the optimized generation of benefits from the information system for the further development of the organization	• Existence of distinct IT governance including formal processes for the (further) development of digitalization strategies (e.g., dynamically adaptable and verifiable strategy, development and transformation plans) • Availability of IT applications to support a learning organization (e.g., process mining tools, BI tools) • Institutionalization (e.g., interprofessional IT steering committee, chief medical information officer, data scientist) • Existence of formal rules and regulations (e.g., for suggestion schemes and data governance)
Digital process quality in the organizational environment: By using existing structures, activities are supported that enable a cooperative and cross-functional dialog for the (further) development of the information system and optimal benefit generation from the information system for further organizational development	• Cooperative and cross-functional exchange activities (especially between clinicians, IT, and administration) to continuously optimize the benefits of the hospital information system • Strong digital leadership on the part of hospital management, chief physicians, and other management functions • Regular communication about developments in the information system to employees (e.g., IT newsletter) • Activities to support user compliance (e.g., incentives to use the system) • Data and process mining activities as well as innovative data analyses for operational management (controlling) to support business intelligence and a learning organization (e.g., analyses of profitability, use of clinical decision support, etc.) • High level of alignment between corporate and IT strategy • Distinct IT innovation and digitalization culture (e.g., user acceptance, intrapreneurship, etc.) • Established business intelligence, i.e. benefits for the organization (time, money, security) and research (data) quality • Proof of conformity with guidelines and laws
Digital outcome quality in the organizational environment: There is a cooperative and cross-functional exchange for the (further) development of the information system. The benefits generated by the information system for further organizational development are optimally utilized in the sense of a learning organization	• Strong digital leadership on the part of hospital management, chief physicians, and other management functions

	Structural quality	Process quality	Outcome quality
Organizational environment	Existence of IT strategies and guidelines, exchange platforms, data governance	Execution of conformity checks, digital participation, data mining	Strategic alignment, regulatory compliance, business intelligence
Information management	Existence of IT governance, digital leadership, resources	Professional activities in IT planning, implementation, operation and evaluation	Economic efficiency, information security, operational security and agility
Field of application	Availability of hardware, software, peripherals, networking	Process-optimizing use of IT applications along the patient journey	Quality of care, patient safety, user satisfaction

Fig. 2 Strategic fields of action and generic target images in the Digital Quality Framework

References

Becker J, Knackstedt R, Pöppelbuß J (2009) Entwicklung von Reifegradmodellen für das IT-Management: Vorgehensmodell und praktische Anwendung. Wirtschaftsinformatik 51(3): 249–260. https://doi.org/10.1007/s11576-009-0167-9

Carvalho JV, Rocha Á, Abreu A (2016) Maturity models of healthcare information systems and technologies: a literature review. J Med Syst 40(6):1–10. https://doi.org/10.1007/s10916-016-0486-5

Cresswell K, Sheikh A, Krasuska M et al (2019) Reconceptualising the digital maturity of health systems. Lancet Digit Heal 1(5):e200–e201. https://doi.org/10.1016/S2589-7500(19)30083-4

Donabedian A (1980) The definition of quality and approaches to its assessment and monitoring. Health Administration Press, Annapolis Junction, p 163

Flott K, Callahan R, Darzi A, Mayer E (2016) A patient-centered framework for evaluating digital maturity of health services: a systematic review. J Med Internet Res 18(4):e75

GMDS e. V. (2020) GMDS-Stellungnahme zur Ausgestaltung von § 14b des Krankenhauszukunftsgesetzes (KHZG): "Evaluierung des Reifegrades der Krankenhäuser hinsichtlich der Digitalisierung". https://www.gmds.de/fileadmin/user_upload/Publikationen/Stellungnahmen/20201007_Stellungnahme_GMDS_KHZG___14b.pdf. Accessed 1 July 2024

Jahn F, Baltschukat K, Buddrus U et al (2015) Benchmarking of hospital information systems - a comparison of benchmarking clusters in German-speaking countries. Dtsch Gesellschaft fur Medizinische Inform Biometrie und Epidemiol eV 11(1). https://doi.org/10.3205/mibe000160

Kücherer C, Liebe J, Schaaf M et al (2016) The status quo of information Management in Hospitals-Results of an online survey. In: Mayr HC, Pinzger M (eds) INFORMATIK 2016. Gesellschaft für Informatik, Bonn, pp 685–698

Liebe JD, Hübner U (2013) Developing and trialling an independent, scalable and repeatable it-benchmarking procedure for healthcare organisations. Methods Inf Med 52(4):360–369. https://doi.org/10.3414/ME12-02-0016

Liebe JD, Jahn F (2019) Adaptierbarkeit und Zielorientierung bei der Anwendung von KIS-Reifegradmodellen. https://www.egms.de/static/de/meetings/gmds2019/19gmds184.shtml. Accessed 1 July 2024

Liebe JD, Buddrus U, Jahn F, Hübner U (2022) Digitalstrategie und Digitalisierungsaktivitäten: Ansätze zur Erfolgsmessung. In: Henke V, Hülsken G, Meier PM, Beß A (eds) Digitalstrategie im Krankenhaus. Springer Gabler, Wiesbaden. https://doi.org/10.1007/978-3-658-36226-3_12

Tarhan AK, Garousi V, Turetken O et al (2020) Maturity assessment and maturity models in health care: A multivocal literature review. Digit Health 6:2055207620914772. https://doi.org/10.1177/2055207620914772

Winter A, Ammenwerth E, Haux R, Marschollek M, Steiner B, Jahn F (2023) Health information systems: technological and management perspectives, 3rd edn. Springer, Cham. https://doi.org/10.1007/978-3-031-12310-8

Digital Maturity Models for Nationwide Healthcare Modernization: Insights from Germany's "DigitalRadar"

Stephan Krumm and Thomas Süptitz

Abstract

The healthcare system continues to undergo digital transformation, necessitating substantial investment decisions. These decisions require a smart strategy and rigorous implementation.

Maturity models can support this process by providing a strategic decision-making framework and making progress transparent and measurable.

Developing a suitable maturity model is complex. It is essential to define appropriate criteria to accurately reflect the healthcare system's unique characteristics. Additionally, implementing a maturity measurement is challenging and requires gaining acceptance from employees and other stakeholders to highlight its necessity.

The Hospital Future Act of 2020 marked the introduction of a maturity model for hospitals at national level in Germany. Initial maturity measurements reveal significant variation in the digitization levels of German hospitals, with digital pioneers and laggards coexisting. These results clearly indicate areas for potential expansion and where political support is needed.

However, an important aspect is often overlooked: maturity models can also serve as an incentive instrument, such as through financing mechanisms. It is possible to promote digital pioneers or target specific gaps. This approach must balance the interests of digital pioneers and laggards, rewarding those who have already invested in digitization while providing incentives for those yet to do so.

Keywords

Digital maturity · National digital assessment · Healthcare policy · Digital transformation · National maturity model

S. Krumm (✉) · T. Süptitz
Cybersecurity and Interoperability, German Federal Ministry of Health, Berlin, Germany
e-mail: Stephan.Krumm@uni-potsdam.de; Thomas.Sueptitz@googlemail.com

A. Scheuer, J. Studzinski (eds.), *Digital Maturity in Hospitals*,
https://doi.org/10.1007/978-3-031-80704-6_3

Abbreviations

BAS	Bundesamt für Soziale Sicherung
DMM	Digital maturity model
DR	DigitalRadar
DRG	Diagnosis-Related group
EMRAM	Electronic Medical Records Adoption Model
HIMSS	Healthcare Information and Management Systems Society
KHZF	Hospital Future Fund
KHZG	Hospital Future Act
KIG	Competence Centre for Interoperability in the Healthcare Sector
PHS	Public Health Service
VAT	Value Added Tax

1 Introduction

Disclaimer: This article reflects exclusively the personal view of the authors.

1.1 The COVID-19 Pandemic and the Hospital Future Act

The year 2020 began with a historic crisis: the COVID-19 pandemic, which reached Europe in the spring. This had a profound impact not only on public health but also on society and the economy. According to the Bundesbank,[1] the economy contracted by over 5% in the first year of the pandemic compared to the previous year (Deutsche Bundesbank 2021).

Recognizing the scale of the crisis, Germany's federal government introduced a one-off economic stimulus package totaling 130 billion euros (Die Bundesregierung 2020a). This package included a temporary reduction in VAT, a child bonus, and additional support for local authorities, as well as a forward-looking investment plan (Die Bundesregierung 2020b). A significant component of this plan was the "Program to Future-Proof Hospitals," which aimed to enhance digital infrastructure and cybersecurity in hospitals. This program allocated 3 billion euros from the EU Program "NextGenerationEU" through federal funds and nearly 1.3 billion euros from state funds. The COVID-19 pandemic thus provided an opportunity to address the investment backlog in digital technologies in hospitals. The stimulus package was designed with the expectation that the measures, including the hospital program, would be rapidly implemented to achieve swift economic benefits.

[1] The Deutsche Bundesbank is Germany's central bank.

To implement the future program legally, the Hospital Future Act (KHZG)[2] was developed and passed by the Bundestag[3] on October 29, 2020—just 5 months after the program's initial announcement. Central to the KHZG is the Hospital Future Fund (KHZF), which provides funding for various investments and includes a digital maturity model to support the funding process.

1.2 The Need for Investments

The relatively short period between the announcement of the future program and the Bundestag's legislative decision was possible because preliminary considerations began as early as 2019.

The historically weak investment financing in the German hospital sector is well-known. This issue stems from the dual hospital financing system, where the 16 *Bundesländer*[4] are responsible for investment costs and the statutory/private insurance funds cover operating costs. According to the German Federal Audit Office (Bundesrechnungshof 2020), "the federal states have inadequately met their investment obligations for years, with funding remaining low or even declining."

It is also important to note that Germany has nearly 1900 hospitals, which are operated either privately, publicly, or as non-profits, and provide various levels of care (Destatis 2023). These hospitals enjoy a high degree of autonomy and operate economically independently, though they are not free to set their own prices. Consequently, market-driven elements exist alongside regulatory constraints.

Additionally, hospitals in Germany have a long history. Except for the former German Democratic Republic, where numerous investment and modernization measures—including new buildings—were undertaken after reunification in the 1990s, there have been few fundamental structural changes in recent decades.

Investment decisions are primarily made by individual hospitals, often in consultation with supervisory authorities at the state level. This is particularly relevant in the area of digitalization, where, except for a few specific areas, there is no comprehensive framework for digital transformation in the hospital sector. As a result, digitalization strategies and goals may vary significantly between facilities. Consequently, some hospitals may excel in certain digital services while lagging in others, such as advanced automated goods transport systems versus more basic tasks like physically printing patient charts for later scanning and archiving.

[2]Gesetzentwurf der Fraktionen der CDU/CSU und SPD Entwurf eines Gesetzes für ein Zukunftsprogramm Krankenhäuser (Krankenhauszukunftsgesetz—KHZG)—https://dserver.bundestag.de/btd/19/221/1922126.pdf.

[3]*Bundestag* is the name of The German Parliament.

[4]Germany has 16 federal states, known as *Bundesländer*. These states have their own governments and responsibilities, including certain aspects of healthcare financing.

Investment decisions are also influenced by the market-regulatory environment. Due to inherent disincentives in models such as the DRG system, decisions often focus on delivering short-term returns on investment.

The consequences of these health system-specific issues for digitalization are frequently highlighted in hospital reports. Germany has been lagging behind in hospital IT digitalization for years. The 2019 Hospital Report (Klauber et al. 2019), which focused on digitalization, revealed that around 40% of German hospitals lack information systems in diagnostic departments, including laboratory, radiology, and pharmacy. An additional 30% do use such systems, but they are not integrated, leading to manual data entry and errors. Consequently, nearly 70% of hospitals did not advance beyond level 2 (out of 8) in the Electronic Medical Records Adoption Model (EMRAM) defined by the Healthcare Information and Management Systems Society (HIMSS) at that time.

1.3 Incentive and Financing Model

The described situation led to the idea of specifying and promoting a limited number of concrete digital services based on demand from hospitals. Funding should be allocated only to those digital services that directly or indirectly enhance patient care. The approach includes promoting internal digitalization within hospitals to build a strong foundation for future digitization measures. Meaningful services for patients, for example digital patient portals/patient interaction platforms, and various internal hospital components must be integrated. When patient-relevant services are prioritized and funded, it often necessitates the creation, adaptation, or modernization of other supporting systems.

The digital services are aligned with the entire patient pathway: from registration and admission, through treatment and operational processes during inpatient care provision, to discharge and transfer to subsequent service providers. The various funding topics enable each hospital to prioritize projects based on its specific needs and level of digital maturity. Despite significant differences among hospitals, this approach aims to achieve a comprehensive impact across the entire hospital sector.

The KHZG therefore integrates various incentive and financing components to maximize the effective use of funding. These include:

- The KHZF provides a total of EUR 3 billion in funding, with an additional EUR 1.3 billion co-financed by the federal states and hospitals. This makes the KHZF the largest of the three healthcare measures in the German Recovery and Resilience Plan.
- Fifteen percent of each individual project budget should be allocated to measures for enhancing cybersecurity.
- A comprehensive funding guideline outlines general and specific conditions and sets technical requirements for digital services. For instance, the funding subject digital patient portal includes 17 mandatory requirements and 16 optional requirements (Bundesamt für Soziale Sicherung [BAS] 2023).

Table 1 Number of applications per subject of application by 31 December 2021 (BAS 2024)

Subject of application	Number of applications
1 Emergency department	395
2 Electronic patient portal	1130
3 Electronic documentation	1533
4 Decision support systems	550
5 Electronic medication management	937
6 Computerized practitioner order entry	553
7 Cloud-computing	151
8 Hospital bed monitoring	26
9 Telemedical networks	270
10 IT security	776
11 Patient rooms	9

- A deduction mechanism has been established to incentivize the nationwide implementation of digital services. Starting in 2025, hospitals must implement and use the funded digital services (subjects 2–6, see Table 1). Failure to do so will result in a deduction of up to 2% from their reimbursement in subsequent years, regardless of whether they have claimed the KHZF funding.
- Correct implementation must be checked and verified by authorized IT service providers.
- All funding recipients are required to participate in the accompanying digital maturity evaluation, known as DigitalRadar (DR).

Already during the application period the numerous structural, organizational, and practical challenges emerged and, despite individual mitigation measures, could not be completely resolved. Thus these challenges continue to influence the success of the program:

- *Funding Responsibilities in Hospital Care in Germany*: As mentioned, hospital planning falls under the jurisdiction of the federal states. While historical and technical reasons support this structure, achieving coherent standardization of administrative processes across Germany is challenging. This difficulty arises because funding policy priorities may vary between the federal government and the federal states, as well as among the states themselves.
- *Clear Future Vision for Integrated Digitization Projects*: Individual hospitals have struggled to develop well-conceived investment projects within the given timeframe. German hospitals face a significant backlog in digitalization, which prompted the launch of the funding program. This backlog makes it challenging for hospitals to create customized funding projects as integrated components of their future IT systems.
- *Scalability of Administrative and Economic Structures*: While the funding program aims to introduce or modernize basic digital services, it must also address

functionalities for which no integrated products were available at the time the funding guidelines were published. Consequently, the industry was tasked with developing and providing products tailored to these needs. Simultaneously, the demand for IT services greatly exceeded the available supply, which also strained the administrative capacities of the authorities, particularly at the program's outset.

By 31 December 2021, the deadline for applications, more than 6000 submissions had been received. The focus was on the following funding areas:

Funding requests prioritized projects that ensure the basic digital infrastructure of hospitals, which might result from the fact that future budget deductions will be applied to funding subjects 2–6 if a hospital neither offers nor uses these designated digital services.

To date, over EUR 2.94 billion in funding has been approved at the federal level. In terms of utilization, the KHZF can be considered highly successful. This success is essential for achieving a nationwide increase in digital maturity (BAS 2024).

2 Digital Maturity Level as an Indicator for the Success of Funding Programs

2.1 Origin of the Digital Maturity Model

Digital maturity models (DMM) for assessing the level of digitalization have a long history in the healthcare sector, with the EMRAM model from HIMSS being particularly notable. In Germany, even before the KHZF, there was debate over how to more accurately measure the digital maturity of medical service providers. However, attempts to use existing models were often met with criticism, as these internationally developed models did not fully account for the unique aspects of the German healthcare system and had specific limitations. This criticism was also evident during discussions about directly applying EMRAM as a measurement tool. Indeed, at the time, there were model-specific limitations that would have complicated its application in Germany.

A comprehensive evaluation of the funding program was incorporated during the parliamentary process of the Hospital Future Act to scientifically assess the improvements achieved through the funding. As part of this evaluation, a tool was to be provided to participating hospitals for structured self-assessment, facilitating a standardized nationwide survey. This tool was also designed to be accessible to hospitals that did not receive funding. The study was initially planned with key measurement dates of 30 June 2021 and 30 June 2023. However, the second measurement was postponed by 12 months due to subsequent amendments to the law, with the possibility of an additional, third, measurement introduced.

The digital maturity model (DMM) as defined in the parliamentary process aims to achieve several objectives, varying in intensity depending on the stakeholder group. It is intended to enhance transparency in multiple ways. Hospitals are given the opportunity to independently assess their level of digitalization, fostering awareness beyond just the IT departments, as future-oriented investment decisions are not made solely within IT. The DMM can also support operational work and decision-making processes.

Hospitals and federal states can also gain insights beyond their own immediate context. Comparing similar hospitals helps identify deficiencies and initiate collaborative improvements. Broad participation enables the identification of best practices and offers insights into how challenges have been addressed. Additionally, hospitals with similar levels of development can be identified, fostering cooperation to achieve mutual benefits.

At the national level, the DR enables an assessment of the funding program's effectiveness. Without the digital maturity model (DMM), demonstrating the concrete impacts of the Program to Future-Proof Hospitals would be challenging.

2.2 Requirements to the Maturity Measurement

The requirements for the maturity level survey were defined in the subsequent implementation of the law. It became evident that merely measuring the success of the progress achieved through the funding would not fully utilize the potential of this tool or justify the effort involved. Consequently, comprehensive requirements were established for the self-assessment tool, particularly concerning its development. The main requirements were:

- *Interdisciplinarity*: The project team responsible for implementation should include a broad range of expertise from various relevant disciplines.
- *Completeness*: The tool should capture structural, process, and outcome quality, covering organizational, technical, and clinical aspects of hospitals. Input from domain experts and the specialist community is essential.
- *International Comparability*: While the model should account for the unique features and complex legal context of the German healthcare system, it must also be based on nationally and internationally recognized maturity models (meta-model) to ensure international comparability.
- *Scientific Criteria*: The model must meet scientific standards, including accuracy, validity, comparability, acceptance, transparency, applicability, effort, traceability, and ongoing development.

3 DigitalRadar Krankenhaus and Beyond

The responsibility to develop and operate such a maturity model was put out to tender to external service providers. The model developed by the consortium that won the tender consists of seven different so-called model dimensions which provide a holistic view of the digital maturity of a hospital (Amelung et al. 2022):

- Structures and Systems: This dimension gives a broad overview on the structure and organizational aspects of the hospital. This includes indicators with respect to capacity, personnel, and finances as well as the availability of selected software applications and allows for contextualization of the data as the type of hospital influences the results.
- Resilience Management and Performance: This dimension covers aspects around organizational and operative, respectively, technical security and the ability to mitigate planned and unplanned disruption of business processes. A large fraction of this dimension targets the area of cybersecurity including measures to ensure confidentiality, integrity, and availability of information.
- Organizational Control and Data Management: This dimension covers aspects of IT infrastructure, data governance, and utilization of data from a clinical but also an administrative perspective.
- Clinical Processes: This dimension captures the extent to which digital resources are integrated into clinical care processes. This covers the entire process from patient admission and treatment to discharge.
- Telehealth: This dimension apprehends capabilities to utilize telemedicine in the medical and nursing services along the entire hospital care spectrum, including intensive acute care, diagnostics, therapy, and rehabilitation.
- Information Exchange: This dimension addresses technical, semantic, and syntactic interoperability. It aims to allow for seamless data exchange on the basis of internationally recognized standards in order to decrease incompatibilities between internal systems within the organization but also with external stakeholders. Ultimately, this shall increase the quality of care, efficiency, and patient safety.
- Patient Participation: This dimension covers the options to which extent patients are able to directly access their own administrative and clinical data and actively interact with the hospital, i.e. with the help of digital patient interaction platforms, digital consent management, or digitally supported anamnesis.

The item collection used in the DR model is a meta-catalog based on several sources, including 65% of the items from EMRAM, requirements from the funding guidelines, and other maturity models like the German KIT-CON model (Dickmann et al. 2020). This catalog consists of 234 items distributed across various dimensions, with each dimension containing a different number of items.

The assessment includes different types of answers: single choice, multiple choice, and, in a few cases, free text. Since the DR is a meta-model, a specific scoring system was developed. Scores range from 0 to 100 points, whereby 100 points indicates full digitalization. Based on their potential impact on patient outcomes, each dimension is weighted differently. Weights were determined through consultations with domain experts. Scores are summed per dimension and adjusted based on their importance to overall digital maturity.

German hospitals offer varying areas of care depending on their official mandate. Some hospitals, for example, are not part of the emergency response system, and others may focus solely on psychiatric and psychosomatic cases. Consequently, certain items may not be relevant for all types of clinics, and the DR score is adjusted accordingly.

An important feature of the model is its international comparability. To this end, an EMRAM indicator score was developed, allowing comparisons with the original full-scale EMRAM model (Amelung et al. 2022). The DR model has been updated for the second measurement in mid-2024 (Konsortium DigitalRadar 2024).

3.1 Measurement and Results

Since the measurement serves both scientific purposes and as a requirement for funding, significant emphasis was placed on ensuring the security of the onboarding process and providing extensive support during data collection. Data collection began on October 19, 2021, via an online platform. By the end of February 2022, nearly 1600 hospitals had completed the questionnaire, including initial quality checks (Amelung et al. 2022).

Ultimately, 1624 hospitals (91% of German hospitals) participated in the first measurement, representing nearly the entire German hospital landscape eligible for funding through the Hospital Future Act. The DR score averaged 33 on a scale from 0 to 100, with scores ranging from a minimum of 2 to a maximum of 64. There is significant variation across dimensions: while the dimensions of Structures & Systems, Resilience Management and Performance, and Organizational & Data Management each scored above 40, the dimensions of Information Exchange, Telehealth, and Patient Participation scored considerably lower. These results correspond to a predicted EMRAM stage of "0" according to the EMRAM indicator score. Among the 1624 participating hospitals, 438 would be classified at EMRAM stage 1, 44 hospitals at stage 2, 14 at stage 3, and 7 at stage 5 (Thun 2024; Table 2).

These results highlight current weaknesses and suggest that the underlying funding framework has prioritized the right areas. The EMRAM indicator score allows for placing the results in an international context. While the German hospital sector has room for improvement, the findings demonstrate that a significant portion of the sector is progressing toward higher digital maturity.

Table 2 DR Score in the model dimensions (Amelung et al. 2022)

Dimension	Mean	SD	Min	Max
Structures and systems	55.3	17.6	4.3	93.9
Resilience management and performance	45.2	16.0	3.8	87.4
Organizational control and data management	40.6	13.4	2.6	84.4
Clinical processes	38.5	14.1	0.7	74.9
Telehealth	18.0	14.1	0.6	76.5
Information exchange	25.1	10.2	0.8	62.2
Patient participation	5.3	9.1	0.3	70.6

SD standard deviation

3.2 Lessons Learned and Transferability to Other Health Care Institutions

With the discussions surrounding a nationwide maturity measurement in the KHZG and the subsequent DR project, Germany has initiated a comprehensive debate on the benefits of DMMs across various areas of the healthcare sector. At the federal level, besides the DR, another maturity model is widely used: the Public Health Service (PHS) maturity model. This model also aims to make progress in digitalization visible.

While extensive knowledge and existing models were available for the hospital sector, this was not the case for the PHS. To address this, a level-based maturity model was developed through a comprehensive stakeholder process. This model defines minimum requirements for the digitalization of PHS facilities. A unique aspect of the PHS is its combination of healthcare and public administration, which have historically operated separately.

The maturity model defines requirements and criteria across five levels (0 to 4) and eight core areas, or "dimensions." Within these dimensions, there are 27 "sub-dimensions." To advance to the next level, an organization must meet at least 80% of the criteria within a dimension. This approach ensures that criteria less relevant to an organization do not hinder progress. Self-assessment is used to evaluate whether an organization meets the defined criteria.

The DR and PHS maturity model are similarly complex and target larger organizations, such as hospitals and PHS institutions, where digitalization involves multiple implementing units. Completing both models requires coordination within the respective organizations to gather the necessary information.

The measurements conducted so far indicate that the extensive criteria catalogs require significant time for institutions to complete the survey. For the DR's first measurement, the survey process generally took between 20 and 40 h per hospital (Amelung et al. 2022). While no data is yet available for the time required for the second measurement carried out in 2024, trends from repeated PHS measurements suggest that subsequent measurements require considerably less time. Specifically, the first PHS measurement averaged around 31 h (Eymann et al. 2022), while the second averaged 10.2 h (Eymann et al. 2023).

A key success factor for the practical application of both models and the usability of the results was making participation mandatory for the respective organizations. In both cases, access to funding was contingent on participation. Another crucial element was providing comprehensive support services, including clarification of questions and information on how to apply the models. Solely relying on written publications without options for operational queries likely reduces acceptance significantly.

Currently, a maturity model for the outpatient practice sector is being tested in Germany. This model is designed for much smaller organizations, typically with fewer employees. It functions primarily as an operational aid to guide the structured digitalization of medical practices (Kassenärztliche Bundesvereinigung Westfalen-Lippe 2024).

The three models demonstrate that the approach of measuring digital maturity can be extended to other areas within the healthcare sector and beyond. However, it is essential to clearly communicate objectives from the outset to ensure necessary acceptance and cooperation among participants and to facilitate broad exchange during model development based on scientific criteria.

From a government perspective, the DR project has provided significant insights and advantages for scaling digital transformation across Germany's hospital sector already. By mandating participation, the project ensured extensive engagement across nearly the entire hospital sector, facilitating comprehensive data collection and analysis that provides valuable insights for future decision-making and policy development. Furthermore, the data is available for scientific research regarding the impact of digital solutions on hospital administration. The cost-effective approach, with an expenditure of approximately €3000 per hospital—including the model and platform development, two rounds of measurement, and scientific evaluation—has strengthened Germany's digital autonomy, equipping the country with the expertise and infrastructure to independently manage and advance its digital healthcare initiatives (Scheuer 2023).

3.3 What Should Happen Next?

Looking ahead, several societal challenges will inevitably impact hospitals. Demographic changes, for instance, will affect hospitals in two principal ways. First, the shortage of skilled labor will make it increasingly difficult for hospitals to recruit suitable employees. Second, the growing prevalence of age-related diseases among an aging population will heighten the significance of addressing these conditions in medical treatment. Technological advancements and innovations, coupled with tailored digital services, present substantial opportunities and considerable potential to address these challenges. These advancements can lead to improved working conditions, more efficient processes, enhanced patient safety, and increased empowerment of citizens. Therefore, it is crucial for hospitals to strategically and effectively navigate technological change and capitalize on the opportunities presented by digitalization. Several key enablers can facilitate this process.

As digitalization progresses, the complexity of IT systems within hospitals also increases. Despite advancements, many hospitals still rely on their internal IT departments to provide a wide range of services. Even though these departments may possess technical expertise, they are often under-resourced and inadequately funded. Consequently, collaboration among hospitals is expected to become increasingly essential. Hospitals will need to specialize in IT services where they excel and procure additional services from partner hospitals or external IT service providers. Such collaborations and IT alliances will require appropriate legal, technical, and organizational frameworks.

In response, the Federal Ministry of Health has introduced a law in 2024 to establish a clear regulatory framework for the secure processing of healthcare data in the cloud (Bundesgesetzblatt 2024). This framework will be further complemented by regulations under the Hospital Care Improvement Act (KHVVG). This act aims to facilitate the formation of care networks among hospitals, enabling them to jointly provide IT services and systems (Bundesministerium für Gesundheit 2024).

Another key enabler is the standardization and harmonization of the IT landscape. Hospitals are intricate organizations with numerous IT systems that have evolved over decades, often lacking interoperability. Consequently, data exchange within a hospital is frequently impeded, leading to situations where information must be re-entered. This redundant data entry introduces several issues, including increased administrative burden, potential errors, and inefficiencies in data management. Addressing these challenges through standardized and interoperable IT systems is essential for improving data accuracy, reducing duplication of effort, and enhancing overall operational efficiency.

To enhance interoperability among IT systems, the Federal Ministry of Health has set a clear objective. Technical standards and norms are developed decentrally by competent stakeholders, based on well-defined requirements and criteria. These standards and norms are centrally reviewed for quality assurance before becoming mandatory for all participants in the healthcare system, ensuring their integration into relevant IT systems. This process of interoperability is overseen by the Competence Centre for Interoperability in the Healthcare Sector (KIG), which is supported by an expert committee known as the Interop Council, along with a broad network of specialists.

This approach is set to be further developed through prospective legislative initiatives. Future legislation should not only establish standards and norms for data storage and transfer but should also mandate reference architectures and functionalities. Reference architectures are essentially "sample models" outlining the IT structure and interactions within information systems. These models are designed to guide manufacturers in developing interoperable and effective systems. Additionally, the legislation should aim to define functionalities essential for all hospitals, insured persons, and patients, such as the ability to access data from an insured person's electronic health record and update treatment information accordingly.

Consequently, DMMs must evolve. Rather than merely verifying the presence of technological artifacts, these models need to measure and transparently demonstrate how digitalization impacts hospital processes. They must also provide a clear pathway for the development and implementation of digitally supported processes.

Ultimately, DMMs also have the potential to become a crucial component in the remuneration framework for hospitals. Currently, the costs associated with digitalization are primarily reflected in reimbursement as either investment or operational expenses. However, these costs' impact on the quality of care is not typically evaluated. To address this gap, it is essential to make the value of digitalization visible and measurable through DMMs and to link this measurement to remuneration.

In practical terms, this could involve providing higher remuneration to hospitals that demonstrate advanced digital maturity compared to those at earlier stages of digitalization. However, implementing this approach presents challenges. It is important to incentivize digital pioneers without penalizing less digitally mature hospitals to the extent that they are discouraged or unable to catch up. Balancing this dynamic will be crucial to ensuring that all hospitals have the opportunity to advance their digital capabilities while recognizing and rewarding those who have made significant strides.

References

Amelung V, Angelkorte M, Augurzky B et al (2022) DigitalRadar. Zwischenbericht. Ergebnisse der ersten nationalen Reifegradmessung deutscher Krankenhäuser. https://www.digitalradar-krankenhaus.de/download/220914_Zwischenbericht_DigitalRadar_Krankenhaus.pdf. Accessed 22 Jul 2024

Bundesamt für Soziale Sicherung (2023) Richtlinie zur Förderung von Vorhaben zur Digitalisierung der Prozesse und Strukturen im Verlauf eines Krankenhausaufenthaltes von Patientinnen und Patienten nach § 21 Absatz 2 KHSFV. https://www.bundesamtsozialesicherung.de/fileadmin/redaktion/Krankenhauszukunftsfonds/20231002_Foerderrichtlinie_Version_04.pdf. Accessed 22 Jul 2024

Bundesamt für Soziale Sicherung (2024) Statistik Krankenhauszukunftsfonds. https://www.bundesamtsozialesicherung.de/fileadmin/redaktion/Krankenhauszukunftsfonds/20240702_Statistik_Krankenhauszukunftsfonds.pdf. Accessed 22 Jul 2024

Bundesgesetzblatt (2024) Gesetz zur Beschleunigung der Digitalisierung des Gesundheitswesens (Digital-Gesetz—DigiG). https://www.recht.bund.de/bgbl/1/2024/101/VO.html. Accessed 22 Jul 2024

Bundesministerium für Gesundheit (2024) Gesetzentwurf der Bundesregierung Entwurf eines Gesetzes zur Verbesserung der Versorgungsqualität im Krankenhaus und zur Reform der Vergütungsstrukturen. https://www.bundesgesundheitsministerium.de/fileadmin/Dateien/3_Downloads/Gesetze_und_Verordnungen/GuV/K/KHVVG_GE_Kabinett.pdf. Accessed 22 Jul 2024

Bundesrechnungshof (2020) Bericht an den Haushaltsausschuss des Deutschen Bundestages nach § 88 Abs. 2 BHO über die Prüfung der Krankenhausfinanzierung durch die gesetzliche Krankenversicherung. https://www.bundesrechnungshof.de/SharedDocs/Downloads/DE/Berichte/2020/krankenhausfinanzierung-gkv-volltext.pdf?__blob=publicationFile&v=1. Accessed 22 Jul 2024

Destatis (2023) Krankenhäuser 2022 nach Trägern und Bundesländern. https://www.destatis.de/DE/Themen/Gesellschaft-Umwelt/Gesundheit/Krankenhaeuser/Tabellen/eckzahlen-krankenhaeuser.html. Accessed 22 Jul 2024

Deutsche Bundesbank (2021) Deutsche Wirtschaft 2020 von Coronavirus-Pandemie schwer getroffen. https://www.bundesbank.de/de/aufgaben/themen/deutsche-wirtschaft-2020-von-coronavirus-pandemie-schwer-getroffen-856036. Accessed 22 Jul 2024

Dickmann F et al (2020) In: Dickmann F, Kümmel K, Oroszi F, Schneider M, Rienhoff O (eds) Der IT-Reifegrad von Krankenhäusern. Verlag Werner Hülsbusch, Fachverlag für Medientechnik und -wirtschaft, Glückstadt

Die Bundesregierung (2020a) Milliardenhilfen beschlossen. https://www.bundesregierung.de/breg-de/aktuelles/konjunkturpaket-geschnuert-1757558. Accessed 22 Jul 2024

Die Bundesregierung (2020b) Besprechung der Bundeskanzlerin mit den Regierungschefinnen und Regierungschefs der Länder am 17. Juni 2020, TOP 3 Bewältigung der Corona-Pandemie. https://www.bundesregierung.de/resource/blob/975226/1761544/f8f0b1e04bb8b8149b332588fd3ba7ab/2020-06-17-bewaeltigung-corona-data.pdf?download=1. Accessed 22 Jul 2024

Eymann T, Heidmann N, Kauffmann AL et al (2022) Ergebnisse der ersten Erhebungswelle zur Erfassung der digitalen Reife der deutschen Gesundheitsämter. https://gesundheitsamt-2025.de/fileadmin/Reifegradmodell/Erster_Zwischenbericht_EvalDiGe_bf.pdf. Accessed 22 Jul 2024

Eymann T, Heidmann N, Kauffmann AL et al (2023) Bericht zu den Ergebnissen der zweiten Erhebungswelle zur Erfassung der digitalen Reife der deutschen Gesundheitsämter und anderer Institutionen des öffentlichen Gesundheitsdienstes. https://gesundheitsamt-2025.de/fileadmin/Reifegradmodell/Zweiter_Zwischenbericht_EvalDiGe_bf.pdf. Accessed 22 Jul 2024

Kassenärztliche Bundesvereinigung Westfalen-Lippe (2024) Digi-Managerin Neue Fortbildung für nicht-ärztliches Praxispersonal. https://www.kvwl.de/themen-a-z/digi-managerin. Accessed 22 Jul 2024

Klauber J, Geraedts M, Friedrich J et al (eds) (2019) Krankenhaus-Report 2019: Das digitale Krankenhaus. Springer, Berlin. https://doi.org/10.1007/978-3-662-58225-1

Konsortium DigitalRadar (2024) Es geht wieder los—Die zweite Erhebungsphase zur Messung des digitalen Reifegrads startet am 15. April 2024. https://www.digitalradar-krankenhaus.de/es-geht-wieder-los-die-zweite-erhebungsphase-zur-messung-des-digitalen-reifegrads-startet-am-15-april-2024/. Accessed 22 Jul 2024

Scheuer A (2023) Germany's Digital Radar: a blueprint for national digital maturity strategies?. https://ictandhealth.com/news/germanys-digital-radar-a-blueprint-for-national-digital-maturity-strategies. Accessed 22 Jul 2024

Thun S (2024) DigitalRadar - Germany's National Digital Maturity Assessment. https://www.himss.org/sites/hde/files/professor-sylvia-thun-digitalradar-germanys-national-digital-maturity-assessment.pdf. Accessed 22 Jul 2024

Digital Maturity Data: Extracting Insights for Health System Management

Justus Vogel, Johannes Hollenbach, Alexander Haering, David Ehlig, and Alexander Geissler

Abstract

With the Hospital Future Act, the German government has initiated a nationwide investment program of more than € 4.3 bn to increase the digital maturity of the country's hospitals. As part of this program, an evaluation project called "DigitalRadar" has been set up. The evaluation goal is to assess whether the funding achieved its goal to increase hospitals' digital maturity. To this end, the DigitalRadar project consortium has developed a digital maturity measure, the DigitalRadar score (DR-score). The DR-score is continuously scaled between 0 (not digitized) and 100 (fully digitized), providing detailed insights into hospitals' digital maturity.

The DR-score allows for detailed analyses yielding different kinds of insights for health care provision and health system management: For instance,

1. transparency of a hospital's digital maturity,
2. the association of hospital profitability and digital maturity, and
3. the relationship between digital maturity and hospital quality.

The original version of the chapter has been revised. A correction to this chapter can be found at https://doi.org/10.1007/978-3-031-80704-6_20

J. Vogel · D. Ehlig · A. Geissler (✉)
Health Economics, Policy and Management, School of Medicine, University of St. Gallen, St. Gallen, Switzerland
e-mail: justus.vogel@unisg.ch; david.ehlig@unisg.ch; alexander.geissler@unisg.ch

J. Hollenbach · A. Haering
RWI – Leibniz-Institut für Wirtschaftsforschung e.V, Essen, Germany
e-mail: johannes.hollenbach@rwi-essen.de; alexander.haering@rwi-essen.de

A. Scheuer, J. Studzinski (eds.), *Digital Maturity in Hospitals*,
https://doi.org/10.1007/978-3-031-80704-6_4

Regarding (1), the transparency of hospitals' digital maturity, the DR-score can be used to understand whether the investment in hospital digitalization from the Hospital Future Act has actually increased digital maturity. Moreover, analyses can reveal what specific aspects of hospital digitalization, e.g., clinical processes or medication management, are more or less developed than others.

To investigate the relationship between digital maturity and (2) profitability or (3) hospital quality, the DR dataset needs to be matched to other datasets, e.g., structured hospital quality report cards. In this article, we describe the commonly used approach for matching hospitals and hospital sites with unique hospital identifiers and site numbers in German secondary datasets. Furthermore, we provide an overview of our findings from two studies investigating the relationship between profitability and digitalization as well as between digitalization and quality.

In the first study, we find a positive, statistically significant relationship between profitability and digital maturity, i.e., more profitable hospitals are digitally more mature. More importantly, membership in a large hospital chain is associated with significantly higher DR-scores. Regarding quality, there is no consistent correlation between digitalization and quality for the two process quality indicators, and only a weak association for the two outcome quality indicators selected for investigation. We plan to challenge these findings and to investigate the effect from within hospital changes in digital maturity on provider quality using longitudinal data once the second measurement period for the DR-score will be completed in the second half of 2024.

To support the generation of insights from digital maturity data for health system management, data quality must improve for matching datasets and for investigating the relationship between digital maturity, hospital profitability, and quality. Researchers must have access to nationwide databases containing financial statement data and quality indicators. Lastly, the digital maturity of hospitals must be measured continuously to monitor whether digitalization will reach its goal in increasing quality and decreasing cost of care.

Keywords

Digital maturity · Hospital Future Act (KHZG) · Hospital quality · Hospital profitability · Digital transformation

Abbreviations

DR-score	DigitalRadar score
KHZG	Krankenhauszukunftsgesetz (German Hospital Future Act)
IK	Institutionskennzeichen (Institution Identification)
SO	Standortnummer (Hospital Site Number)
EMRAM	Electronic Medical Record Adoption Model
EBITDA	Earnings Before Interest, Taxes, Depreciation, and Amortization

1 Background and Political Context in Germany

Digital transformation is a key enabler of a patient-centered, integrated healthcare system. It helps eliminate communication and care disruptions, supports holistic treatment decisions, and connects stakeholders. In this way, digital solutions actively contribute to strengthening patient safety and improving the quality of care.

German hospitals are early adopters of new technologies and have witnessed all stages of digitalization in the past decades, from early lab test protocols through digital imaging archives up to AI-based clinical decision support systems. This history led to a multitude of hard- and software systems of different medical specialties and from different development stages and thus a fragmented system architecture in many hospitals. Moreover, various studies in recent years have highlighted the challenges of digitalization in the healthcare sector and have shown that there is considerable potential for development in many areas in Germany (e.g., standardization, interoperability, patient participation). In particular, the level of digitalization in German hospitals appears to have room for improvement when compared internationally (Stephani et al. 2019).

The German Hospital Future Act (Krankenhauszukunftsgesetz, KHZG) came into effect in October 2020 with the aim of improving the digital infrastructure of German hospitals (Federal Ministry of Health 2024). Among its key features is a €4.3 billion endowment of federal and state funds to invest in digital infrastructure of hospitals. Hospitals were able to apply for support from this fund for various types of investments. Fundable topics include, for instance, digital medication management, (semi-)automated clinical decision support systems, and the digitalization of emergency care.

As part of the KHZG, the German Federal Ministry of Health has commissioned the DigitalRadar Hospital consortium to carry out the evaluation of the funding program.[1] The overall goal is to assess the status quo of the digital maturity of German hospitals in an initial survey and to measure the progress made as a result of the investments initiated by the funding program in follow-up surveys (Amelung et al. 2022).

2 The DigitalRadar Score

To evaluate the funding program, a digital maturity model tailored to the German hospital landscape was developed and transformed into a tool that enables the standardized measurement of digital maturity of German hospitals based on surveys answered directly by hospitals themselves. This tool is called the "DigitalRadar score" or simply DR-score.

[1] See project partners on the DigitalRadar Hospital consortium website http://www.digitalradar-krankenhaus.de.

The development of the digital maturity model and the DR-score started with in-depth research on digital maturity models used in Germany and in other countries. Moreover, key factors linked to hospital digitalization were compiled in a criteria catalog. These criteria were organized into seven dimensions for a holistic view of digital maturity: Structures and Systems, Resilience Management and Performance, Organizational Governance and Data Management, Clinical Processes, Telehealth, Information Sharing, and Patient Participation. Each dimension is structured into two to eight subdimensions, and overall, there are 27 subdimensions. An expert advisory board, including representatives from hospitals, statutory health insurances, and medical associations, oversaw all development steps. After compiling the criteria, a pilot phase began on August 19, 2021, involving 12 hospitals. Weekly webinars, along with mail and phone support, were provided to address questions and offer feedback. Post-pilot, participants also tested the accompanying evaluation of the DigitalRadar survey. A standardized questionnaire and interviews with representatives from four hospitals gathered feedback, which was used to finalize the main survey.

The DR-score ranges from 0 (not digitized) to 100 (fully digitized). Hospitals were required to complete the survey used to calculate the DR-score in order to receive funding from the endowment established through the KHZG. The initial survey was completed by a total of 1624 hospitals between October 5, 2021, and December 17, 2021. The survey was conducted on a custom platform. During survey completion, the DigitalRadar Hospital Consortium offered technical and content support via phone, email, and biweekly video conferences. FAQs, recordings, and presentations were also made available and regularly updated online.

After data collection, quality control was conducted, and hospitals were consulted to correct any implausible information. The results were made available to the hospitals via a dashboard that allows hospitals to benchmark themselves with their peers.

Following the publication of an interim report on the status quo of digital maturity in German hospitals (Amelung et al. 2022), the survey was expanded to include several new questions that reflect current trends, but the core of the survey remained unchanged to allow for a valid comparison between the two measurements. Data collection for the second survey (243 questions) began in April 2024.

Regarding the calculation of the DR-score, points are awarded for each answer selected. A maximum of one point can be awarded per question. In order to calculate the DR-scores, first the degree of fulfillment is calculated in each of the seven dimensions identified in the criteria catalog. This is done by dividing the sum of points received by the number of highest possible points in each dimension. To reflect differences in the importance of each of the dimensions for hospital digitalization, the degree of fulfillment is given an additional weighting before they are aggregated to form the final DR-score. The weights were determined by the expert advisory board described above.

Germany is characterized by a very heterogeneous hospital landscape. To ensure comparability between hospitals, the calculation distinguishes between several hospital types. On the one hand, a distinction is made between somatic and psychiatric/

psychosomatic hospitals, as these have very different requirements, structures, and processes. For example, some questions cannot be applied to psychiatric/psychosomatic hospitals and are therefore excluded in the calculation of their DR-score. Furthermore, questions regarding emergency care are marked as irrelevant for hospitals without an emergency care unit. Hospitals also have the option to mark some questions related to special circumstances (e.g., network integration of robotic assistance systems) as not applicable to their operation, which is also considered in the calculation. Still, even if irrelevant questions are excluded, the DR-score of each hospital is normalized to the DR-score range of 0 to 100 points. As a result, flexible scoring is possible while still allowing for comparability across the heterogeneous hospital sector.

The DR-score is a measure of the degree to which a broad set of normative criteria for a "digitally mature hospital" is met. However, it works differently than tools such as EMRAM,[2] which measure digital maturity using eight levels (0 to 7), each of which specifies a narrow set of normative criteria. To reach the next EMRAM level, a set of criteria must be met (Pettit 2013). As a result, there is much less variation and bunching at certain (higher) levels.

The DR-score, on the other hand, can capture nuanced differences and provides a more natural representation of digital maturity. As a result, hospitals are quasi-normally distributed across the DR-score range, with an average of 33 out of 100 points and roughly 70% between 23 and 44 points (Fig. 1). Figure 1 also illustrates the distribution of DR-scores by type of care, ownership, and size. Psychiatric hospitals on average tend to score only slightly higher than acute somatic care hospitals. Public hospitals tend to have higher DR-scores than private and nonprofit hospitals. Lastly, large hospitals are more digitally mature than small hospitals.

Hospitals that underperform in one dimension can compensate and achieve a good DR-score by performing well in another. However, the DR-score is not a black box, as the degree of fulfillment in each of its dimensions and subdimensions can be used to identify strengths and weaknesses, both at the hospital level and at the aggregate (system) level.

Overall, 1624 hospitals participated in the initial survey, representing 91% of all German hospitals, thus providing a comprehensive picture of the German hospital landscape. Sample characteristics in terms of hospitals' type of care, teaching status, ownership, size, and level of emergency care are presented in Fig. 2.

3 Matching of Digital Maturity Data with Other Data Sources

With the DR-score collected for more than 90% of all hospitals, Germany stands out as one of few countries where hospitals' digital maturity can be investigated in depth at a national level.

[2] Electronic Medical Record Adoption Model provided by HIMSS.

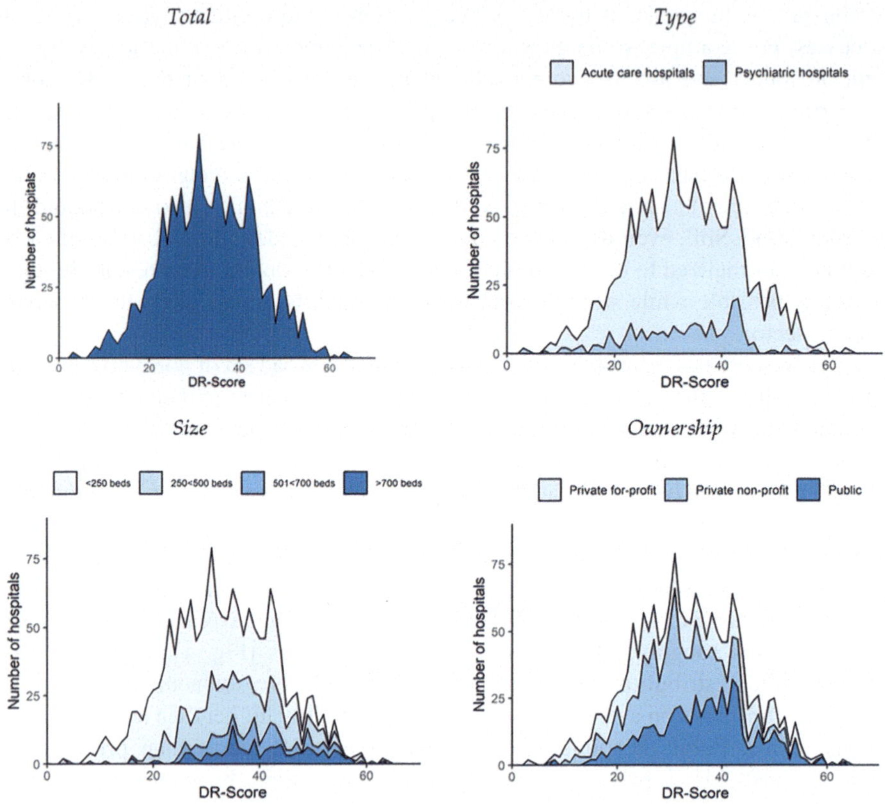

Fig. 1 DR-Score histograms total and by hospital type, ownership and size. *Note*: 1624 observations in each panel

However, research questions relating digital maturity to other outcomes, such as hospital profitability or quality, require additional data sources. In Germany, such data sources exist in the form of financial statements data from the DAFNE database (Bureau van Dijk 2021), used, for instance, for an annual scientific report on the financial state of the German hospital landscape, called the Hospital Rating Report (Augurzky et al. 2022), or the structured hospital report cards as required by the External Inpatient Quality Assurance program (Freistühler et al. 2018).

These data sources can be matched using unique hospital identifiers, the so-called Institution Identification (Institutionskennzeichen, IK) and the hospital site number of the German Hospital Directory (Standortnummer, SO). A hospital or (small) hospital group can have multiple sites, which in many cases are separate hospitals operating under the same IK number (e.g., in the case of hospitals owned by a county or in municipal ownership). Each hospital site is given an SO. Matching hospital sites with the IK and the SO across different datasets works well for most hospitals. A considerable share, however, cannot be matched using IK and the SO alone, e.g.,

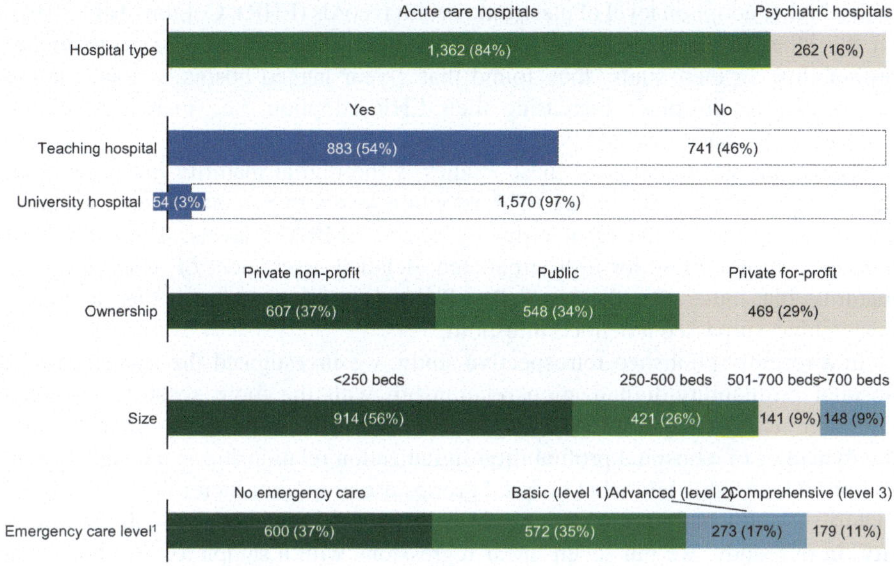

Fig. 2 Hospital sample characteristics of the DR sample ($n = 1624$)

some data are reported on IK level and some data are reported on SO level. Hence, a decision must be made whether these hospitals are excluded, if the data reported on IK level are used for all sites of the IK, or if the values of the sites of an IK are averaged. Another problem arises because SO number entries are sometimes missing, or they are mistyped. In these cases, additional information such as address data needs to be used for matching.

We performed matching as described above to investigate the relationship between profitability and digital maturity as well as digital maturity and quality. In both cases, we needed to address additional challenges unique to the matched datasets which we describe in the following two chapters.

4 Association of Hospital Profitability and Digital Maturity

When considering the association of hospital profitability and digital maturity, both a (1) positive profitability-digitalization and a (2) positive digitalization-profitability relationship seem plausible. That is because (1) hospitals might use profits to invest in digitalization, e.g., to increase the efficiency and quality of their processes and to cut costs in the mid- to long-run. This might, in turn, lead to (2) higher profits for more digitized hospitals.

Both a profitability-digitalization (Shen and Ginn 2012; Ginn et al. 2011) and a digitalization-profitability (van Poelgeest et al. 2015; Collum et al. 2016) relationship have been investigated in the literature. Shen and Ginn (2012), for instance, found that higher hospital profitability, measured as total margin, was associated with a higher adoption level of electronic health records (EHR). Collum et al. (2016), on the other hand, investigated whether stronger EHR adoption leads to higher profitability. In their study, they found that 2-year lagged hospitals' total margin was higher for hospitals increasing their EHR adoption, i.e., increasing digital maturity in the base year had positive financial effects 2 years later.

One common limitation of these studies is the digital maturity indicator used. Studies rely on ordinally scaled EHR adoption levels (e.g., comprehensive, basic, and no adoption) or the eight ordinally scaled EMRAM levels. These types of measures do not allow for a differentiated, detailed assessment of hospital digital maturity but rather—as the term "level" implies—showcase whether a certain hospital has surpassed a defined threshold.

In a recently published retrospective study, we investigated the dynamics of a hospital profitability-digitalization relationship with the more sensitive, detailed DR-score as digitalization measure (cf. above) (Vogel et al. 2024b). Investigating the dynamics of a hospital profitability-digitalization relationship in a health system can provide insights into what hospital groups use profits for increasing their digital maturity—and what other factors might be associated with hospitals' digital maturity. In our study, we ran seven linear regressions with a sample of 860 hospitals, providing both DR and profitability data. In each new regression, we added a different control variable or changed control variables altogether to analyze the association of hospitals' profitability with the DR-score and to identify other hospital characteristics associated with digital maturity. In addition, we ran a sub-group analysis stratifying hospitals by their chain size.

Our main findings were:

- Profitability is positively and statistically significantly associated with digital maturity. We found this relationship to be relatively small in magnitude, however: In our first regression, for an average hospital, an increase of its EBITDA-margin[3] from 4.6% to 5.6% (i.e., a 21.7% increase) was associated with a 1.0% increase of its DR-score, namely by 0.36 points from on average 34.90 to 35.26. Depending on the set of independent variables considered, this magnitude changed only slightly.
- Chain membership and especially membership in a large chain had the strongest positive association with digital maturity. Belonging to one of the largest three hospital chains was associated with a 11.24 points higher DR-score, for instance.
- Within large hospital chains, profitability is more strongly associated with digital maturity.
- In an additional analysis of DR-score items we saw that hospitals belonging to large chains more frequently have a (central) IT strategy. We concluded that this might explain why large chains are digitally more mature and additionally hypothesized that they might benefit from cross-financing between hospital sites and central project management resources and structures.

[3] Earnings before Interest, Taxes, Depreciation, and Amortization.

In the initial DR-score sample, there are 1624 hospital sites (cf. above). Our final sample "only" contained 860 hospital sites, however. Indeed, for our study, one main challenge was to find and match DR-scores and EBITDA-margins at hospital site level. While there was no financial statement data available for a large part of the DR-score sample, for the analyzed 860 hospital sites, we could only exploit 510 financial statements. For hospital chains, we could only access financial statement data at chain level and therefore had to use the same EBITDA-margin for all hospitals that belonged to the same chain. While this did not make our analysis infeasible, it limits our study and makes interpretation of results more complex (Vogel et al. 2024b).

To comprehensively assess the dynamics of the association of hospital profitability and digital maturity, detailed financial statement data must be made available to research. Such data could also be used to investigate a potential digitalization-profitability relationship, as done by Wang et al. (2018) and Collum et al. (2016) using large hospital samples from the United States, for instance.

5 The Association of Hospital (Process) Digitalization and Quality of Care

Hospital digitalization is expected to increase the efficiency of health service delivery, reduce costs, and/or enhance the quality of care. Especially with the promise that more digitized processes improve the delivered quality in hospitals, several countries set up legislations to fund digitization efforts (e.g., the Hospital Future Act in Germany or the HITECH Act in the United States of America). While there is evidence that specific digitization initiatives such as the implementation of clinical decision support systems for specific medical conditions and in single center settings have improved quality of care (Bright et al. 2012), empirical results on large-scale retrospective observational studies on the relationship between digitalization and quality have shown inconclusive or insignificant results (van Poelgeest et al. 2015; Martin et al. 2019; Agha 2014). These latter results could potentially be explained by (1) no clear logical association between the digital maturity indicator and the used quality indicator, (2) the sensitivity of the digital maturity indicator to detect changes in digitalization, and (3) the sensitivity of the quality indicator to detect quality differences and to pick up the effects from the digital maturity changes at hospital level.

To address the aforementioned issues, we used the data from the DigitalRadar project from 2021 and linked it with selected quality indicators from the German structured quality reports. The selection of quality indicators, as well as the relevant (sub-)dimensions of digitalization, was based on a logical construct, i.e., which digitalization measure is likely to influence process or outcome quality for selected medical conditions. To increase the robustness of the measured quality indicators per hospital, we averaged quality indicator values over 2020 and 2021. We isolated two processes—pre-operative waiting time for osteosynthesis and hip replacement surgery after femur fracture—and two risk-adjusted outcome quality indicators—inpatient mortality ratio of patients hospitalized for outpatient-acquired pneumonia, ratio of new cases with decubitus—out of a comprehensive set of 155 (2020) and 158 (2021) published quality indicators (Institut für Qualitätssicherung und Transparenz im Gesundheitswesen (IQTIG) 2021, 2022). Sample sizes ranged between 665 and 1566 hospital sites, depending on the indication. For each indication, we ran a set of multivariate linear regressions with the DR-score, or relevant subdimensions as explanatory variables and the respective quality indicator as dependent variable (cf. preliminary results in our working paper (Vogel et al. 2024a)).

From our preliminary results, we can deduce two main take-aways:

- We did not find a consistent digitalization-quality relationship across several DR-score subdimensions for any of the selected quality indicators. For the two process quality indicators for femur fracture, all associations were insignificant. For the two outcome quality indicators, selected subdimensions were (weakly) associated with better (or worse) quality.
- More concretely, in the case of patients hospitalized for outpatient-acquired pneumonia, we found that higher (process) digitalization is associated with better outcomes. With regard to the ratio of new cases with decubitus/ulcers, greater process digitization was associated with worse quality, albeit rather weak in magnitude. This negative association could also be due to better documentation quality of decubitus/ulcer cases due to more digitized processes—and thus, these results have to be interpreted with care.

Our inconclusive results might be due to the insufficient statistical sensitivity of the (process) quality indicators and their inability to reflect digitalization effects, leading to the conclusion that it is necessary to refine quality indicators to be able to detect effects from higher digital maturity on quality. For instance, pre-operative waiting time for hip replacement/osteosynthesis after femur fracture is assessed on how many cases fail to be operated within 24 h after admission and not on a continuous variable such as pre-operative waiting time in hours or minutes. Accordingly, a large share of hospitals in our sample has "perfect" quality of zero cases operated after 24 h after admission.

Limitations in Germany arise due to the difficulty to match quality indicators from the structured quality reports with data from the DigitalRadar project due to occasional measurements on different levels (hospital vs. hospital site). In the future, these sources should be harmonized on the level of measurement and include the same unique identifier.

Lastly, we might see effects on quality from digitalization changes within hospitals, which cannot be investigated with our cross-sectional dataset but are only detectable using a longitudinal dataset. The second measurement of the DigitalRadar project in 2024 and its evaluation might thus yield different results.

6 Conclusion

With the DR-score, researchers and practitioners can utilize a very detailed, continuously scaled digital maturity indicator to investigate and understand the digital maturity of all care-relevant hospitals in Germany. While this dataset is unique in its breadth (details on digital maturity) and length (number of observations), it needs to be matched to other datasets to investigate research questions linked to the goals of hospital digitalization, e.g., improving quality.

So far, we could only investigate associations of digital maturity with profitability and quality with cross-sectional data. As a research outlook, investigating causal relationships will become possible once panel data is available. For the DR-score, the second data measurement takes place in 2024 and a third one is planned (cf. above). We plan to use these data to test the positive profitability-digitalization relationship we found (Vogel et al. 2024b), to analyze a digitalization-profitability relationship as in Collum (2016) and to further investigate a potential digitalization-quality relationship (Vogel et al. 2024a).

In conclusion, we argue for a continuous measurement of hospital digital maturity. This is needed to monitor the development of hospitals' digital maturity in Germany and digitalization goals at system level such as higher quality and lower costs of care.

Acknowledgments We thank all our colleagues of the Konsortium DigitalRadar Krankenhaus for the constructive project work, and the German Federal Ministry of Health (BMG) as well as all hospitals participating in the evaluation project. Special thanks to Astrid Wering.

Conflict of Interest Both the Chair of Health Economics, Policy and Management of the University of St. Gallen and the RWI—Leibniz-Institut für Wirtschaftsforschung e.V. received funding for their work in the DigitalRadar evaluation project from the German Federal Ministry of Health. The digitalization funding program of the Hospital Future Act is part of the EU Program "NextGeneration."

References

Agha L (2014) The effects of health information technology on the costs and quality of medical care. J Health Econ 34:19–30. https://doi.org/10.1016/j.jhealeco.2013.12.005

Amelung V, Angelkorte M, Augurzky B et al (2022) Zwischenbericht - Ergebnisse der ersten nationalen Reifegradmessung deutscher Krankenhäuser. https://www.bundesgesundheitsministerium.de/service/publikationen/details/digitalradar-zwischenbericht. Accessed 1 July 2024

Augurzky B, Krolop S, Hollenbach J et al (2022) Krankenhaus Rating Report 2022: Vom Krankenhaus zum Geisterhaus? Heidelberg

Bright TJ, Wong A, Dhurjati R et al (2012) Effect of clinical decision-support systems: a systematic review. Ann Intern Med 157:29–43. https://doi.org/10.7326/0003-4819-157-1-201207030-00450

Bureau van Dijk (2021) DAFNE database - German company data & financials. https://www.bvdinfo.com/en-gb/our-products/data/national/dafne. Accessed 17 Mar 2022

Collum TH, Menachemi N, Sen B (2016) Does electronic health record use improve hospital financial performance? Evidence from panel data. Health Care Manag Rev 41(3):267–274. https://doi.org/10.1097/HMR.0000000000000068

Federal Ministry of Health (2024) Krankenhauszukunftsgesetz für die Digitalisierung von Krankenhäusern. https://www.bundesgesundheitsministerium.de/krankenhauszukunftsgesetz. Accessed 1 July 2024

Freistühler M, Roeder N, Franz D (2018) Institute for quality assurance and transparency in the healthcare system. Z Herz Thorax Gefasschir 32:331–338. https://doi.org/10.1007/S00398-018-0231-0/TABLES/1

Ginn GO, Shen JJ, Moseley CB (2011) Hospital financial position and the adoption of electronic health records. J Healthc Manag 56(5):337–352. https://doi.org/10.1097/00115514-201109000-00009

Institut für Qualitätssicherung und Transparenz im Gesundheitswesen (IQTIG) (2021) Öffentliche Berichterstattung von Ergebnissen der externen stationären Qualitätssicherung in den Qualitätsberichten der Krankenhäuser. Empfehlungen zum Erfassungsjahr 2020. Berlin

Institut für Qualitätssicherung und Transparenz im Gesundheitswesen (IQTIG) (2022) Öffentliche Berichterstattung von Ergebnissen der externen stationären Qualitätssicherung in den Qualitätsberichten der Krankenhäuser. Empfehlungen zum Erfassungsjahr 2021. Berlin

Martin G, Clarke J, Liew F et al (2019) Evaluating the impact of organisational digital maturity on clinical outcomes in secondary care in England. NPJ Digit Med 2:41. https://doi.org/10.1038/S41746-019-0118-9

Pettit L (2013) Understanding EMRAM and how it can be used by policy-makers, hospital CIOs and their IT teams. World Hosp Health Serv 49(3):7–9

Shen JJ, Ginn GO (2012) Financial position and adoption of electronic health records: a retrospective longitudinal study. J Health Care Finance 38(3):61–77

Stephani V, Busse R, Geissler A (2019) Benchmarking der Krankenhaus-IT: Deutschland im internationalen Vergleich. In: Klauber J, Geraedts M, Friedrich J, Wasem J (eds) Krankenhaus-Report 2019. Springer, Berlin, pp 17–32. https://doi.org/10.1007/978-3-662-58225-1_2

van Poelgeest R, Heida JP, Pettit L et al (2015) The association between eHealth capabilities and the quality and safety of health Care in The Netherlands: comparison of HIMSS analytics EMRAM data with Elsevier's "the best hospitals" data. J Med Syst 39:90. https://doi.org/10.1007/S10916-015-0274-7

Vogel J, Haering A, Kuklinski D et al (2024a) Is there a relationship between hospital process digitalization and quality of care?—Observational study using data from hospital quality report cards and the DigitalRadar Project. Essen. https://doi.org/10.4419/96973260

Vogel J, Hollenbach J, Haering A et al (2024b) The association of hospital profitability and digital maturity—an explorative study using data from the German DigitalRadar project. Health Policy (New York) 142. https://doi.org/10.1016/J.HEALTHPOL.2024.105012

Wang T, Wang Y, McLeod A (2018) Do health information technology investments impact hospital financial performance and productivity? Int J Account Inf Syst 28:1–13. https://doi.org/10.1016/J.ACCINF.2017.12.002

Digital Maturity Strategy for Modernizing Dubai's Healthcare Ecosystem

Osama El-Hassan, Mahera AbdulRahman, Mohammed Al Redha, and Armin Scheuer

Abstract

This paper reflects the 10-year journey of the Emirate Dubai in utilizing a digital maturity model to assess the adoption of electronic medical record (EMR) in healthcare facilities in the Emirate. The adoption model has been exploited as a planning tool to assess the impact of the various digital health investments made by these facilities to improve efficiency, quality, and patient outcomes. The maturity assessment was also utilized to inform on the readiness of these healthcare facilities for the implementation of the Emirate's Health Information Exchange Platform. The research study provides a thorough analysis of the Dubai EMRAM survey results between 2011 and 2022 and highlights the rationale behind the fluctuation of EMRAM average scores across the different years of implementation.

Keywords

Electronic medical record · Maturity models · Interoperability · Benchmarking

Abbreviations

CDS	Clinical Decision Support
CPOE	Computerized Provider Order Entry
EHR	Electronic Health Record

O. El-Hassan (✉) · M. AbdulRahman · M. Al Redha
Health Informatics & Smart Health Department, Dubai Health Authority, Dubai, UAE
e-mail: oeelhassan@dha.gov.ae; mrad@dha.gov.ae; maalreda@dha.gov.ae

A. Scheuer
Lemonmint HealthTech Communications, Berlin, Germany
e-mail: armin.scheuer@lemonmint.org

A. Scheuer, J. Studzinski (eds.), *Digital Maturity in Hospitals*,
https://doi.org/10.1007/978-3-031-80704-6_5

eMAR	Electronic Medicines Administration Records
EMR	Electronic Medical Record
EMRAM	EMR Adoption Model
HIE	Health Information Exchange
HIT	Health Information Technology
HITECH	Health Information Technology for Economic and Clinical Health
NABIDH	Network Analysis Backbone for Integrated Dubai Health

1 Introduction

A distinct feature of health systems worldwide over the past 30 years has been the increased utilization and reliance on Information and Communications Technology (ICT). Also called Health Information Technology (HIT) or digital health, these developments are being pursued because of their potential to transform the delivery of health care by improving efficiency, safety as well as the experience of both physicians and patients. Digital health applications can be seen in various settings but one of the most effort-consuming and impactful is the Electronic Medical Record (EMR).

An EMR is a digital version of a patient's paper medical chart and personal information that contains a patient's medical history, diagnoses, medications, treatment plans, immunization dates, allergies, radiology images, laboratory test results, etc.

The goal of the EMR is to facilitate a clinical environment that is paper-free yet include all clinical workflows across all departments and secure and efficient communication between clinicians and patients. Successful inclusion of clinical workflows and supporting collaborations between clinical teams and patients lays the basis for improving efficiency and leveraging healthcare outcomes.

Despite the clear benefits of EMR systems for healthcare facilities, the procurement and the implementation of such systems are always problematic. EMR systems usually require a substantial investment. Both the solution provider and the healthcare facility have to work hand in hand to tailor and customize the solution to the existing clinical workflows and to address key obstacles to enabling the smooth transition to the new system such as data migration, training and compliance to enterprise and regulatory requirements. Additional aspects may exacerbate the cost such as maintenance and system upgrade.

Many researchers have reported a net return of EMR investment in terms of cost saving, efficiency, and improved outcomes (Delpierre et al. 2004). A number of systematic reviews have been accumulated and published. However, these reviews presented mixed-messages on the actual benefits of EMR implementation compared to their expenses. One review found that while EMR may improve user and patient satisfaction rates, their impact on patient outcomes and quality of care was inconclusive (Delpierre et al. 2004). A flagship research study following a comprehensive review of the published literature on the benefits and costs of EMR concluded that, while in individual pilot sites there was clear evidence that health care had been made safer, more effective, and more efficient, the findings were not generalizable to other settings (Shekelle et al. 2006). In particular, there was a lack of knowledge about what specific systems and implementation methods should be adopted. In particular, much of the positive evidence came from bespoke systems developed at academic medical centers or in very large health systems (Garrido et al. 2004). These insights clearly show that the positive impact of EMR implementations depends heavily on the accumulated experience of the implementers and organizational digital maturity within the healthcare facility.

The adoption of EMR in healthcare facilities has significantly increased in the last decade (Kanakubo and Kharrazi 2019). In the Gulf Cooperation Countries (GCC) region, the adoption of EMR solution has been initiated many years ago although the pace of change was not expeditiously flat-out (AlSadrah 2020; Musa et al. 2023). In many cases, the decision to move to EMR solutions was a response to regulatory requirements such as subscribing to insurance claim clearing-houses, or public health reporting. Alternatively, in some other cases, the EMR implementation was a corner-stone of an internal digital health strategy of healthcare facilities to achieve excellence in operations and care delivery processes.

Conversely, in other parts of the world such as the USA, regulators were more proactive in disseminating EMR adoption. The implementation of EMRs is considered as an instrument to tame the increasing costs of healthcare and as such has become a strategic goal (Payne et al. 2013). In the USA, the 2009 Health Information Technology for Economic and Clinical Health (HITECH) Act allows the use of tax-payers' money to provide incentive payments to early adopters of EMR solutions with providers (Burke 2010). To qualify for these payments, physicians must have an operationally "certified" or approved EMR system that is capable of collecting, managing, and sharing specified data sets and performance measures which together constitute a well-defined criterion that is known as "meaningful use."[1] It includes as a minimum e-prescribing and electronic exchange of health information to improve quality. This requirement has led to considerable new research efforts to assess the proportion of providers that are achieving meaningful use and the characteristics of those that have done so (Patel et al. 2013).

The prevalence of such nation-wide incentive programs for digital health transformation developed the necessity for digital maturity assessment methodologies to set and assess improvement goals and priorities for these transformations. Additionally, these methodologies should provide guidance for step-wise improvements that comprises all ancillaries and clinical departments of a health facility. Last but not the least, they should provide a benchmark for appraising current outcomes of these transformations.

[1] https://www.ama-assn.org/practice-management/medicare-medicaid/meaningful-use-electronic-health-record-ehr-incentive.

Digital health maturity is defined as the level to which digital health systems are alleviated to deliver high-quality health care, resulting in improved services for an enhanced patient experience (Martin et al. 2019). Assessing digital maturity is of paramount importance in hospital settings due to the complexity and cost of healthcare delivery that includes multidisciplinary aspects both on acute and chronic cases (Eden et al. 2019). To evaluate digital maturity, a maturity model-based assessment methodology can be used to allow an organization to evaluate its current digital status across a chain of dimensions (Johnston 2017). Nevertheless, limitations exist in current approaches for assessing digital maturity in hospitals, as there is a lack of agreement over which capabilities should be assessed (Krasuska et al. 2020). There are a number of adoption and maturity models available to measure the level of implementation of Health Information Exchange (HIE) (Carvalho et al. 2016; Duncan et al. 2022).

This research study investigates a 10-year city-wide implementation carried by the Dubai Health Authority (DHA), which is the healthcare regulator of the Emirate of Dubai, using a digital health maturity model assessment methodology developed by the Healthcare Information and Management Systems Society (HIMSS). The Electronic Medical Record Maturity Model (EMRAM) from HIMSS defines various stages of maturity in terms of EMR's adoption level by healthcare providers. These stages are rated on a scale of 0 to 7, with 0 being the least mature stage and 7 being the most mature. EMRAM comprises five focus areas: Data Capture & Health Information Exchange (HIE), Patient Engagement, Healthcare Analytics & Outcomes Measurement, Resilience Management, and Clinical User Adoption. According to the HIMSS EMRAM website,[2] more than 100 indicators are used by the EMRAM to assess an organization's progress within these five focus areas.

Electronic Medical Records (EMRs) have been adopted at various healthcare facilities in Dubai since the early 1990s. Until 2011, there were no policies or regulations that nudged healthcare facilities to adopt automation within their healthcare services. The introduction of automated Emirate-wide healthcare insurance claim clearing-house in 2013 was a game changer and forced healthcare facilities to adapt their plans to enable digital submission of claims which entails having EMRs that are capable of proper standardized clinical documentations. The EMRAM survey methodology, initially developed in 2007 in the USA and then provided to healthcare facilities internationally, provided a consensus-based platform for benchmarking improvements in EMR adoption and exchanging best practice in the digital health field.

In Dubai, a policy was released by Dubai Health Authority in January 2021 mandating EMR implementation in all healthcare facilities including hospitals, outpatient clinics, dental clinics, pharmacies, labs, and rehabilitation facilities to facilitate the roll-out of the city-wide Health Information Exchange platform. The COVID-19 pandemic put a tremendous pressure on healthcare facilities to fast-track

[2] https://www.himss.org/what-we-do-solutions/maturity-models-emram.

their automation plans to address the pressing needs for public health notification and extending healthcare services beyond hospital walls (Levene et al. 2020).

The city-wide Health Information Exchange, namely NABIDH which stands for the Network Analysis Backbone of Integrated Dubai Health, went through a long journey that was initiated during a ministerial retreat in 2013. NABIDH is a healthcare platform that aims to establish a unified longitudinal record for all patients in Dubai and to securely exchange trusted healthcare information across public and private facilities. The objective of NABIDH is to allow healthcare providers to construct a 360° view of patient information to support clinical decisions and improve healthcare outcomes.

2 Methods

This study reflects the 10-year journey of the Emirate-wide use of the EMR Adoption Model (EMRAM), a self-reported digital maturity assessment tool that has been developed by HIMSS and deployed in a partnership with Dubai Health Authority across the Emirate's hospitals. Participating healthcare providers, comprising public and private hospitals in Dubai, submit their data through the EMRAM web-based tool and are then ranked at one of eight stages which cover the spectrum of EMR implementation from stage zero (a paper-based environment) to stage seven (an environment that is highly interconnected, orchestrated, and paper-free). Participants are also given an overall EMRAM Score which combines the current stage with credit for partially achieving higher stage applications. Healthcare facilities that achieve EMRAM Stage 6 and higher are validated via an on-site assessment and therefore usually exempted from the Dubai EMRAM online survey as long as their assessment is valid—which is typically for a period of 3 years.

The requirements of each stage are shown below in Table 1. HIMSS and DHA teams are responsible for setting up the EMRAM cycle or data collection round. They also follow up with participants to make sure that they complete the survey on-time and to the required standards. The teams also conduct validation procedures to ensure the eligibility of participants who are ranked at stage five or above; these include site visits and telephone contact with chief information officers (CIO). Occasionally contacting software vendors, who are involved in establishing or upgrading participants' EMR systems, is required. Data collection rounds of the EMRAM survey have been executed on an annual basis between 2011 and 2022 apart from the year of 2018 during which the survey was suspended due to some administrative challenges.

3 Results

Between 2011 and 2017, the average EMRAM score among Dubai hospitals has shown a steady increase from (1.9: 2011) to (4.0: 2017). Nevertheless, with major EMRAM criteria updates in January 2018, HIMSS significantly raised the

Table 1 Situational analysis of EMR maturity model within hospitals in Dubai; from 2011 to 2022

		2011	2012	2013	2014	2015	2016	2017	2019[a]	2020	2021	2022
EMRAM	Cumulative capabilities	n (%)										
Stage 7	Complete EMR, external health information exchange (HIE), data analytics, governance, disaster recovery, privacy and security	0	0	0	0	0	0	0	0	0	0	0
Stage 6	Technology-enabled medication, blood products and human milk administration, risk reporting, full clinical decision support (CDS)	0	0	0	0	0	0	2 (6)	4 (12)	4 (10)	6 (15)	0
Stage 5	Full physician documentation, using structured templates, intrusion/device protection	0	2 (7.5)	6 (21)	6 (21)	10 (31)	11 (33)	14 (42)	6 (18)	11 (28)	11 (28)	9 (21)
Stage 4	50% of all medical orders are placed using CPOE and is supported by a CDS, Nursing/ allied health professional documentation has reached 90%, and basic business continuity	1 (4)	2 (7.5)	1 (3.5)	1 (3.5)	1 (3)	1 (3)	0	1 (3)	3 (8)	5 (13)	13 (30)
Stage 3	50% of nursing and allied health documentation, The eMAR application is implemented, and role-based security	1 (4)	1 (4)	2 (7)	1 (3.5)	1 (3)	1 (3)	1 (3)	0	0	0	1 (2)
Stage 2	Major ancillary clinical systems are enabled with internal interoperability feeding data to a single clinical data repository (CDR) and basic security	7 (30)	10 (37)	10 (36)	12 (41)	12 (38)	17 (52)	15 (46)	0	0	0	2 (5)
Stage 1	All major ancillary clinical systems are installed (laboratory, pharmacy, radiology, and cardiology).	9 (39)	9 (33)	8 (29)	6 (21)	6 (19)	1 (3)	0	10 (29)	7 (18)	5 (13)	0
Stage 0	All three ancillaries are not installed	5 (22)	3 (11)	1 (3.5)	3 (10)	2 (6)	2 (6)	1 (3)	13 (38)	14 (36)	12 (31)	18 (42)
	No. of participated hospitals	**23**	**27**	**28**	**29**	**32**	**33**	**33**	**34**	**39**	**39**	**43**

[a] In 2018, the EMRAM criteria were updated. This is the reason for the drop of EMRAM scores between 2017 and 2019. The annual survey was temporarily suspended in 2018

automation bar to advance the technology adoption and to enable data-informed clinical processes, which incurs a substantial drop in EMRAM score to (2.2: 2019). Between 2019 and 2021, average EMRAM scores of Dubai hospitals increased from (2.2: 2019) to (2.6: 2022) as shown in Table 1. However, it is not possible to directly compare the current EMRAM scores with the results of the evaluation from the previous year (average 3.0) due to modified criteria of the EMRAM assessment methodology in the year 2022.

More than 50% of hospitals in Dubai were compliant with EMRAM Stage 4 and above standards (#21) as shown in Table 1. These hospitals employ Computerized Provider Order Entry (CPOE) and Electronic Medicines Administration Records (eMAR); have a clear clinical and information governance, and keep track of both clinical results and patient satisfaction goals. On the other hand, around 42% of the EMRAM Stage 0 surveyed hospitals need to intensify their digital transformation initiatives in order to maximize their organizational performance and patient health outcomes. About 7% (#3) of the hospitals that achieved Stage 2 or Stage 3 on the EMRAM showed that they have already started the process of going digital.

In 2011, the majority of hospitals in Dubai (21 out of 23, or 91%) were in EMRAM Stages 0, 1, or 2 when the survey was first conducted. Over time, more newly opened healthcare facilities joined the survey, increasing the number of participating hospitals from 23 in 2011 to 43 in 2022.

Between 2020 and 2022, 19 healthcare facilities improved their compliance with EMRAM criteria and advanced to a higher EMRAM stage. Improvements were observed in seven hospitals in Dubai, resulting in their progression to higher EMRAM stages. Notably, 9 out of 21 hospitals successfully met all the criteria for Stage 5 enrollment.

To assess IT spending, hospitals in Dubai are required to submit their expenditure data. Of the 36 hospitals in Dubai, 16 provided IT spending figures. Among these, seven hospitals (44%) reported operational IT expenditures of less than 1%, comparable to 2020, where 40% (6 out of 15 hospitals) reported similar spending levels. This expenditure is typically lower than what is observed in hospitals meeting EMRAM Stages 6 and 7.

3.1 Development of Digital Maturity in Dubai Over the Assessment Period

The long-term tracking of digital maturity in Dubai over a period of eleven years (2011–2022; no assessment in 2018) allows us to observe the steady progression and evolution of digital capabilities within the Emirate. Table 2 summarizes the average scores over the assessment period:

From 2011 to 2022, the number of hospitals participating in the assessment of digital maturity in Dubai more than doubled, increasing from 21 hospitals in 2011 to 43 in 2022. More research needs to be done to understand the factors of how the increasing number of participating hospitals influences the digital maturity scores. For instance, new hospitals might achieve higher digital scores more quickly

Table 2 Progression of digital maturity in Dubai hospitals

Year	2011	2012	2013	2014	2015	2016	2017	2019	2020	2021	2022
Average score	1.3	1.8	2.4	2.3	2.7	2.9	3.4	2	2.5	2.9	2.4
(n) hospitals	23	27	28	29	32	33	33	34	39	39	43

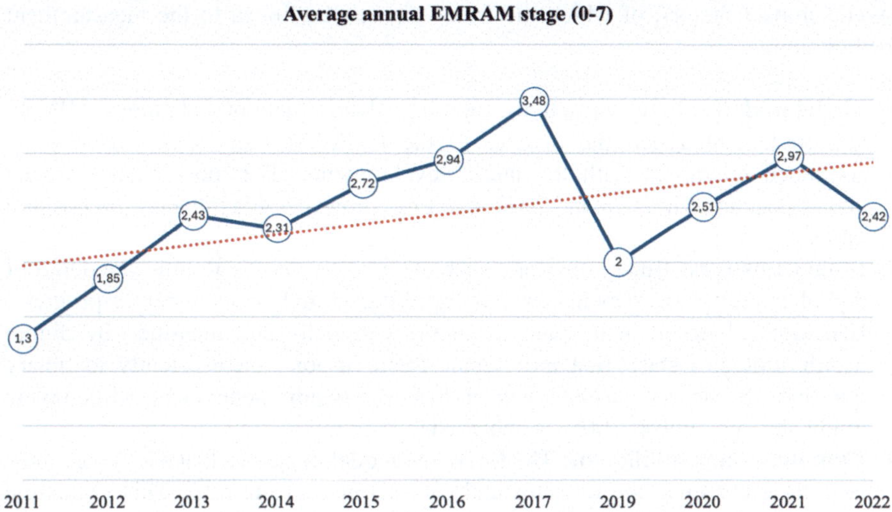

Fig. 1 Average annual EMRAM stage scores (0–7) of hospitals in Dubai, illustrating a positive trend line. The figure also highlights the impact of EMRAM criteria updates in 2018 and 2022

as they start with state-of-the-art equipment and systems, while older hospitals may need more time to transform legacy systems and processes. However, the following analysis of the average scores considers only the overall trends in digital maturity without delving into these specific influences.

From 2011 to 2022, the average annual EMRAM scores of hospitals in Dubai show a general upward trajectory with some fluctuations (Fig. 1). Starting at an average score of 1.3 in 2011, the digital maturity of hospitals steadily increased over the next few years, reaching 2.43 in 2013 and 2.31 in 2014. A continued rise was observed with scores moving to 2.72 in 2015 and 2.94 in 2016, showing gradual advancements in digital adoption.

The peak average score was achieved in 2017, where hospitals reached an average EMRAM stage of 3.48. However, a notable drop occurred in 2019, with the average score falling to 2.0, which is attributed to changes in assessment criteria by HIMSS, the developer of the assessment model. The 2018 update to EMRAM introduced tighter criteria and new requirements, making it more challenging for healthcare organizations to achieve higher levels within the model.

Following this dip, scores began to recover with a rise to 2.51 in 2020 and further improvement to 2.97 in 2021. However, due to the updating of EMRAM criteria in 2022, the average score slightly declined again in the same year, ending at 2.42. Despite these fluctuations, the overall trend line reveals a steady, though modest, upward trajectory in the digital maturity of hospitals in Dubai. With a calculated slope of approximately 0.075, the trend line indicates that, on average, the digital maturity stage of hospitals has increased by about 0.075 stages per year. At this rate of progress, moving from one full stage to the next (e.g., from Stage 2 to Stage 3) would take approximately 13.3 years. This slow gradual progress can be attributed to several impact factors, of which the following can be linked to the measurement methodology:

1. **Model updates:** In response to the ongoing advancement of technology, HIMSS has implemented periodic updates to the EMRAM framework, ensuring its continued alignment with the latest developments. The most notable model updates occurred in 2018 and 2022, exerting a considerable influence on hospital ratings.
2. **Binary answers:** Binary answers, such as "yes" or "no," can limit the depth of digital maturity assessments by not capturing partial or incremental progress. Behaviors observed in hospital evaluations globally and identified by digital health strategies show that individuals often respond inconsistently to binary questions. Some may answer "yes" if 90% of a requirement is met, while others might say "no" unless 100% is achieved.
3. **Complete stage fulfillment:** The EMRAM model mandates that each stage must be fully completed before a hospital can advance to the next. This sequential approach can delay progression, as hospitals cannot move forward incrementally within a stage. For example, a hospital with a fully integrated IT-supported medication management loop but without a digitalized lab would remain at the same stage as a hospital with no digital capabilities at all, both being certified at Stage 0 (Stephani et al. 2019).
4. **Alignment with priorities:** EMRAM's reliance on US-best practices can clash with market specifics, national health policies, and local priorities in other parts of the world. As hospitals will always focus on meeting specific local healthcare needs and regulations first, they may not fully align with EMRAM's staged requirements (Liebe et al. 2022).

To address these challenges and better reflect the evolving digital landscape, HIMSS "modernized" the EMRAM measurement in 2022, moving away from a strict linear model to a more nuanced assessment that utilizes a Likert scale. The Likert scale provides a flexible approach that measures the extent to which hospitals have implemented various digital capabilities, rather than requiring full completion of each stage in a sequential manner. Additionally, the introduction of a threshold of 70% completion within each stage enables hospitals to progress to the next stage without having to meet all requirements fully, thereby streamlining the advancement process and better reflecting incremental progress (Torres 2022).

3.2 International Comparison of Digital Maturity in Dubai

EMRAM scores offer insights into individual hospital performance and facilitate benchmarking against international peers, provided sufficient and timely data is available from a statistically relevant number of hospitals (Stephani et al. 2019) (Table 3).

Key observations and interpretation of results:

Table 3 Percentage of hospitals in each EMRAM stage

EMRAM stages in %	0	1	2	3	4	5	6	7	Participating hospitals
Dubai (2021)	42	0	5	2	30	21	0	0	39 (87%) out of 45
Germany (2021)	69	27	3	1	0	0	0	0	1.624 (90%) out of 1893
USA	78	2	0	1	0	1	11	7	3.690 (60%) out of 6120
Ontario	76	7	1	1	0	1	12	3	144 (36%) out of 400
Australia	96	1	0	0	0	1	1	0	299 (43%) out of 697

This comparison is based on data used for the international comparison of Germany's[3] DigitalRadar results (Amelung et al. 2022) and includes Dubai. In both regions, the EMRAM data is robust, covering almost all hospitals at a specific point in time (2021). In other regions, data may be older and less comprehensive, often collected over years, which could limit benchmarking accuracy and fail to reflect the impact of the 2018 EMRAM update. Therefore, while the findings are accurate for the data presented, they may not fully correspond to the actual situation. This comparison, given the available data, is the most balanced approach we could take under the circumstances.

1. **Lower Concentration in Stage 0**: Dubai has significantly fewer hospitals at Stage 0, with 42% of its hospitals in this initial stage. This is notably lower compared to Australia (96%), and even Germany (69%), Ontario (76%), and the USA (78%). This indicates that a considerable portion of Dubai's hospitals have already moved beyond the initial stage and are advancing toward higher levels of digital maturity.
2. **Lack of Hospitals in Early Stages (1 and 2):** Dubai has no hospitals in Stage 1 and only 5% in Stage 2, which is markedly different from Germany, where 27% of hospitals are in Stage 1. This indicates that Dubai's hospitals, when progressing beyond Stage 0, tend to leap toward higher stages (like Stages 4 and 5), skipping the early stages that are more populated in other regions.
3. **Significant Representation at Advanced Stages**: Dubai shows a substantial presence in advanced stages of digital maturity, particularly at Stages 4 and 5. A combined total of 51% of Dubai's hospitals are in these stages (30% in Stage 4 and 21% in Stage 5), which is significantly higher than other regions. In comparison, the representation in these stages is minimal in other regions—Germany, USA, Ontario, and Australia each have 1% or fewer hospitals in Stage 4, and similar low percentages at Stage 5 (Germany: 0%, USA: 1%, Ontario: 1%, and Australia: 1%). This strong presence in Stages 4 and 5 highlights Dubai's determination to achieve mid-level to advanced digital maturity relative.
4. **No Hospitals in Stages 6 and 7**: Dubai, same as Germany and Australia, has no hospitals in the highest stages. This does not necessarily mean that advanced hospitals do not exist in these regions; rather, it may only indicate that no hospital

[3] In Germany, the scores are considered "indicators" derived from the DigitalRadar scores, which may not fully reflect the actual EMRAM stage in a full-scale assessment.

has applied for validation at the top-2 stages. This could be due to a variety of reasons, such as a complex onsite validation process or other obstacles involved in achieving and maintaining these validations. In contrast, the USA and Ontario have a notable percentage of hospitals at these higher stages (USA: 11% in Stage 6, 7% in Stage 7; Ontario: 12% in Stage 6, 3% in Stage 7). This could suggest that the EMRAM methodology, which was developed in North America, might be more closely aligned with the requirements and specifics of the health systems in these regions, potentially making the validation process more accessible or prioritized there.

3.3 Conclusions from International Benchmarking

Dubai's hospitals have demonstrated a remarkable level of advancement in digital maturity compared to the benchmarking data from Germany, the USA, Ontario, and Australia. With significantly fewer hospitals in the initial stages and a strong presence in advanced stages, Dubai clearly leads in achieving a higher level of digital maturity across relatively more hospitals than any of its peers in this comparison.

Dubai, along with Germany, stands out as one of the few regions with a comprehensive database, boasting around 90% of hospitals assessed for digital maturity. The key success factor behind this high participation rate is the strong support from health authorities in each region—Dubai Health Authority and Germany's Ministry of Health (Scheuer 2023). However, Dubai has clearly outpaced Germany in advancing beyond the early stages and reaching higher levels of digital maturity. A significant factor contributing to Dubai's strong performance in digital maturity could be its substantial financial investments in digital hospital transformation, coupled with the absence of legacy systems that often complicate the transformation process (i.e., the challenge of upgrading versus replacing versus implementing new systems). Additionally, Dubai's long-term commitment to continuously measuring digital maturity over an extended period has supported informed decision-making and facilitated the closing of gaps in digital transformation. In contrast, Germany only initiated its first comprehensive digital maturity measurement in 2021, which may explain the differences observed between the two regions.

What sets Dubai apart in the international landscape of digital maturity is the availability of long-term, comprehensive data that tracks the progression of its hospitals over time. Unlike many regions where data is sporadic or limited, Dubai has consistently monitored and assessed its hospitals' digital maturity, allowing for a clear and sustained view of progress. This long-term tracking has enabled Dubai to make informed, strategic decisions about digital transformation in healthcare and measure their impact and progress.

The commitment to gathering detailed digital maturity data has been reinforced by strong governmental support, specifically from the Dubai Health Authority, which actively promotes and sustains these assessments. This long-term data availability provides Dubai with a unique advantage, offering a clear, actionable roadmap for advancing digital transformation in healthcare. As a result, Dubai has been able to outpace peers, including Germany and the United States, not just in advancing to higher stages of digital maturity on the EMRAM scale but also in maintaining a consistent, data-driven approach to achieving and sustaining these advancements over time.

4 Discussion

We have presented the journey of EMR adoption and the transformation from paper to electronic data in Dubai since 2011. The adoption of an EMR was seen in terms of its capacity to support particular tasks, like integrating clinical decision support into clinical platforms, automating the collection of patient-reported outcomes, and integrating population health management initiatives as the adoption of a basic EMR became conventional. Thus, EMRAM is used as an instrument to measure the maturity level of EMR implementation in Dubai.

The reported average score for Dubai EMRAM survey is 2.6, which is slightly lower than the comparable figure from the previous year's assessment (3.0). The drop can be attributed to maturity model adjustments made in 2018 versus 2022, as well as a change in the sample's makeup from 2021 to 2022. One of the major challenges faced is that many hospitals in Dubai had no EMRs. Four hospitals that already had an EMR were only at the Stage 0 or 1 according to the EMRAM assessment.

Although there is a decrease in the average EMRAM score compared to the previous year, the level of digitalization in Dubai's healthcare facilities is increasing (e.g., increased technology availability and integration with EMR). The current EMRAM criteria (version 2022) raise the bar compared to EMRAM version 2018, not least due to the addition of further focus areas or minimum requirements. However, despite the additional requirements, the hospitals achieved a higher average score (2.6: 2022) than when EMRAM 2018 was introduced (2.2: 2019), which shows the increasing level of digitization.

The implementation of EMR systems within healthcare facilities has always been a huge challenge from both investment and change management perspectives. The degree of success of these implementations is highly dependable on healthcare provider's ability to tailor the system to clinical workflows and to ensure the engagement of clinical teams. Another challenge was finding qualified personnel to train medical professionals and clinicians on using EMR. This is in line with other studies reporting the obstacles and challenges they faced in implementing EMR within the health sector (HIMSS 2017). DHA facilitates the help of training materials like videos and virtual events. There are also technological challenges like different interoperability standards for data exchange. In 2017, different interoperability standards were in use by different EMR vendors. Hence, DHA tried to unify these standards gradually; through different regulatory actions starting by giving options until making it mandatory to use standards that are accepted by DHA.

We have noticed that governmental hospitals in the Emirate of Dubai were a good example of how healthcare providers use technology to support optimal patient care, reduce medical errors, and improve operational efficiency. Government healthcare facilities steadily improved their digital maturity from 2012 to 2017 and after the implementation of new EMRAM criteria in 2018. One of the core stumbling blocks in Stages 1 and 2 is managing basic security/privacy requirements. Gaps in EMRs as well as IT security prevent a large number of hospitals in Dubai from reaching EMRAM Stage 2; in particular, these include technological/functional gaps in cardiology and annual review of Physical Access, Acceptable Use, and Data Destruction policies. High levels of EMR maturity are typically associated with high levels of IT investment. Whilst the volume of IT spend is important, most often it is the strategic nature of the spending that plays a more significant role in the overall level of EMR maturity and adoption (e.g., spending the right amount of money on the right thing at the right time is often the key to success). This makes it more difficult to implement investment projects.

Another major milestone was achieving EMRAM Stage 2 or higher in the maturation of EMRs, which would facilitate the interoperability and homogeneity of EMRs among healthcare providers, as well as the future integration of EMRs with local and regional health information exchanges. The DHA has extensively persuaded healthcare providers to adopt system-wide interoperability by providing relevant roadmaps for technology introduction and facilitating the development and adoption of new information exchange standards; however, this will take some time. The continued lack of interoperability may adversely affect patient safety, clinical outcomes, and population health management efforts (AbdulRahman et al. 2022). In addition, hospitals have variable levels of capital assets, resources, and IT-driven mindsets that may lead to different adoption patterns of EMR functionalities. Nevertheless, many milestones were achieved with our healthcare reform project that is related to the use of information and technology standards such as the International Classification of Diseases, 10th Revision (ICD-10) (WHO 2021) Health Level Seven International (HL7)[4] (HL7 2023) Health Insurance and Claims Management System, and telehealth systems.

One aspect that is not adequately addressed in the current EMRAM models is the impact of EMR implementation on the workforce. The models could be enhanced by incorporating factors such as workforce readiness, development, and the ongoing training required to maintain and operate the EMR systems effectively. Addressing these factors in future versions of the model would ensure that the human elements of technology adoption, including the preparedness and continuous development of healthcare professionals, are given due consideration, which is essential for the sustainable success of EMR implementations.

In the Emirate of Dubai, the EMR implementation in healthcare facilities is progressing in maturity. The study investigated the situation and challenges associated with implementing EMR. Several challenges were found, including the

[4]HL7 Standards. 2023; Available at https://www.hl7.org/.

healthcare facility's lack of an EMR, an EMR with a low EMRAM score, or an EMR that was not compliant to the necessary standards for the exchange of health information due to the lack of awareness about available vendors and related interoperability standards (AbdulRahman et al. 2024). Funding, policymaking, and organizational resistance can all seriously impede the implementation of EMR. Evidence, however, indicated that governmental support and national standards are the most important factors influencing EMR adoption in hospitals (Liang et al. 2021). As a result, promoting interoperability programs and pertinent policies and standards are put into practice in Dubai. However, hospitals may be discouraged from using EMR and reaping its benefits due to the lack of EMR incentives by DHA. We believe that other Gulf countries will benefit from our experience in using EMR technologies and EMRAM to improve their health information systems' maturity level.

5 Conclusion

Carrying out EMRAM surveys within the active social network of CIOs from both public and private hospitals in Dubai has fostered a constructive environment for interaction between all in-patient healthcare providers and the primary healthcare regulator in the Emirate, the DHA.

The EMRAM assessments in Dubai have been smartly exploited by DHA to set standards for digital health and gradually calibrate them to serve its long-term digital health strategy. For example, cherishing the achievements of the best performers and highest score jumpers of the EMRAM survey with DHA's annual awards ceremony developed a healthy competitive spirit between healthcare facilities and cultivated the discussion between CEO and CIOs on annual digital health investments which became EMRAM-score centric. More importantly, DHA's technical team exploited the EMRAM assessments to monitor and calibrate healthcare facilities' readiness to participate in the Dubai Health Information Exchange. The reported EMRAM scores were the main differentiating factors in prioritizing healthcare facilities which were selected for wave 1 and 2 of the HIE onboarding.

Based on our experience, we strongly advocate the use of digital maturity models of EMR implementation as a strategic planning tool by healthcare regulators to support nation-wide or state-wide digital health initiatives such as the implementation of Health Information Exchange platforms.

6 Research Limitations

This study has several limitations: it was conducted in health facilities under jurisdiction of DHA, which only oversees the Emirate of Dubai. A larger study containing all health facilities in the UAE might be more comprehensive. The annual survey was temporarily suspended in 2018 due to some administrative obstacles which entailed a one-year gap of data and contributed to the substantial drop in the overall score of Dubai EMRAM which was mainly affected by the applied changes in the assessment criteria in 2018. Last but not the least, although the EMRAM methodology comprises an embedded verification mechanism which is mainly anchored on phone call interviews with the CIOs of participating healthcare facilities, the reported EMRAM scores do not reach the accuracy level or counterpart EMRAM surveys that are crowned with site visits.

Acknowledgments The authors would like to thank the HIMSS team for their continuous support over the years to the Dubai EMRAM Program. Research collaborators such as Manal Al Malki from Jazan University, KSA and Ahmed Balkhair have contributed significantly to this research.

References

AbdulRahman M, Al-Tahri F, AlMehairi MK et al (2022) Digital health technology for remote care in primary care during the COVID-19 pandemic: experience from Dubai. Telemed J E Health 28(8):1100–1108. https://doi.org/10.1089/tmj.2021.0459

Abdulrahman M, El-Hassan O, Redha MAA, Almalki M (2024) Adoption of electronic medical records in healthcare facilities in the Emirate of Dubai. Healthc Inform Res 30(2):154–161. https://doi.org/10.4258/hir.2024.30.2.154

AlSadrah SA (2020) Electronic medical records and health care promotion in Saudi Arabia. Saudi Med J 41(6):583–589. https://doi.org/10.15537/smj.2020.6.25115

Amelung V, Angelkorte M, Augurzky B et al (2022) Ergebnisse der ersten nationalen Reifegradmessung deutscher Krankenhäuser [Zwischenbericht]. https://www.digitalradar-krankenhaus.de/download/220914_Zwischenbericht_DigitalRadar_Krankenhaus.pdf. Accessed 13 Sep 2024

Burke T (2010) The health information technology provisions in the American Recovery and Reinvestment Act of 2009: implications for public health policy and practice. Pub Health Rep 125(1):141–145. https://doi.org/10.1177/003335491012500119

Carvalho JV, Rocha A, Abreu A (2016) Maturity models of healthcare information systems and technologies: a literature review article. J Med Syst 40(6):131. https://doi.org/10.1007/s10916-016-0486-5

Delpierre C, Cuzin L, Fillaux J, Alvarez M, Massip P, Lang T (2004) A systematic review of computer-based patient record systems and quality of care: more randomized clinical trials or a broader approach?. Int J Qual Health Care 16(5):407–416

Duncan R, Eden R, Woods L et al (2022) Synthesizing dimensions of digital maturity in hospitals: systematic review. J Med Internet Res 24(3):e32994. https://doi.org/10.2196/32994

Eden R, Burton-Jones A, Grant J et al (2019) Digitising an Australian university hospital: qualitative analysis of staff-reported impacts. Aust Health Rev 44(5):677–689. https://doi.org/10.1071/ah18218

Garrido T, Raymond B, Jamieson L et al (2004) Making the business case for hospital information systems--a Kaiser Permanente investment decision. J Health Care Finance 31(2):16–25

HIMSS (2017) Electronic Medical Record Adoption Model (EMRAM) Burlington, VT, USA: HIMSS Analytics. http://www.himssanalytics.org/emram. Accessed 22 Mar 2024

Johnston DS (2017) Digital maturity: are we ready to use technology in the NHS? Future Health J 4(3):189–192. https://doi.org/10.7861/futurehosp.4-3-189

Kanakubo T, Kharrazi H (2019) Comparing the trends of electronic health record adoption among hospitals of the United States and Japan. J Med Syst 43(7):224. https://doi.org/10.1007/s10916-019-1361-y

Krasuska M, Williams R, Sheikh A et al (2020) Technological capabilities to assess digital excellence in hospitals in high performing health care systems: international eDelphi exercise. J Med Internet Res 22(8):e17022. https://doi.org/10.2196/17022

Levene LS, Seidu S, Greenhalgh T, Khunti K (2020) Pandemic threatens primary care for long term conditions. BMJ 371:m3793. https://doi.org/10.1136/bmj.m3793

Liang J, Li Y, Zhang Z et al (2021) Adoption of Electronic Health Records (EHRs) in China during the past 10 years: consecutive survey data analysis and comparison of Sino-American challenges and experiences. J Med Internet Res 23(2):e24813. https://doi.org/10.2196/24813

Liebe J, Jahn F, Buddrus U, Hübner U (2022) Können Reifegradmodelle die Umsetzung von Digitalisierungsstrategien unterstützen? Eine vergleichende Analyse von 42 Modellen. mdi Forum der Medizin_Dokumentation und Medizin_Informatik 2_2022 24:54–59. https://www.gmds.de/fileadmin/user_upload/AG_mwmKIS/Vergleich_Reifegradmodelle_mdi_2_2022_web.pdf

Martin G, Clarke J, Liew F et al (2019) Evaluating the impact of organisational digital maturity on clinical outcomes in secondary care in England. NPJ Digit Med 2:41. https://doi.org/10.1038/s41746-019-0118-9

Musa S, Dergaa I, Al Shekh Yasin R, Singh R (2023) The impact of training on electronic health records related knowledge, practical competencies, and staff satisfaction: a pre-post intervention study among wellness center providers in a primary health-care facility. J Multidiscip Healthc 16:1551–1563. https://doi.org/10.2147/JMDH.S414200

Patel V, Jamoom E, Hsiao JC et al (2013) Variation in electronic health record adoption and readiness for meaningful use: 2008-2011. J Gen Intern Med 28(7):957–964. https://doi.org/10.1007/s11606-012-2324-x

Payne T, Bates D, Berner E et al (2013) Healthcare information technology and economics. J Am Med Inform Assoc 20(2):212–217. https://doi.org/10.1136/amiajnl-2012-000821

Scheuer A (2023) Germany's digital radar: a blueprint for national digital maturity strategies? https://ictandhealth.com/news/germanys-digital-radar-a-blueprint-for-national-digital-maturity-strategies. Accessed 22 Mar 2024

Shekelle P, Morton S, Keeler E (2006) Costs and benefits of health information technology. Evid Rep Technol Assess 132:1–71. https://doi.org/10.23970/ahrqepcerta132

Stephani V, Busse R, Geissler A (2019) Benchmarking der Krankenhaus-IT: Deutschland im internationalen Vergleich. In: Klauber J, Geraedts M, Friedrich J, Wasem J (eds) Krankenhaus Report 2019. Springer, Berlin, Heidelberg. https://doi.org/10.1007/978-3-662-58225-1_2

Torres A (2022) Role of digital health post Covid – EMRAM in 2022. https://hm.fnol.cz/uploads/composer/mnu5e5pgqf-Torres.pdf. Accessed 22 Mar 2024

World Health Organization (2021) International statistical classification of disease and related health problems, Tenth Revision (ICD-10). World Health Organization, Geneva. https://www.who.int/standards/classifications/classification-of-diseases. Accessed 22 Mar 2024

Driving EMR Adoption to Transform Türkiye's Healthcare System

Gurhan Zincircioglu and Ilker Kose

Abstract

The adoption of Electronic Health Records (EHRs) is vital for digital transformation in healthcare, enhancing patient care, efficiency, data-driven decision-making, and patient engagement. Despite challenges such as cost, interoperability, data security, and user resistance, the long-term benefits make EHRs essential for modern healthcare. Türkiye faced issues with healthcare accessibility, insurance coverage, quality, efficiency, and equity before 2003.

A Health Transformation Program (HTP) was launched in 2003 by the Ministry of Health (MoH) and Ministry of Labor and Social Security (MoLSS). The program aimed to improve healthcare through radical reforms, including restructuring the MoH, building health insurance capacity, reorganizing healthcare delivery, strengthening human resources, building health infrastructure, and project management. Since 2003, some improvements such as enhanced patient access, health outcomes, and satisfaction have been achieved, though regional disparities and financial sustainability issues remain. Efforts to improve efficiency and input quality in public hospitals have been noted, with a focus on financial sustainability. To standardize the Nationwide Health Information System (HIS), National Health Data Dictionary (NHDD), Health Coding Reference Server (HCRS), Minimum Health Data Sets (MHDSs), and guidelines for IT procurement and data security have been developed.

G. Zincircioglu (✉)
Business Engineering, Tire State Hospital, Tire/Izmir, Turkey

I. Kose
Engineering Faculty of Alanya University, CEO of Health 4.0 Co (Sağlık 4.0 A.S), Alanya/Antalya, Turkey
e-mail: ilker.kose@alanyauniversity.edu.tr

A. Scheuer, J. Studzinski (eds.), *Digital Maturity in Hospitals*,
https://doi.org/10.1007/978-3-031-80704-6_6

For digitalization and integration, Family Physician Information System (FPIS), Sağlık-Net (Health-Net), e-Nabız (e-Pulse), Central Physician Booking System (CFBS), and other applications like teleradiology and e-prescription have been developed. For guiding and measuring adoption, the MoH used HIMSS EMRAM and O-EMRAM digital maturity models.

Significant progress in EHR adoption has been achieved in most public hospitals and oral and dental health hospitals in Türkiye, with all hospitals using at least basic EHR functions since 2021. The HTP in Türkiye has notably improved healthcare quality indicators and EHR adoption, contributing to a more integrated, efficient, and patient-centered healthcare system.

Keywords

Digital transformation · Transformation in health · Health informatics · HER · Adoption

Abbreviations

CDSS	Clinical Decision Support System
CFBS	Central Physician Booking System
EHR	Electronic Health Record
EMRAM	Electronic Medical Record Adoption Model
FPIS	Family Physician Information System
HCRS	Health Coding Reference Server
HIMSS	Healthcare Information and Management Systems Society
HIS	Hospital Information System
HL7	Health Level Seven
HTP	Health Transformation Program
IT	Information Technology
MHDS	Minimum Health Data Sets
MoH	Ministry of Health
MoLSS	Ministry of Labor and Social Security
O-EMRAM	Outpatient Electronic Medical Record Adoption Model
ODHH	Oral and Dental Health Hospital
NHDD	National Health Data Dictionary
NHIS	National Health Information System
WB	World Bank
WHO	World Health Organization

1 Foreword

The healthcare landscape in Türkiye has been undergoing a radical transformation in recent years, and the rapid integration of digital technologies is reshaping our approach to health and medical care. One of the cornerstones of the digital health revolution is the widespread adoption of Electronic Health Records (EHRs). Today, these digital repositories of patient information have become more comprehensive, interoperable, and secure than ever before. By seamlessly sharing patient data, healthcare providers can provide more coordinated and efficient care (Ratwani 2017).

EHRs, telemedicine, wearable devices, big data intelligence (AI), and other advancements in healthcare refer to the integration of digital technology into all aspects of healthcare. This transformation is fundamentally changing how healthcare services are delivered and how healthcare organizations operate. This transformation involves the adoption of various analytical tools and artificial technologies. With the potential to fundamentally transform the practice of medicine and the delivery of healthcare, artificial intelligence (AI) in particular is a powerful and disruptive area of computer science (Bajwa et al. 2021).

The rapid advancement of technology has ushered in an era of digital transformation, revolutionizing various sectors, including healthcare. At the heart of this transformation in health lies the field of health informatics, which leverages information technology to enhance healthcare delivery, improve patient outcomes, and streamline operations. One of the pivotal elements driving this change is the widespread adoption of Electronic Health Records (EHRs).

These digital systems not only centralize patient information but also facilitate real-time data access and interoperability among healthcare providers. As health informatics continues to evolve, the integration of EHR into everyday medical practice exemplifies the profound impact of digital transformation on the healthcare landscape.

2 The Health Transformation Program of Türkiye

Türkiye[1] had a population of nearly 65 million in 2003, which has grown to 85 million in 2024, and has a surface area of 785,562 km^2. These numbers make Türkiye the European country with the highest population and largest surface area. Despite its size, Türkiye has a unitary state and central government structure, which can be advantageous for driving large and long-term programs.

The HTP of Türkiye was initiated in 2003 by the Ministry of Health (MoH) and Ministry of Labor and Social Security (MoLSS). The program was 10 years long, very comprehensive, and included radical steps that required strong administrative support and political stability. Due to the advantages of procurement procedures, a

[1]The United Nations announced the name change of the country from Turkey to Türkiye in 2022.

robust management infrastructure, and a government-independent approach, the relevant ministries prefer to plan these long-term and integrated projects as a World Bank program.

2.1 The Situation Before the Health Transformation Program

Regarding the literature and reports, the situation of Turkish Healthcare system before the HTP can be described in four dimensions, as follows:

2.1.1 Accessibility and Coverage of Healthcare Services

- **Access to Healthcare Facilities**: Prior to the HTP, the accessibility of healthcare facilities and services in Türkiye was characterized by complexity and fragmentation. Access to public health facilities was not universally streamlined, leading to potential disparities in healthcare access across different regions of Türkiye (Ali Jadoo et al. 2014).
- **Health Insurance Coverage**: The percentage of the population with health insurance coverage or access to subsidized healthcare was limited before the reforms. The existing healthcare system featured inequalities in access to healthcare services, indicating that a significant portion of the population did not have adequate health insurance coverage or access to subsidized healthcare (Ali Jadoo et al. 2014; Gürsoy 2015).

2.1.2 Quality and Efficiency of Healthcare Delivery

- **Overall Quality of Medical Care:** The quality of medical care before the reforms was marked by challenges such as inadequate availability of modern equipment, potential shortages of trained healthcare professionals, and varying levels of patient satisfaction. This scenario indicated the need for improvements in the quality of healthcare services to ensure better patient outcomes (Gürsoy 2015).
- **Efficiency and Coordination:** The healthcare system's efficiency in terms of waiting times, referral processes, and resource management was suboptimal. Factors such as overcrowding in public hospitals, long waiting times, and poor responsiveness contributed to inefficiencies in the system (Gürsoy 2015).

2.1.3 Financing and Affordability of Healthcare

- **Primary Sources of Healthcare Funding**: The healthcare system relied on a mix of funding sources, including government allocations, private insurance, and out-of-pocket payments. However, the financial burden on individuals and families was significant, especially for those without adequate health insurance coverage (Bener et al. 2019; Gürsoy 2015).
- **Affordability Challenges:** The population faced challenges in affording necessary medical treatments and services, particularly due to the prevalence of out-of-pocket payments and potential disparities in healthcare coverage and financial accessibility (Bener et al. 2019; Gürsoy 2015).

2.1.4 Equity and Disparities in Healthcare

- **Regional and Socioeconomic Disparities:** Disparities in healthcare access and outcomes existed between urban and rural areas, as well as across socioeconomic groups. Urban areas might have had better access to healthcare facilities and services compared to rural areas, contributing to disparities in healthcare provision (Bener et al. 2019; Gürsoy 2015).
- **Inequalities in Healthcare Provision:** The healthcare system prior to the reforms exhibited significant gaps and inequalities in the provision of healthcare services. These disparities highlighted the need for reforms to address equity issues and ensure more uniform access to healthcare services across different segments of the population (Bener et al. 2019; Gürsoy 2015).

In summary, prior to the implementation of the HTP in 2003, the Turkish healthcare system faced challenges related to accessibility, quality, financing, and equity. These challenges necessitated comprehensive reforms to enhance the overall effectiveness and inclusivity of the healthcare system.

2.2 The Components of the Program

The main components of Turkish HTP are expressed in the Word Bank project appraisal document exactly as follows (Baris 2004):

"**Component A: Restructuring the MOH for Effective Stewardship.** This component supports MOH's institutional transition from a mainly provider of services to a policy maker and regulator of service provision, while retaining key public health functions.

Component B: Building Capacity of the Health Insurance and Social Security Institutions. This component supports the establishment of a single health insurance fund.

Component C: Reorganizing the Delivery of Healthcare Services. This component has four subcomponents: (i) adoption of family medicine as an organizational model for the provision of outpatient or primary health care services; (ii) integration and harmonization of MOH and SSK hospitals towards great autonomy; and (iii) building a new patient referral mechanism to re-enforce system hierarchy.

Component D: Strengthening Human Resources Capacity aims at developing skills and competence of the health workforce.

Component E: Building Infrastructure for Health and Social Security. This component supports the development of national standards in line with the realignment of institutional roles and responsibilities.

Component F: **Project Management.** This component provides support for project coordination between the two ministries, and overall project governance."

2.3 Achievements

The achievements of the HTP in Türkiye in terms of healthcare quality indicators include:

- **Improved Patient Access and Health Outcomes**: The comprehensive health reforms in Türkiye have succeeded in improving patient access, health outcomes, and patient satisfaction (Ökem and Çakar 2015). The program has contributed to enhancing the overall quality of healthcare services, leading to positive changes in patient experiences and health outcomes (Bener et al. 2019; Ökem and Çakar 2015).
- **Uneven Distribution Across Regions**: While significant progress has been made, these achievements are not evenly distributed across regions in Türkiye. This suggests that there may still be variations in healthcare quality and outcomes based on geographical locations within the country (Bener et al. 2019; Ökem and Çakar, 2015).
- **Long-Term Financial Sustainability**: The long-term financial sustainability of the healthcare system remains a work in progress for policymakers. This indicates that while improvements have been made, there are ongoing efforts to ensure the sustainability of the healthcare system in the future (Ökem and Çakar 2015).
- **Efficiency and Input Quality in Public Hospitals**: Efforts to improve efficiency and input quality in public hospitals have been noted as part of the achievements of the HTP. However, the need to improve allocative efficiency has been emphasized as a crucial aspect for achieving financial sustainability (Ökem and Çakar 2015).

There are more achievements of Turkish HTP reported by the WB (Bump and Sparkes 2014; Aran and Rokx 2014), the World Health Organization (WHO) (Kyalwazi et al. 2018), and in the literature expressing the achievement in different aspects, such as patient satisfaction (Baltaci et al. 2012), universal health coverage (Atun et al. 2013), and workforce policy (Agartan 2015).

In summary, the Health Transformation Program in Türkiye has achieved notable improvements in healthcare quality indicators, including enhanced patient access, health outcomes, and patient satisfaction. However, there is a continued focus on addressing regional variations and ensuring the long-term financial sustainability of the healthcare system.

3 Building a Nationwide, Standardized, and Integrated HIS

3.1 Standardization

As a key component of the HTP, building a National Health Information System (NHIS) first required relevant standardization studies in data, coding, and communication between healthcare facilities and the MoH. The main standardization studies are listed as below:

1. **National Health Data Dictionary (NHDD)**: The MoH started with data standardization and developed the NHDD, publishing it in 2007 as one of the first dictionaries in the healthcare domain (Kose 2007). The latest version of the NHDD includes around 500 data elements defined by the ISO/IEC 11179-4 standard (General Directorate of Health Information Systems 2024b).
2. **Health Coding Reference Server (HCRS)**: The HCRS has been operational since 2007 as a central repository, sharing all structured data elements with all healthcare facilities. Since all data elements have a unique ID and update date, the HCRS enables all HISs to automatically update all structured data elements whenever the MoH makes a change (Kose 2017; Kose et al. 2008; General Directorate of Health Information Systems 2024a).
3. **Minimum Health Data Sets (MHDSs)**: MHDSs are the data sets including the basic data element for each medical event, such as admission, examination, prescription, laboratory test, and discharge (Kose et al. 2008).
4. **Health Level Seven (HL7)**: As an ANSI standard, HL7 has carried an essential role in enabling interoperability within and between hospitals for many years. Türkiye was one of the first countries to use the HL7 version 3, which is web services-based and enables developing the clinical document architecture. HL7 has been used for the transfer of MHDSs from all healthcare facilities to the MoH. By means of the central EHR repository, the hospitals can also retrieve the patient data if there is a consent (Kabak et al. 2008; Namli et al. 2008; Yuksel et al. 2009; Turhan et al. 2008).
5. **Guidelines for Procurement of the IT systems**: Procuring the appropriate hardware, technology, and information system for a public hospital can easily become a problem, especially when considering the challenge of using it properly and safely without high-capacity technical staff. In the early stages of the HTP, the MoH's technical capacity in Turkish public hospitals was notably lacking. The MoH tried to solve this bottleneck by standardizing the procurement documents. First the MoH standardized the functional and user requirements of HISs and published a guideline to help the hospitals prepare the respective tender document. The very early version of that guideline was published in 2003, and then regularly updated due to new requirements (Ülgü 2008). Additionally, the MoH has also published guidelines for hardware and infrastructure procurements (Sağlık Bilgi Sistemleri Genel Müdürlüğü 2024).
6. **Regulations on Privacy and Data Security of HISs**: In parallel to digitalization, data privacy and security issues have been a top priority since the beginning of the

HTP. In 2003, the MoH issued one of the first circulars about data security among all public institutions in Türkiye. Over time, the MoH updated this policy by adding relevant standards and rules (General Directorate of Health Information Systems 2019).

3.2 Digitalization and Integration

Family Physician Information System (FPIS): As a part of the Component C of the HTP, *Reorganizing the Delivery of Healthcare Services* (Baris 2004), the MoH initiated a structural shift in the primary care from health centers to family physicians, so that every citizen will be registered with a specific family physician. This change brought an important opportunity in the digitalization of the healthcare system, which was within the scope of another component of the HTP program, *Component E: Building Infrastructure for Health and Social Security* (Baris 2004). Regarding Component E, The MoH first developed its own FPIS for the pilot application in 2005 for ten provinces, which enabled the first central EHR to be stored in the MoH database. The MoH then developed data and communication standards using the NHDD, MHDS, and HCRS in 2006 and 2007, putting vendors in charge of developing and distributing standardized FPISs for all family physicians. Currently, the MoH functions as a policymaker, while vendors provide relevant services to nearly 23,000 family physicians across the country. Since 2005, thousands of family physicians have been using a standard application for their clinical functions such as vaccination, mother and baby monitoring, chronic disease management, examination, and for exchanging the relevant EHR data with the MoH. Since there have been some other structural changes in the healthcare system in Türkiye such as developing a universal health insurance coverage, FPIS was not only an information system, but an important supporter and integrator of the structural change of the healthcare system in primary care (Tatar et al. 2011).

Sağlık-Net (Health-Net) and e-Nabız (e-Pulse): In parallel to the digitalization of primary care, the usage of standardized HIS by the state hospitals was approximately 90% in 2008. It was a good time for integrating the state hospitals with the central EHR which have been used for family physicians. Then, the Sağlık-Net project was finalized and first deployed in 2008. Sağlık-Net was the complementary part of the centralized EHR, which would be called as e-Nabız (e-Pulse) (Birinci 2023) latterly inspired from its mobile application name (Eşiyok et al. 2023; Dogac et al. 2011; Dogac 2012).

Central Physician Booking System (CFBS): The widespread digitalization process enables added-value applications like the Centralized Family Budget System (CFBS), which integrate and bind different nationwide health information systems, such as Sağlık-Net and FPIS. CFBS has been operational since 2009 and includes all family physicians and state hospitals with mobile and web applications which can also be accessed through e-government authentication system and call centers with hundreds of desks in different locations (Bucak et al. 2018; Küçük et al. 2021).

Others: Since 2009, the year Sağlık-Net became operational, the MoH has launched many applications, such as teleradiology (Aydogdu and Aydogdu 2013; Gultekin et al. 2018), e-prescription (Bulut et al. 2019), and Life is Safe at Home (Hayat Eve Sığar) which is a mobile application including geographical information system during Covid 19 pandemic (Tükenmez and Karakaş 2022).

3.3 Adoption

After standardization, digitalization, and integration, the last and perhaps most important phase of the HTP was the adoption of the EHR to increase healthcare quality and patient safety. While the MoH had nearly completed the overall digitalization and integration of many added-value applications by the early 2010s, in other words, in the first ten years of the HTP, healthcare staff, including physicians, pharmacists, and nurses, were still acclimating to using information systems properly. As the second phase of the HTP, the MoH declared a new target in their Strategic Plan for 2013–2018: "defining and disseminating the digital hospital concepts across the country" (MoH 2013).

4 Adopting and Using the EHR for Increasing Healthcare Quality

4.1 Digital Health Progress Monitoring

As the MoH sets a country-wide target for digital hospitals, measuring the status quo and progress became an important issue. The MoH addressed this through a long-term collaboration with the Healthcare Information and Management Systems Society (HIMSS), which began in 2013. The MoH and HIMSS first signed a five-year protocol to measure the usage and adoption of health information systems by healthcare staff in all public hospitals using the Electronic Medical Record Adoption Model (EMRAM).

This protocol was subsequently updated in 2018 for another five years and in 2024 for three years. The Turkish MoH was aware that the EHR itself was not directly beneficial for healthcare quality and patient safety, but rather the EHR functions—meaningful usage of EHR such as e-order, clinical decision support systems (CDSS), digital imaging, etc.—were directly impactful (Köse et al. 2022). Thus, the collaboration between HIMSS and the Turkish MoH aligned well in vision and timing, as both parties aimed to increase the adoption level of EHR among healthcare staff to enhance healthcare quality and patient safety.

4.2 Methodology

The first phase of the HTP was finalized in 2013, along with the completion of structural reforms. As a result of this transformation, Türkiye established around 75 oral and dental health hospitals (ODHH), which are mostly outpatient facilities, in addition to secondary and tertiary public healthcare facilities. Thus, the Turkish MoH used EMRAM for inpatient hospitals and O-EMRAM for outpatient hospitals as tools for measuring the status and progress in terms of EHR adoption.

Although the essential target was measuring the adoption level, the real goal was the widespread adoption of EHR among healthcare staff. To achieve this, numerous studies were conducted by both HIMSS and the MoH. HIMSS worked with local partners, including academicians, to conduct studies such as training, applying the digital maturity surveys, guiding hospitals for stage 6 and stage 7 validations, and organizing gap analysis and regional workshops. Additionally, the MoH and HIMSS have organized an annual international event called HIMSS Eurasia.[2]

While EMRAM and O-EMRAM have been used as tools, the MoH has simultaneously pursued its own strategies and investments in digitalization, IT infrastructure, and EHR adoption for all hospitals, taking into account the priorities and progress levels of the hospitals year by year.

4.3 Achievements of 10 Years Studies

The results of the digital maturity surveys have been published in three studies in literature.

The first paper, including the EMRAM surveys conducted between 2014 and 2017, showed that the bottom-up approach to adopting the EHR, which was successful in the USA (Sheikh et al. 2014), was also successful in Türkiye. The study also indicated that smaller hospitals were more successful in adopting the EHR compared to larger hospitals, which was the opposite of the trend observed in the USA. Additionally, the study measured that 63.1% of Turkish public hospitals had at least basic EHR functions, and 36% had comprehensive EHR functions. These results were slightly higher than those of Korean hospitals in 2017 but slightly lower than those of US hospitals in 2015 and 2017 (Kose et al. 2020).

The second paper based on EMRAM surveys conducted in 2021 was published 2023. This study showed that 33.7% of public hospitals in Türkiye have only basic EHR functions, while most of them, 66.3%, have extensive EHR functions. In other words, all hospitals (100%) have at least basic EHR functions. This result points to remarkable progress since 2017. This level also indicates that Türkiye has slightly better adoption than the USA (96%) and much better than China (85.3%) and Korea (58.1%) (Köse et al. 2023).

[2]HIMSS Eurasia- https://himsseurasia.com. Organization- Ekspoturk. Accessed 1 July 2024.

The only study based on O-EMRAM surveys conducted in 2021 at the ODHH was published in 2024. Very similar adoption levels were observed as in public hospitals, with 59 (40.7%) of ODHCs/ODHHs having comprehensive EHR functions, while the remaining 86 (59.3%) had basic EHR functions. Thus, it can be stated that all ODHHs in Türkiye use at least basic EHR functions (Köse et al. 2024).

5 Conclusions and Perspectives

The adoption of EHRs is a critical component of digital transformation in healthcare. It enhances patient care, improves efficiency, supports data-driven decision-making, and engages patients in their healthcare. While challenges such as cost, interoperability, data security, and user resistance exist, the long-term benefits of EHR adoption make it an essential investment for modern healthcare systems. This transformation, supported by health informatics, represents a significant shift toward a more integrated, efficient, and patient-centered healthcare system.

The HTP in Türkiye has significantly improved healthcare quality indicators and EHR adoption, contributing to a more integrated, efficient, and patient-centered healthcare system. Continued efforts are required to address regional disparities, ensure financial sustainability, and leverage emerging technologies like AI and big data analytics to further enhance healthcare delivery and outcomes.

References

Agartan TI (2015) Health workforce policy and Türkiye's health care reform. Health Policy 119(12):1621–1626. https://doi.org/10.1016/J.HEALTHPOL.2015.09.008

Ali Jadoo SA, Aljunid SM, Sulku SN et al (2014) Turkish health system reform from the people's perspective: a cross sectional study. BMC Health Serv Res 14:30. https://doi.org/10.1186/1472-6963-14-30

Aran M, Rokx C (2014) Türkiye on the way of universal health coverage through the health transformation program (2003-13). https://documents1.worldbank.org/curated/en/243161468308038702/pdf/913260WP0UHC0C00Box385331B00PUBLIC0.pdf. Accessed 1 July 2024

Atun R, Aydin S, Chakraborty S et al (2013) Universal health coverage in Türkiye: enhancement of equity. Lancet 382(9886):65–99. https://doi.org/10.1016/S0140-6736(13)61051-X

Aydogdu C, Aydogdu Ö (2013) An overview of telemedicine in Türkiye. BTAIJ 7(3):89–101. https://www.tsijournals.com/articles/an-overview-of-telemedicine-in-turkey.pdf

Bajwa J, Munir U, Nori A et al (2021) Artificial Intelligence in healthcare: transforming the practice of medicine. Future Health J 8(2):e188–e194. https://doi.org/10.7861/fhj.2021-0095

Baltaci D, Celesun T, Eroz R et al (2012) Evaluation of patient satisfaction with family physicians after implementation of family medicine in Türkiye. HealthMED 6(2):469–475. http://pdf.healthmed.ba/HealthMED_6_2_stampa.pdf

Baris E (2004) Türkiye Health Transition Project Project Appraisal Document. https://documents1.worldbank.org/curated/en/657501468764374311/pdf/Project0Inform1nt010Appraisal0Stage.pdf. Accessed 1 July 2024

Bener A, Alayoglu N, Çatan F et al (2019) Health services management in Türkiye: failure or success? Int J Prev Med 10(1):30. https://doi.org/10.4103/IJPVM.IJPVM_422_17

Birinci Ş (2023) A digital opportunity for patients to manage their health: Türkiye National Personal Health Record System (The e-Nabız). Balkan Med J 40(3):215–221. https://doi.org/10.4274/BALKANMEDJ.GALENOS.2023.2023-2-77

Bucak IH, Almis H, Dogan F et al (2018) A retrospective analysis of central physician appointment system data in a Tertiary Health Center in Türkiye. Telemed J E Health 24(3):216–221. https://doi.org/10.1089/TMJ.2017.0109

Bulut S, Yıldız A, Kaya S (2019) Evaluation of transition to electronic prescriptions in Türkiye: perspective of family physicians. Int J Health Policy Manag 8(1):40–48. https://doi.org/10.15171/IJHPM.2018.89

Bump JB, Sparkes SP (2014) A political economy analysis of Türkiye's Health Transformation Program. https://documents1.worldbank.org/curated/en/254431468101383013/pdf/886000REPLACEM0PolEcon0Turkey0Final.pdf. Accessed 1 July 2024

Dogac A (2012) Interoperability in eHealth systems. Proc VLDB Endow 5(12):2026–2027. https://doi.org/10.14778/2367502.2367568

Dogac A, Yuksel M, Avcl A et al (2011) Electronic health record interoperability as realized in the Turkish Health Information System. Methods Inf Med 50(02):140–149. https://doi.org/10.3414/ME10-01-0022

Eşiyok A, Uslu Divanoğlu S, Çelik R (2023) Digitalization in Healthcare - Mobile Health (M-Health) Applications. Aksaray Üniversitesi İktisadi ve İdari Bilimler Fakültesi Dergisi 15(2):165–174. https://doi.org/10.52791/aksarayiibd.1241287

General Directorate of Health Information Systems (2019) The Guideline for Information Security and Privacy V 2.1.

General Directorate of Health Information Systems (2024a) Web Site of Health Coding Reference Server. Ministry of Health of Türkiye. Accessed 1 July 2024

General Directorate of Health Information Systems (2024b) Web Site of National Health Data Dictionary. Ministry of Health of Türkiye. https://e-saglik.gov.tr/TR,7141/usvs.html. Accessed 1 July 2024

Gultekin M, Ozturk C, Karaca S et al (2018) Centralization of mammography reporting with mobile trucks: Turkish experience. Prev Med Rep 10:317–322. https://doi.org/10.1016/J.PMEDR.2018.04.008

Gürsoy K (2015) An overview of Turkish Healthcare System after Health Transformation Program: main successes performance assessment further challenges and policy options. Sosyal Güvence Dergisi 7:83–112. https://dergipark.org.tr/tr/download/article-file/152166

Kabak Y, Dogac A, Kose I et al (2008) The use of HL7 CDA in the National Health Information System (NHIS) of Türkiye. 9th International HL7 Interoperability Conference, pp 49–55

Kose I (2007) National Health Data Dictionary a way towards to terminology union (Sağlık bilişiminde terminoloji birliğine doğru: Ulusal sağlık veri sözlüğü). SD Platform, pp 8–9. http://www.sdplatform.com/Yazilar/Konuk-Yazilar/464/Saglik-bilisiminde-terminoloji-birligine-dogru-Ulusal-saglik-veri-sozlugu.aspx. Accessed 1 July 2024

Kose I (2017) Health information systems in Türkiye (Sağlık Bilgi Sistemleri: Sağlık-Net e-Nabız vb.). In: Özen O (ed) Sağlık Diplomasisi, 1st edn. İstanbul Medipol Üniversitesi Yayınları, pp 221–252

Kose I, Akpinar N, Gurel M et al (2008) Türkiye's National Health Information System (NHIS). Proceeding of the eChallenges Conference; E-Challenges Conference, pp 170–177. https://www.srdc.com.tr/share/publications/2008/9.pdf

Kose I, Rayner J, Birinci S et al (2020) Adoption rates of electronic health records in Turkish Hospitals and the relation with hospital sizes. BMC Health Serv Res 20(1). https://doi.org/10.1186/s12913-020-05767-5

Köse İ, Cece S, Bayraktar E (2022) Are electronic health records beneficial or not? J Health Syst Policies 4(1):1–20. https://doi.org/10.52675/jhesp.1082576

Köse İ, Cece S, Yener S et al (2023) Basic electronic health record (EHR) adoption in Türkiye is nearly complete but challenges persist. BMC Health Serv Res 23(1). https://doi.org/10.1186/s12913-023-09859-w

Köse İ, Cece S, Yener S et al (2024) Electronic Health Record Adoption in Oral and Dental Health Centres in Türkiye. J Health Manag. https://doi.org/10.1177/09720634241237596

Küçük A, Demirci M, Kerman G et al (2021) Evaluating hospital appointment systems in Türkiye: challenges and opportunities. Health Policy Technol 10(1):69–74. https://doi.org/10.1016/J.HLPT.2020.11.008

Kyalwazi M, Subramanian L, Schwarz D et al (2018) Country case studies on primary health care Türkiye family practice for quality in universal health coverage. https://iris.who.int/bitstream/handle/10665/326253/WHO-HIS-SDS-2018.35-eng.pdf?sequence=1. Accessed 1 July 2024

MoH (2013) The strategic plan of Turkish MoH 2013-2018. Accessed 1 July 2024

Namli T, Aluc G, Sinaci AA et al (2008) Testing the Conformance and Interoperability of NHIS to Türkiye's HL7 Profile. 9th International HL7 Interoperability Conference, pp 63–68. http://www.srdc.metu.edu.tr/webpage/publications/2008/11.pdf.

Ökem ZG, Çakar M (2015) What have health care reforms achieved in Türkiye? An appraisal of the "Health Transformation Programme". Health Policy 119(9):1153–1163. https://doi.org/10.1016/J.HEALTHPOL.2015.06.003

Ratwani RM (2017) Electronic health records and improved patient care: opportunities for applied psychology. Curr Dir Psychol Sci 26(4):359–365. https://doi.org/10.1177/0963721417700691

Sağlık Bilgi Sistemleri Genel Müdürlüğü (2024) Saglik Bilgi Yonetim Sistemi Alim Kilavuzu. T.C. Sağlık Bakanlığı. https://sbsgm.saglik.gov.tr/TR,43797/saglik-bilgi-yonetim-sistemi-alim-kilavuzu-sbys-ak.htm l. Accessed 1 July 2024

Sheikh A, Jha A, Cresswell K et al (2014) Adoption of electronic health records in UK hospitals: lessons from the USA. Lancet 384(9937):8–9. https://doi.org/10.1016/S0140-6736(13)61841-3

Tatar M, Mollahaliloglu S, Sahin B et al (2011) Türkiye Health system review 13(6). Health Systems in Transition. https://iris.who.int/bitstream/handle/10665/330325/HiT-13-6-2011-eng.pdf. Accessed 1 July 2024

Tükenmez B, Karakaş B (2022) A virtual panopticon in Türkiye during the COVID-19 pandemic: the case of location-based mobile application "Hayat Eve Sığar." Urban Assemblage: The City as Architecture Media AI and Big Data. https://hdl.handle.net/11511/94553. Accessed 1 July 2024

Turhan K, Kurt B, Uzun E (2008) SAĞLIK-NET Project in Türkiye and HL7 v3 Implementation. World Acad Sci Eng Technol. Int Sch Sci Res Innov 2(8):812–816. https://zenodo.org/record/1070519/files/8238.pdf

Ülgü MM (2008) Hastane Bilgi Sistemleri Alim Çerçeve Ilkeleri. TC Sağlk Bakanlğ Bilgi .Işlem Daire Başkanlğ Ankara:5. Accessed 1 July 2024

Yuksel M, Dogac A, Akdag R et al (2009) Automatic transformation of local EHRs to the international standards: application in Türkiye's NHIS. EChallenges Conference. Accessed 1 July 2024

Physician-Led Innovation: Digital Maturity at Johns Hopkins Aramco Healthcare

Tamara Sunbul

Abstract

This chapter delves into the digital transformation journey of Johns Hopkins Aramco Healthcare (JHAH) over a decade, focusing on physician-led innovations that have significantly elevated digital maturity within the organization. By centering their strategy on physician engagement and empowerment, JHAH has seamlessly integrated advanced digital technologies into their healthcare delivery system, thereby enhancing patient care standards and operational efficiency. The global context of digital healthcare is explored, highlighting the rapid adoption of AI, IoT, and Big Data analytics, which are revolutionizing healthcare through improved diagnostic accuracy, personalized treatment plans, and predictive analytics.

Despite the challenges of integrating these technologies, JHAH's collaboration with Johns Hopkins Medicine and their robust governance framework have ensured successful implementation. The chapter also underscores the critical role of physician engagement throughout the EHR implementation phases, transitioning physicians from users to builders and trainers, thus fostering a culture of continuous improvement. Through strategic governance, collaborative leadership, and championing physicians as change agents, JHAH has not only enhanced clinical outcomes and user satisfaction but also achieved significant milestones in digital healthcare. This comprehensive analysis provides valuable insights and recommendations for healthcare organizations striving for digital maturity and aiming to navigate the complexities of digital transformation successfully.

T. Sunbul (✉)
Clinical Informatics, Johns Hopkins Aramco Healthcare, Dhahran, Saudi Arabia
e-mail: tamara.sunbul@jhah.com

A. Scheuer, J. Studzinski (eds.), *Digital Maturity in Hospitals*,
https://doi.org/10.1007/978-3-031-80704-6_7

Keywords

Digital maturity · Change management · Digital transformation · Physician engagement · Electronic health records (EHRs) · Electronic medical records (EMRs)

Abbreviations

AI	Artificial Intelligence
IoT	Internet of Things
EHR	Electronic Health Record
EMR	Electronic Medical Record
mHealth	Mobile Health
NLP	Natural Language Processing

1 Introduction

In the rapidly evolving landscape of healthcare, the integration of digital technologies has become crucial for enhancing patient care standards and improving operational efficiency. This chapter provides an in-depth examination of a meticulously designed and implemented digital transformation strategy centered on physician engagement and empowerment. Emphasizing their critical role, this strategy has significantly advanced digital maturity within hospital environments. The collaborative partnership between healthcare organizations and physicians serves as the foundation for the seamless integration and optimization of digital technologies, thereby redefining and enhancing the entire spectrum of the healthcare delivery system.

2 Global Context of Digital Healthcare Digitization

The global adoption of healthcare digitization is accelerating, driven by technological advancements, the demand for improved patient outcomes, and the need for cost-effective healthcare delivery (McKinney et al. 2020). Technologies such as artificial intelligence (AI), the Internet of Things (IoT), and Big Data analytics are revolutionizing healthcare systems worldwide (Ting et al. 2017). These advancements enable more accurate diagnoses, personalized treatment plans, and predictive analytics for proactive patient care (Obermeyer and Emanuel 2016).

Moreover, digital health technologies like telemedicine, mobile health (mHealth) apps, and wearable devices are increasingly being adopted to enhance patient engagement and provide remote monitoring solutions (Topol 2019). These technologies are especially crucial in addressing healthcare access challenges in rural and underserved areas, where traditional healthcare infrastructure may be lacking (Davenport and Kalakota 2019).

However, healthcare systems globally are grappling with significant digital transformation challenges. Integrating new technologies with existing systems, ensuring data interoperability, managing financial constraints, and overcoming resistance to change among healthcare professionals are major hurdles (Islam et al. 2015). Regulatory compliance and maintaining patient data privacy add additional layers of complexity to the digital transformation process (Heintzman 2015).

A. **Artificial Intelligence (AI):** AI has become a cornerstone of modern healthcare, offering transformative potential across various domains. AI technologies, including machine learning (ML) and natural language processing (NLP), are being utilized to enhance diagnostic accuracy, streamline administrative processes, and improve patient outcomes (Yu et al. 2012).
   - **Diagnostic Accuracy:** AI algorithms can analyze medical images with a level of precision that rivals human experts. For example, AI-driven tools can detect early signs of diseases such as cancer, diabetic retinopathy, and cardiovascular conditions by examining imaging data from X-rays, MRIs, and CT scans (McKinney et al. 2020). For instance, AI systems have achieved diagnostic accuracy rates of up to 94.5% for breast cancer screening, surpassing the average radiologist (McKinney et al. 2020).
   - **Predictive Analytics:** AI models can predict patient outcomes based on historical data, helping healthcare providers to identify high-risk patients and intervene early. Obermeyer and Emanuel's (2016) study found that using predictive models reduced hospital readmissions by 20%. Predictive analytics are particularly useful in managing chronic diseases and preventing hospital readmissions.
   - **Personalized Medicine:** AI facilitates the development of personalized treatment plans by analyzing genetic information and patient data. This approach enables healthcare providers to tailor treatments to individual patients, improving efficacy and reducing adverse effects (Ashley 2016).
   - **Administrative Efficiency:** AI-powered chatbots, ambient clinical documentation, NLP, and virtual assistants are being deployed to handle routine administrative tasks such as appointment scheduling, patient inquiries, and billing. This reduces the workload on healthcare staff and enhances operational efficiency (Khanra et al. 2020).

B. **Internet of Things (IoT)**: IoT is another technology that is significantly impacting healthcare. IoT involves the interconnection of devices, sensors, and systems to collect and exchange data, enabling real-time monitoring and remote healthcare delivery (Islam et al. 2015).
   - **Remote Patient Monitoring**: IoT devices such as wearable fitness trackers, smartwatches, and implantable sensors allow continuous monitoring of patients' vital signs. This data is transmitted to healthcare providers in real-time, enabling timely interventions and reducing the need for frequent hospital visits (Islam et al. 2015; Vest and Gamm 2010)). Studies have shown that

remote patient monitoring can reduce hospital readmissions by up to 25% (Islam et al. 2015).

- **Chronic Disease Management**: IoT devices are particularly beneficial for managing chronic diseases such as diabetes, hypertension, and heart disease. For instance, continuous glucose monitors (CGMs) help diabetic patients maintain optimal blood sugar levels by providing real-time data and alerts (Heintzman 2015).
- **Smart Hospitals**: IoT technology is transforming hospital operations by enabling smart systems for asset tracking, environmental monitoring, and patient flow management. These systems enhance operational efficiency and improve patient safety by ensuring that medical equipment is always available and properly maintained (Yu et al. 2012).
- **Telehealth and Telemedicine**: IoT devices facilitate telehealth services by enabling remote consultations and diagnostics. Patients can receive medical advice and treatment from the comfort of their homes, which is particularly beneficial for those living in remote or underserved areas (Kruse et al. 2018).

C. **Big Data Analytics**: Big Data analytics in healthcare involves the processing and analysis of vast amounts of data generated from various sources, including electronic health records (EHRs), medical imaging, genomic data, and wearable devices (Raghupathi and Raghupathi 2014).

- **Population Health Management**: Big Data analytics helps healthcare providers manage population health by identifying trends and patterns in large datasets. This enables the development of targeted interventions and preventive measures to improve public health outcomes (Raghupathi and Raghupathi 2014). For example, analyzing population data helped reduce the incidence of heart disease in a community by 15% (Raghupathi and Raghupathi 2014).
- **Clinical Decision Support**: By analyzing clinical data, Big Data tools provide healthcare professionals with evidence-based insights and recommendations. This enhances clinical decision-making and improves patient care quality (Bates et al. 2014).
- **Research and Development**: Big Data is revolutionizing medical research by enabling the analysis of complex datasets to discover new treatments and understand disease mechanisms. For example, genomic data analysis is advancing the field of precision medicine by identifying genetic factors that influence disease susceptibility and treatment response (Ashley 2016).
- **Operational Efficiency**: Big Data analytics improves operational efficiency by optimizing resource allocation, reducing costs, and identifying areas for process improvement. For instance, predictive analytics can forecast patient admissions, allowing hospitals to better manage staffing and resources (Bates et al. 2014).

2.1 The Evolution of Electronic Health Records (EHRs)

EHRs have evolved significantly from simple digital versions of paper charts to comprehensive, integrated systems that support clinical decision-making, streamline workflows, and enhance patient safety. The adoption of EHRs has been driven by the need to improve care coordination, reduce medical errors, and facilitate data sharing among healthcare providers (McKinney et al. 2020; Ting et al. 2017).

A. **Historical Context and Evolution:** Historically, EHRs were primarily used for administrative purposes, such as billing and scheduling. Early EHR systems were limited in scope and functionality, mainly serving as digital repositories for patient information. However, as technology advanced and the healthcare industry recognized the potential benefits of EHRs, these systems began to evolve (Obermeyer and Emanuel 2016).
   The initial adoption of EHRs focused on digitizing paper records to improve efficiency and reduce the physical space required for storing paper charts. This transition laid the foundation for more sophisticated systems that could offer greater value beyond simple record-keeping. As healthcare providers began to see the potential for EHRs to enhance patient care, the focus shifted toward developing systems that could support clinical decision-making and improve patient outcomes (Topol 2019)
B. **Modern EHR Systems and Their Functionalities:** Modern EHR systems now offer advanced functionalities, including real-time access to patient data, electronic prescribing, and clinical decision support. These systems facilitate seamless communication among healthcare providers, thereby enhancing the overall quality of patient care (Davenport and Kalakota 2019).
   - **Real-Time Access to Patient Data**: One of the most significant advancements in EHR systems is the ability to access patient data in real-time. This feature ensures that healthcare providers have up-to-date information about their patients from different sources, which is crucial for making informed clinical decisions. Real-time access also allows for better coordination of care, as multiple providers can view and update patient records simultaneously (Islam et al. 2015).
   - **Electronic Prescribing (e-Prescribing)**: Electronic prescribing has transformed the way medications are prescribed and dispensed. By using EHR systems for e-prescribing, healthcare providers can send prescriptions directly to pharmacies electronically, reducing the risk of errors associated with handwritten prescriptions. Studies have shown that e-prescribing can reduce medication errors by up to 85% (Heintzman 2015).
   - **Clinical Decision Support**: Modern EHR systems incorporate clinical decision support tools that provide evidence-based recommendations to clinicians at the point of care. These tools can alert providers to potential drug interactions, suggest alternative treatments, and provide reminders for preventive care. For instance, clinical decision support systems have been shown to

improve adherence to clinical guidelines and reduce the incidence of adverse drug events (Yu et al. 2012; Kruse et al. 2018).

- **Comprehensive Documentation**: Advanced EHR systems offer comprehensive documentation capabilities, allowing providers to capture detailed patient information efficiently. This includes structured templates for documenting clinical encounters, automated coding for billing purposes, and the ability to attach diagnostic images and test results to patient records. Enhanced documentation improves the accuracy and completeness of patient records, which is essential for continuity of care (Raghupathi and Raghupathi 2014).
- **Digital Front Door and Patient Engagement**: Modern EHR systems serve as a digital front door, enhancing patient engagement by providing easy access to health information, appointment scheduling, and communication with healthcare providers. Patient portals and mhealth apps enable patients to actively participate in their care, schedule appointments, and access test results, fostering a partnership between patients and providers in managing health (Bates et al. 2014).
- **Care Gaps and Social Determinants of Health**: EHR systems are equipped to identify care gaps by capturing data on social determinants of health. This capability allows for personalized interventions and a more holistic approach to patient care. By integrating social, economic, and environmental factors, EHR systems help providers address the root causes of health disparities and deliver tailored care plans (Ashley 2016).
- **Enabling Value-Based Care**: EHR systems support value-based care by tracking clinical outcomes, patient satisfaction, and cost efficiency. These systems enable healthcare organizations to shift from volume-based to value-based care models, improving quality while controlling costs. Metrics such as readmission rates, patient outcomes, and preventive care measures are monitored to ensure value-driven healthcare delivery (Khanra et al. 2020).
- **Clinical Pathways and Optimal Care Paths**: EHR systems facilitate the development and adherence to clinical pathways and optimal care paths. These standardized protocols guide providers in delivering evidence-based care, reducing variability, and improving outcomes. Clinical pathways ensure that patients receive consistent and high-quality care across different settings and providers (Adler-Milstein and Huckman 2013).
- **Dashboards and Point-of-Care Actionable Data**: EHR systems provide dashboards that offer real-time, actionable data at the point of care. These dashboards present key metrics and clinical information in an easily digestible format, enabling healthcare providers to make informed decisions quickly. Point-of-care data improves clinical efficiency and supports timely interventions (Vest and Gamm 2010).

2.2 The Critical Role of Physicians in EHR Implementation

- **Importance of Physician Engagement:** Physician engagement is pivotal in the successful implementation of EHR systems. Physicians play a crucial role in shaping EHR functionalities to align with clinical workflows and patient care needs. Their involvement ensures that the system is user-friendly, efficient, and enhances patient safety. Engaging physicians from the outset helps to address potential concerns, gather valuable insights, and foster a sense of ownership and commitment to the new system (McKinney et al. 2020; Ting et al. 2017).
  Studies have shown that physician engagement significantly influences the acceptance and effective use of healthcare technology. When physicians are actively involved in the implementation process, they are more likely to embrace the technology and advocate for its use among their peers. This leads to higher adoption rates, improved data quality, and better patient outcomes (Adler-Milstein and Huckman 2013; Sheikh et al. 2015).
  For instance, healthcare facilities with high physician engagement reported a 30% faster implementation timeline and a 20% increase in user satisfaction compared to facilities with lower engagement levels (Adler-Milstein and Huckman 2013; Sheikh et al. 2015). Additionally, the involvement of physicians in decision-making processes ensures that EHR systems are designed to meet clinical needs, which can reduce the time required to reach full implementation and operational efficiency (Khanra et al. 2020; Grol and Grimshaw 2003).
- **People, Process, and Technology Aspects:** The successful implementation of EHR systems requires a comprehensive approach that integrates people, process, and technology.
  - **People:** The engagement of physicians and other healthcare professionals is critical. Studies have shown that healthcare organizations with high levels of physician engagement experience smoother transitions to digital systems, higher user satisfaction, and improved patient outcomes. For example, facilities with strong physician engagement reported a 20% increase in user satisfaction and a 15% reduction in the time required to reach full implementation (Obermeyer and Emanuel 2016; Topol 2019).
    Training and Education: Continuous training and education are essential to ensure that physicians are comfortable and proficient with the new technology. Hands-on training sessions, simulations, and ongoing technical support help mitigate resistance and build confidence among physicians (Davenport and Kalakota 2019).
    Leadership and Governance: Establishing a governance framework that includes physician leaders in decision-making roles ensures that the EHR system meets clinical needs. Physician champions can gather input from their peers, test system functionalities, and provide feedback to the implementation team (Islam et al. 2015).

- **Process**: Effective process management is crucial for the seamless integration of EHR systems into existing workflows. This includes designing workflows that minimize disruption to clinical practices and ensure that the EHR system supports, rather than hinders, patient care.
  - **Workflow Integration**: Aligning EHR functionalities with clinical workflows is essential for maximizing efficiency and effectiveness. For instance, integrating electronic prescribing and clinical decision support tools into daily routines can reduce medication errors and improve patient outcomes. In one study, electronic prescribing reduced medication errors by up to 85% (Heintzman 2015).
  - **Change Management**: Managing change is a significant aspect of the EHR implementation process. This involves clear communication, addressing concerns, and demonstrating the tangible benefits of the new system. Effective change management strategies can reduce resistance and foster a culture of innovation (Yu et al. 2012).
- **Technology**: The technical aspects of EHR implementation involve selecting the right system, ensuring data interoperability, and maintaining cybersecurity.
  - **System Selection and Customization**: Choosing an EHR system that is customizable and scalable is vital for meeting the specific needs of the healthcare organization. Customization allows the system to adapt to the unique workflows and requirements of different departments (Kruse et al. 2018).
  - **Data Interoperability**: Ensuring that the EHR system can seamlessly exchange data with other digital health tools is crucial for achieving a cohesive healthcare ecosystem. Interoperability enables better care coordination and improves the accuracy and completeness of patient information (Raghupathi and Raghupathi 2014).
  - **Cybersecurity**: Protecting patient data from cyber threats is a critical concern. Implementing robust cybersecurity measures, such as encryption, multi-factor authentication, and continuous monitoring, helps safeguard sensitive information and maintain compliance with data privacy regulations (Bates et al. 2014).

3 Case Study: Johns Hopkins Aramco Healthcare (JHAH), EHR Implementation

Successful EHR implementations showcase the transformative potential of digital technologies and IT systems in improving healthcare delivery. Multiple studies demonstrate significant improvements in patient safety and operational efficiency (Adler-Milstein and Huckman 2013; Vest and Gamm 2010).

For instance, improvements in patient safety and operational efficiency with the system's ability to provide real-time alerts for potential medication errors and its comprehensive documentation capabilities have greatly reduced the incidence of adverse events. Additionally, the integration of clinical decision support tools within the EHR enables healthcare providers to make more informed and timely clinical decisions (Adler-Milstein and Huckman 2013; Vest and Gamm 2010).

Specifically, features such as automated alerts for drug interactions, streamlined workflows for order entry, and access to up-to-date patient records across multiple departments. These enhancements have not only improved clinical outcomes but also increased the efficiency of healthcare delivery, reducing the time healthcare providers spend on administrative tasks and allowing more time for patient care (Sheikh et al. 2015).

A. **Contextualizing JHAH's Visionary Approach:** Situated in the innovative terrain of Saudi Arabia's healthcare landscape, Johns Hopkins Aramco Healthcare (JHAH) represents a paradigm shift in digital healthcare innovation. As the first organization in the country to implement a comprehensive electronic health record (EHR) system, JHAH has achieved notable milestones in digital healthcare advancement.

B. **Specific Challenges Faced by JHAH**

   a. **Cultural Considerations:** When commencing to install its EHR ten years ago, one of the challenges JHAH faced was integrating technology within the traditional healthcare practices in Saudi Arabia. The cultural emphasis on personal interactions and face-to-face consultations posed a significant barrier to the adoption of digital health tools like Portals, Digital Patient engagement tools, EHRs, and telemedicine. At the time the EHR was being installed, patients and healthcare providers often preferred in-person consultations due to trust and established practices.

   To address these cultural nuances, JHAH implemented strategies that balanced the benefits of digital technologies with the cultural preferences of both patients and healthcare providers. For instance, JHAH focused on educating both groups about the advantages of EHRs, such as improved patient safety, better care coordination, and enhanced accessibility to health records as well as an extensive marketing campaign for its patient portal, showcasing all features and associated benefits. By conducting workshops, training sessions, and awareness campaigns, JHAH aimed to build trust and demonstrate the tangible benefits of the new system.

   b. **Regulatory Landscape:** The healthcare system in Saudi Arabia is governed by specific guidelines and standards set by local health authorities, which require planning and execution to ensure compliance. JHAH had to align the EHR implementation with these regulatory requirements to ensure data privacy, security, and patient safety.

   JHAH worked closely with regulatory bodies to ensure that their EHR system met all necessary compliance standards. This involved regular consultations with regulatory authorities. By maintaining a proactive approach to compliance, JHAH was able to mitigate risks and ensure a smooth implementation process.

Unique Patient Care Requirements: Saudi Arabia's healthcare needs differ from those in other regions, necessitating customized solutions. For example, managing chronic diseases and ensuring access to healthcare services in remote areas are critical issues that JHAH's EHR system needed to address. Additionally, the prevalence of chronic conditions such as diabetes and cardiovascular diseases requires a robust system capable of providing comprehensive care management and remote monitoring capabilities.

To meet these unique patient care requirements, JHAH tailored its EHR functionalities to support chronic disease management and telehealth services. The EHR system included features such as remote patient monitoring, automated alerts for chronic disease management, and integrated telemedicine capabilities. These enhancements enabled JHAH to provide high-quality care to patients in remote areas, improving access to healthcare services and overall patient outcomes.

c. **Risk of a Big Bang Go-Live:** One of the most significant risks JHAH undertook was executing a Big Bang go-live approach across all its hospitals and clinics with multiple EHR modules simultaneously. This ambitious strategy involved the rapid implementation of various EHR components, which heightened the potential for disruptions and required meticulous coordination and planning.
To manage this risk, JHAH employed a comprehensive project management approach, ensuring that all stakeholders were adequately prepared and that contingency plans were in place. Extensive training programs were conducted to ensure that staff were proficient in using the new system from day one. Despite the challenges, the successful execution of the Big Bang go-live demonstrated JHAH's commitment to a swift and efficient digital transformation, minimizing prolonged disruptions and quickly reaping the benefits of the new EHR system.

C. **Collaborative Framework with Johns Hopkins Medicine:** The collaborative framework established between JHAH and Johns Hopkins involved regular consultation meetings, joint decision-making processes, and the establishment of specialized committees to oversee various aspects of the EHR implementation. These committees, comprising members from both JHAH and Johns Hopkins, ensured that all decisions were well-informed and aligned with the strategic goals of both institutions.

a. **Partnership and Decision-Making Process:** The partnership between JHAH and Johns Hopkins was instrumental in guiding the EHR implementation process. Johns Hopkins provided expertise in clinical workflows, technology integration, and change management, helping JHAH navigate the complexities of digital transformation. This collaboration facilitated the adoption of best practices and ensured that the EHR system met the highest standards of patient care and operational efficiency.

b. **Adoption of Best Practices:** JHAH adopted several best practices from Johns Hopkins, including robust training programs, continuous system evaluation, and iterative improvements. For example, Johns Hopkins' approach to

continuous system evaluation and iterative improvements was integrated into JHAH's EHR strategy. This involved regular assessments of the EHR system's performance, gathering feedback from users, and implementing necessary changes to enhance system functionality and user experience.
The partnership also facilitated knowledge transfer and capacity building. Johns Hopkins provided observerships to JHAH EHR leadership staff, ensuring they were well-equipped to handle the new system. This included JHAH EHR leadership staff participating in a go-live at one of the Johns Hopkins facilities that went live. These hands-on training sessions, simulations, and continuous technical support helped build confidence and proficiency among JHAH EHR leadership staff.

3.1 Strategic Imperative: Physician Empowerment

Empowering physicians through governance structures, collaborative leadership, and championing them as change agents were pivotal strategies in the successful EHR implementation at JHAH. These approaches not only enhanced physician engagement and system adoption but also contributed to better clinical outcomes and higher user satisfaction. By fostering a collaborative leadership environment and leveraging the influence of physician champions, JHAH was able to navigate the complexities of digital transformation effectively (Vest and Gamm 2010; Sheikh et al. 2015).

A. ***Governance Framework Established by JHAH:*** Understanding that the successful integration of EHR transcended the realm of mere software implementation, JHAH aimed at sculpting the landscape of long-term patient care. Leveraging its strategic partnership with Johns Hopkins, JHAH erected a robust governance framework that intricately valued and incorporated insights from both clinical and non-clinical staff. This innovative structure ensured a holistic and inclusive approach, garnering comprehensive support from every specialized domain (Sheikh et al. 2015; Grol and Grimshaw 2003; Kruse et al. 2016).
JHAH's governance framework, both during the project mode and post-stabilization, played a critical role in the successful implementation and ongoing optimization of the EHR system, ensuring that the project stayed aligned with organizational goals and regulatory requirements. By establishing several specialized committees, each with specific responsibilities and decision-making authority, and ensuring efficient decision-making processes, JHAH was able to address challenges effectively and achieve significant improvements in patient care and operational efficiency.
Project Mode Governance Framework: In the project mode, JHAH established a robust governance framework to oversee the EHR implementation, ensuring alignment with organizational goals and regulatory requirements (Fig. 1). This structured governance framework ensured focused decision-

Fig. 1 Project mode governance framework

making and effective project management. Decisions were made at the lowest level possible and only escalated to higher levels when necessary, allowing for efficient resolution of issues and minimizing delays. This framework included several specialized committees, each with specific responsibilities and decision-making authority.

- **Executive Management Team (EMT)**: This team provided overall strategic direction and ensured that the project remained aligned with the organization's mission and goals.
- **EHR Steering Committee**: Responsible for overseeing the entire EHR project, this committee included senior leaders from various departments. They were tasked with making high-level decisions, addressing major project issues, and ensuring resource allocation.
- **EHR Ambulatory Steering Subcommittee**: Focused on the implementation of EHR systems in outpatient settings, this subcommittee addressed specific challenges and requirements related to ambulatory care.
- **EHR Inpatient Steering Subcommittee**: This subcommittee managed the implementation of EHR systems in inpatient settings, ensuring that the system met the needs of hospital-based care.
- **EHR MyChart Steering Subcommittee**: Dedicated to the patient portal, this subcommittee ensured that the MyChart system was user-friendly and provided patients with seamless access to their health information.
- **EHR Cadence Steering Subcommittee**: Focused on the scheduling aspects of the EHR system, this subcommittee worked on optimizing appointment scheduling and related workflows.
- **EHR Patient Advisory Subcommittee**: Included patient representatives who provided input on patient-facing aspects of the EHR system, ensuring that the system met patient needs and expectations.

Fig. 2 Post-stabilization governance framework

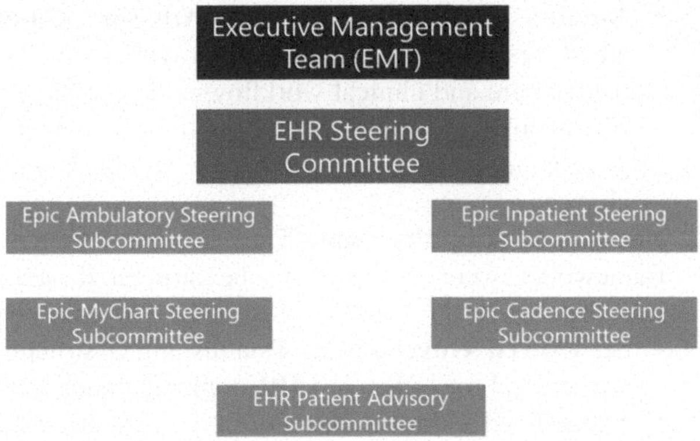

This structured governance framework ensured focused decision-making and effective project management. Decisions were made at the lowest level possible and only escalated to higher levels when necessary, allowing for efficient resolution of issues and minimizing delays.

1. **Post-Stabilization Governance Framework:** After the EHR implementation reached stabilization, JHAH transitioned to a governance framework that supported ongoing optimization and continuous improvement (Fig. 2). This framework included:
   - **Board of Managers**: Provided overall oversight and strategic direction for the EHR system post-implementation.
   - **Executive Committee**: Ensured that the EHR system continued to align with the organization's strategic objectives and addressed high-level issues.
   - **EHR Steering Committee**: Continued to oversee the EHR system, focusing on ongoing improvements and addressing any systemic issues that arose.
   - **Healthcare IT Committee**: Managed the technical aspects of the EHR system, ensuring system stability, security, and integration with other digital health tools.
   - **Clinical Leadership Council**: Comprised of clinical leaders, this council provided continuous feedback on the system's performance and recommended enhancements to meet evolving clinical needs.
   - **Cross Council Advisory Group**: Included representatives from various stakeholder groups, ensuring that diverse perspectives were considered in decision-making processes.
   - **Data & Documentation Governance Group**: Focused on data governance, ensuring data quality, integrity, and compliance with regulatory requirements.
   - **Epic@JHAH Project Management Office**: Managed ongoing projects related to the EHR system, coordinating efforts to implement new features and enhancements.
   - **Clinical Chiefs**: Provided leadership and guidance on clinical matters, ensuring that the EHR system supported high-quality patient care.

- **Nursing and Clinical Services Advisory Council**: Included nursing and allied health professionals who provided input on the system's impact on patient care and clinical workflows.
- **Physician Champions**: Acted as liaisons between the clinical staff and the governance committees, gathering feedback and advocating for necessary changes.

2. **Decision-Making Processes**: The decision-making processes in both governance frameworks were designed to be structured and inclusive. Key principles included:
   - **Structured Governance**: Establishing a structured governance framework was crucial for effective EHR implementation and ongoing management.
   - **Specialized Committees**: Forming specialized committees ensured focused decision-making, with each committee addressing specific aspects of the EHR system.
   - **Decisions at the Lowest Level**: Decisions were made at the lowest possible level to ensure efficiency. Issues were only escalated to higher levels for resolution if necessary and the higher-up committees were utilized to inform leadership about progress.

B. ***Empowering Through Collaborative Leadership:*** Empowering physicians through collaborative leadership was a cornerstone of the EHR implementation strategy at JHAH. This approach ensured that physicians were not only involved in decision-making processes but also felt a sense of ownership and responsibility toward the success of the system.

- **Involvement in Decision-Making**: Physicians were actively involved in the governance committees, such as the Clinical Advisory Committee and the EHR Steering Committee. Their participation in these committees ensured that their insights and expertise were integral to the decision-making process. This collaborative leadership model fostered a sense of inclusion and empowerment among the physicians (McKinney et al. 2020; Ting et al. 2017).
- **Leadership Training Programs**: JHAH implemented leadership training programs specifically designed for physicians. These programs aimed to equip physicians with the necessary skills to lead their peers and contribute effectively to the EHR implementation. Training covered areas such as change management, communication strategies, and technical proficiency with the EHR system (Obermeyer and Emanuel 2016).
- **Mentorship and Support**: Senior physicians and clinical leaders acted as mentors to their junior colleagues, providing guidance and support throughout the implementation process. This mentorship model helped to build a strong leadership pipeline and ensured continuity in collaborative leadership efforts (Topol 2019).
- **Regular Feedback and Iteration**: Regular feedback sessions were conducted to gather input from physicians on the functionality and usability of the EHR system. This feedback was used to make iterative improvements, ensuring that the system evolved in line with clinical needs. The collaborative approach to

feedback fostered a culture of continuous improvement and empowered physicians to drive meaningful changes.

C. ***Championing Physicians as Change Agents:*** At the core of JHAH's transformative journey lay the identification and empowerment of revered champions—individuals revered for their expertise and profound knowledge within their respective domains. These champions emerged as linchpins, steering pivotal decision-making processes, spearheading training initiatives, and championing system customization. Their influence engendered a profound sense of ownership among peers, fostering a collective dedication to system enhancement and refinement (Adler-Milstein and Huckman 2013). By positioning physicians as advocates for the new system, JHAH was able to facilitate smoother transitions and higher acceptance rates among clinical staff.
   - **Identification and Training of Champions**: JHAH identified physicians who were enthusiastic about the EHR system and had a natural ability to influence their peers. These physicians were given additional training to become EHR champions. They played a crucial role in advocating for the system, addressing concerns, and providing peer-to-peer support
   - **Role Modeling and Advocacy**: Physician champions acted as role models for their colleagues, demonstrating the benefits and effective use of the EHR system. Their advocacy helped to build trust and reduce resistance among other clinical staff. By sharing their positive experiences and success stories, physician champions encouraged their peers to embrace the new system.
   - **Peer Support Networks**: JHAH established peer support networks led by physician champions. These networks provided a platform for clinicians to share their experiences, seek advice, and collaboratively solve problems related to the EHR system. The peer support model facilitated knowledge sharing and fostered a supportive environment for change.
   - **Incentives and Recognition**: To further motivate physician champions, JHAH implemented a recognition program that acknowledged their contributions to the EHR implementation. Recognitions and awards were given to those who demonstrated exceptional leadership and made significant impacts on the adoption of the system. This recognition not only honored the champions but also inspired others to take on similar roles.

The impact of championing physicians as change agents was evident in several key areas:

- **Increased Adoption Rates**: The active involvement of physician champions led to higher adoption rates of the EHR system. Data indicated a 30% increase in adoption rates in departments where physician champions were actively involved compared to those without.
- **Improved User Satisfaction**: The support and advocacy of physician champions contributed to higher levels of user satisfaction. Surveys showed a 20% increase in satisfaction among clinical staff who had access to peer support from physician champions.

- **Reduced Resistance to Change**: The presence of physician champions helped to mitigate resistance to change. By addressing concerns and providing reassurance, champions played a critical role in easing the transition to the new system. Reports indicated a 15% reduction in resistance-related issues in areas with active champion involvement.
- **Enhanced Clinical Outcomes**: The effective use of the EHR system, driven by physician champions, led to improved clinical outcomes. For example, departments with strong champion leadership saw a 25% improvement in clinical documentation accuracy and a 10% reduction in adverse events.

D. ***Pivotal Role of Physician Engagement Throughout Phases—Transition from Users to Builders and Trainers*:** A key factor in the successful implementation of EHR systems at JHAH was the transition of physicians from mere users to active builders and trainers. This approach not only enhanced the system's functionality but also ensured greater acceptance and enthusiasm among the clinical staff. By leveraging their clinical expertise and providing specialized training programs, JHAH ensured that the EHR system was user-friendly, efficient, and aligned with clinical workflows. This approach not only improved staff proficiency and satisfaction but also enhanced patient safety and care quality, demonstrating the importance of physician engagement in healthcare technology adoption.
   - ***Validation and Implementation:*** Physicians were engaged in the early stages, participating in validation sessions during EHR implementation of EHR implementation. Their clinical expertise was invaluable in customizing the system to meet the specific needs of different departments and in choosing essential features and assessing content, ensuring alignment with clinical workflows and patient safety protocol. For instance, physicians provided input on the design of order entry workflows, clinical documentation templates, and decision support tools. Their involvement ensured that the EHR system was intuitive and aligned with clinical practices, leading to a more efficient and user-friendly system.
   - ***Building Phase and Physician Builders:*** JHAH's approach transcended traditional training paradigms by empowering physicians as builders, enabling them to take ownership of their respective areas and navigate complexities within the EHR system. This phase was crucial in fostering a sense of ownership and deep understanding among physicians, ensuring they were not just users but architects of the system.

- ***Transition to STS:*** After the initial implementation, many of these physician builders transitioned to roles as trainers, and Specialists Training Specialists (STS), for their peers. They conducted specialized training programs designed to equip their peers with the skills needed to effectively use the EHR system. These training sessions included hands-on workshops, simulations, and continuous education modules. The impact of these training programs was significant, as evidenced by increased staff proficiency and satisfaction. For example, feedback from staff indicated a 25% increase in confidence and competence in using the EHR system after undergoing training conducted by physician trainers. This cascading training model ensured a more organic and effective dissemination of knowledge and expertise among the medical staff. The specialized training programs developed by these physician trainers were tailored to address the unique needs of different clinical roles. These programs included:
  - **Hands-On Workshops**: Practical, scenario-based workshops where physicians could practice using the EHR system in a controlled environment. These workshops focused on common tasks such as order entry, patient documentation, and using clinical decision support tools.
  - **Simulations and Playground Environment Access**: High-fidelity simulations that replicated real-world clinical scenarios. These simulations allowed physicians to experience the EHR system's functionality in a realistic setting, helping them to understand how it would integrate into their daily workflows.
  - **Continuous Education**: Ongoing education sessions that provided updates on new features, system enhancements, and best practices. These sessions ensured that physicians remained proficient in using the EHR system and could take full advantage of its capabilities.
  - **Key Training Points:** By implementing these key points, JHAH ensured that their physicians were well-prepared and confident in using the EHR system, leading to better overall system adoption, higher user satisfaction, and improved patient care outcomes.

    Don't stop training at go-live, or new hires/new functionality (Onboarding education): Training should not end at the go-live phase. Continuous onboarding education is crucial for new hires and for introducing new functionalities to existing staff.

    Thrive training is needed with ongoing training programs tailored to different roles to maximize comfort and efficiency: Tailored training programs should be continuously provided to ensure all roles are comfortable and efficient in using the system.

    Role-specific, continuous training curriculum. Internally sourced & delivered by like roles. E.g., STS trainers: Continuous training should be role-specific and delivered by trainers who share similar roles to those being trained, such as STS trainers.

    And not to forget, workflow education, as important as digital education: Workflow education is just as important as digital education to ensure that staff understand how to integrate the EHR system into their daily workflows.

Hardwire training using a blended learning approach (e.g., classroom, virtual/e-learning, and playground): Employ a blended learning approach that includes classroom instruction, virtual/e-learning, and access to a playground environment for practical application.

Don't underestimate the value of personalization sessions and optimization rounds: Personalization sessions and optimization rounds are valuable for tailoring the system to individual needs and continuously improving its use.

- ***Stewards of the EHR:*** Post-implementation, Physicians at JHAH played a crucial role in the continuous improvement of the EHR system and assumed the role of stewards of the EHR, overseeing enhancements and system upgrades... Their ongoing involvement in recommending and implementing system enhancements ensured that the EHR system evolved to meet industry best practices as well as the changing clinical needs and incorporated feedback from frontline users.
  - **System Enhancements**: Physicians provided valuable insights into areas where the EHR system could be improved. For example, they identified the need for a more robust clinical decision support tool to assist with medication management. In response, JHAH integrated a real-time alert system for drug interactions, allergies, and contraindications. This enhancement led to a 30% reduction in adverse drug events, significantly improving patient safety
  - **Feedback Mechanism**: JHAH established a formal feedback mechanism where physicians could submit suggestions for system improvements. These suggestions were reviewed by the EHR Steering Committee and prioritized based on their potential impact on patient care and clinical workflows. This feedback loop ensured that the system remained responsive to the needs of its users and continuously improved over time.

E. ***Success Stories and Impact:*** The continuous involvement of physicians in recommending and implementing system enhancements resulted in several notable success stories:

- ***Enhanced Clinical Workflows:*** Physicians identified inefficiencies in the clinical documentation process and recommended changes to streamline workflows. As a result, JHAH implemented templates and auto-fill features that reduced the time required for documentation by 20%, allowing physicians to spend more time on direct patient care.
- ***Improved Patient Outcomes:*** The integration of advanced clinical decision support tools, based on physician recommendations, led to better clinical decision-making and improved patient outcomes. For instance, the addition of predictive analytics tools helped identify high-risk patients, enabling early interventions and reducing hospital readmissions by 15%.
- ***Increased User Satisfaction:*** Continuous enhancements and the active involvement of physicians in the EHR system's evolution contributed to

high levels of user satisfaction. Surveys conducted among clinical staff indicated a 20% increase in overall satisfaction with the EHR system, highlighting the positive impact of physician engagement on system usability and acceptance.

F. ***Measuring Success and Adoption*:** JHAH's monumental accomplishments, including the attainment of HIMSS Stage 6 shortly after implementation, the prestigious MyChart Diamond status, and the remarkable surge in patient adoption rates, aptly showcased the institution's fervent embrace and resounding success with EHR. This unwavering focus on engagement and empowerment substantiated high patient adoption rates, solidifying a steadfast commitment to patient-centric care (Duncan et al. 2022).

4 Outcomes, Insights, and Recommendations for Healthcare Organizations

The journey of implementing EHRs at Johns Hopkins Aramco Healthcare (JHAH) offers valuable lessons and insights for other healthcare organizations embarking on similar transformation endeavors aimed at elevating the digital maturity at enterprise level. Here are the key outcomes, insights, and recommendations derived from this comprehensive experience:

- ***Communication and Collaboration as Key Pillars:*** The success of EHR implementation underscores the critical importance of open communication channels and collaborative relationships. Active involvement of all stakeholders, including physicians, nurses, administrative staff, and IT professionals, in the decision-making processes is essential. This inclusive approach ensures that technological solutions are meticulously tailored to meet the intricate demands of clinical practice, fostering a sense of ownership and commitment across the organization.
- ***Tailored Training Programs as Catalysts*:** Tailored training programs that address specific workflow intricacies and incorporate real-life scenarios are indispensable. These initiatives help to bridge the gap between theoretical knowledge and practical application, facilitating the seamless integration of technology into clinical practice. By providing hands-on workshops, high-fidelity simulations, and continuous education sessions, healthcare organizations can significantly enhance staff proficiency and confidence in using advanced digital tools.
- ***Iterative Implementation for Continuous Improvement:*** Adopting a phased rollout and refining the system based on real-time feedback is crucial for successful digital transformation. This adaptive and iterative strategy allows for incremental enhancements, making it easier for stakeholders to accept and integrate new technologies. Continuous feedback mechanisms should be established to gather input from end-users, enabling ongoing improvements that align with the evolving needs of healthcare providers and patients.

- ***Sustainable Engagement and Roadmaps for Growth*:** The future of healthcare digitization promises significant advancements, and healthcare systems must be prepared to adapt to these changes while maintaining physician engagement. Sustainable engagement can be achieved by prioritizing continuous training, involving physicians in decision-making processes, and adopting scalable and secure technologies. This approach ensures that healthcare organizations can navigate the evolving digital landscape successfully, fostering a culture of continuous learning and innovation.
- ***Adapting to Evolving Digital Landscapes—Maintaining Physician Engagement:*** As healthcare systems adapt to continuously evolving digital landscapes, maintaining physician engagement remains crucial. Here are some strategies to ensure sustained engagement:
  1. **Continuous Training and Education**: Regular training programs should be conducted to keep physicians updated on new technologies and system enhancements. Continuous education ensures that physicians remain proficient and confident in using advanced digital tools.
  2. **Involvement in Decision-Making**: Physicians should be actively involved in the decision-making processes related to technology adoption and implementation. Their input is invaluable in ensuring that digital tools align with clinical workflows and patient care needs.
  3. **Feedback Mechanisms**: Establish robust feedback mechanisms to gather input from physicians on the usability and functionality of digital tools. This feedback should be used to make iterative improvements, ensuring that the systems evolve to meet the changing needs of healthcare providers.
- ***Adapting to Evolving Digital Landscapes—Adapting to Technological Changes:*** Healthcare organizations must remain agile and adaptable to leverage new technologies effectively. Key considerations include:
  1. **Scalability and Flexibility**: Digital systems should be scalable and flexible to accommodate future technological advancements and changing healthcare needs. This involves choosing platforms that can integrate with new tools and technologies as they emerge.
  2. **Patient-Centric Approaches**: Digital transformation efforts should always prioritize patient care. Technologies that enhance patient engagement, improve outcomes, and streamline care delivery should be at the forefront of adoption strategies.
  3. **Data Security and Privacy**: As digital tools become more pervasive, ensuring data security and patient privacy is paramount. Implementing robust cybersecurity measures and adhering to privacy regulations will build trust among patients and providers.
  4. **Collaborative Efforts**: Collaboration between healthcare providers, technology vendors, and regulatory bodies is essential for successful digital transformation. Shared goals and collaborative efforts will drive innovation and improve healthcare delivery.
- ***Measuring Digital Maturity:*** To ensure the effectiveness and impact of digital transformation projects, healthcare organizations must measure their digital maturity. In 2023, JHAH's achievement of HIMSS EMRAM Stage 7 and CHIME's

Most Wired Stage 9 for both acute and ambulatory care highlights the importance of this assessment. Digital maturity assessments provide a comprehensive evaluation of an organization's technological capabilities and readiness. Available tools and models such as the HIMSS EMRAM (Electronic Medical Record Adoption Model) and CHIME's Digital Health Most Wired survey offer standardized metrics to gauge an organization's digital maturity. These assessments help hospital management understand their current digital maturity level, identify gaps, and prioritize areas for improvement. Knowing the digital maturity score benefits hospital management by highlighting areas for improvement, benchmarking against peers, and guiding strategic planning to enhance patient care and operational efficiency. Measuring digital maturity is crucial for drawing a line in the sand regarding where an organization currently stands and where it needs to go. This process provides insights into digital strengths and weaknesses, supports strategic planning, and enhances overall organizational performance.

5 Conclusion

As JHAH commemorates their illustrious seven-year odyssey with EHR, their resounding success echoes the importance of fostering shared system ownership between IT infrastructure and end-users. Fueled by the influential voices of champions, the governance framework fostered an ingrained culture of continual enhancement and unparalleled patient care delivery. Their journey serves as an unparalleled testimony, illustrating how the empowerment and engagement of champions, along with alignment with industry benchmarks, served as pivotal catalysts propelling an organization's digital maturity toward international acclaim.

In summation, the ascension of digital maturity in healthcare pivots significantly upon the strategic emphasis on engaging and empowering physicians as collaborative partners. This harmonious fusion of their unparalleled expertise with innovative digital solutions not only elevates patient care standards and operational efficiency but also propels the entire healthcare landscape into an era brimming with technological empowerment and unparalleled possibilities.

JHAH's major accomplishments include obtaining HIMSS Stage 6 shortly after implementation and recently achieving HIMSS EMRAM Stage 7 and CHIME's Most Wired Stage 9 for both acute and ambulatory care. Additionally, JHAH received the KLAS Arch Collaborative EHR Experience Pinnacle for Physicians and Nursing, which is awarded to Arch Collaborative members who have attained the highest Net EHR Experience Scores (NEES) of all Arch Collaborative organizations. The institution also achieved Epic Gold Stars Level 10, became a four-time awardee of the Epic Honor Roll Cum Laude, and adopted the prestigious MyChart Diamond status. These accolades, along with the remarkable surge in patient adoption rates, aptly showcase JHAH's fervent embrace and resounding success with its EHR. This unwavering focus on engagement and empowerment substantiated high patient adoption rates, solidifying a steadfast commitment to patient-centric care.

References

Adler-Milstein J, Huckman RS (2013) The impact of electronic health record use on physician productivity. Am J Managed Care 19(SP 10):345–352. https://www.ajmc.com/view/the-impact-of-electronic-health-record-use-on-physician-productivity

Ashley EA (2016) Towards precision medicine. Nat Rev Genet 17:507–522. https://doi.org/10.1038/nrg.2016.86

Bates DW, Saria S, Ohno-Machado L et al (2014) Big data in health care: using analytics to identify and manage high-risk and high-cost patients. Health Aff (Project Hope) 33(7):1123–1131. https://doi.org/10.1377/hlthaff.2014.0041

Davenport T, Kalakota R (2019) The potential for artificial intelligence in healthcare. Future Healthc J 6(2):94–98. https://doi.org/10.7861/futurehosp.6-2-94

Duncan R, Eden R, Woods L et al (2022) Synthesizing dimensions of digital maturity in hospitals: systematic review. J Med Internet Res 24(3):e32994. https://doi.org/10.2196/32994

Grol R, Grimshaw J (2003) From best evidence to best practice: effective implementation of change in patients' care. Lancet 362(9391):1225–1230

Heintzman ND (2015) A digital ecosystem of diabetes data and technology: services, systems, and tools enabled by wearables, sensors, and apps. J Diabetes Sci Technol 10(1):35–41. https://doi.org/10.1177/1932296815622453. https://www.thelancet.com/journals/lancet/article/PIIS0140-6736(03)14546-1/abstract

Islam SMR, Kwak D, Kabir MH et al (2015) The internet of things for health care: a comprehensive survey. IEEE Access 3:678–708. https://doi.org/10.1109/ACCESS.2015.2437951

Khanra S, Dhir A, Islam AKMN et al (2020) Big data analytics in healthcare: a systematic literature review. Enterp Inf Syst 14(7):878–912. https://www.tandfonline.com/doi/full/10.1080/17517575.2020.1812005

Kruse CS, Kothman K, Anerobi K et al (2016) Adoption factors of the electronic health record: a systematic review. JMIR Med Inform 4(2):e19. https://doi.org/10.2196/medinform.5525

Kruse SC, Karem P, Shifflett K et al (2018) Evaluating barriers to adopting telemedicine worldwide: a systematic review. J Telemed Telecare 24(1):4–12. https://doi.org/10.1177/1357633X16674087

McKinney SM, Sieniek M, Godbole V et al (2020) International evaluation of an AI system for breast cancer screening. Nature 577:89–94. https://doi.org/10.1038/s41586-019-1799-6

Obermeyer Z, Emanuel EJ (2016) Predicting the future—big data, machine learning, and clinical medicine. N Engl J Med 375(13):1216–1219. https://doi.org/10.1056/NEJMp1606181

Raghupathi W, Raghupathi V (2014) Big data analytics in healthcare: promise and potential. Health Inf Sci Syst 2:3. https://doi.org/10.1186/2047-2501-2-3

Sheikh A, Sood HS, Bates DW (2015) Leveraging health information technology to achieve the "triple aim" of healthcare reform. J Am Med Inform Assoc 22(4):849–856. https://academic.oup.com/jamia/article/22/4/849/1746364

Ting DSW, Cheung CY, Lim G et al (2017) Development and validation of a deep learning system for diabetic retinopathy and related eye diseases using retinal images from multiethnic populations with diabetes. JAMA 318(22):2211–2223. https://doi.org/10.1001/jama.2017.18152

Topol EJ (2019) High-performance medicine: the convergence of human and artificial intelligence. Nat Med 25:44–56. https://doi.org/10.1038/s41591-018-0300-7

Vest JR, Gamm LD (2010) Health information exchange: persistent challenges and new strategies. J Am Med Inform Assoc 17(3):288–294. https://academic.oup.com/jamia/article/17/3/288/740136

Yu L, Lu Y, Zhu X (2012) Smart hospital based on internet of things. J Netw 7(10):1654–1661. https://www.semanticscholar.org/paper/Smart-Hospital-based-on-Internet-of-Things-Yu-Lu/9ee56096afa6ca1c28fac41ce3577ada09b6a2a3

Part II

Strategic Initiatives to Raise Digital Maturity

Catalonia's Journey to the Open Platform Paradigm in Healthcare

Jordi Piera-Jiménez and Gerard Carot-Sans

Abstract

The autonomous Spanish region of Catalonia is undergoing a significant transformation in its healthcare system through the adoption of an open platform paradigm, focusing on people-centred care, interoperability, and the strategic use of digital technologies. This approach aims to advance healthcare delivery by enhancing integration, flexibility, and user-friendliness within the digital health ecosystem. Key initiatives include the development of the Catalan eHealth Platform, the implementation of the Shared Electronic Health Record (HC3), and the deployment of various eHealth solutions such as ePrescription and the My Health portal (LMS). These efforts are driven by the region's commitment to improving health outcomes, fostering innovation, and ensuring the sustainability of the healthcare system.

Catalonia's digital health journey reflects its high digital maturity, marked by advanced Health information technologies (HIT) infrastructure and strategic use of health data to build value-added services. Despite challenges such as system fragmentation, data security, and the need for continuous stakeholder engagement, Catalonia's strategic planning and robust governance model have laid a strong foundation for this transition. The region's experience offers valuable insights for global healthcare systems aiming to leverage digital technologies for better patient care. Catalonia's journey towards an open platform paradigm highlights the critical role of collaboration, open standards, and a people-centric approach in achieving digital health transformation.

J. Piera-Jiménez (✉) · G. Carot-Sans
Catalan Health Service, Barcelona, Spain

Digitalization for the Sustainability of the Healthcare System (DS3) Research Group, IDIBELL Biomedical Research Institute, L'Hospitalet de Llobregat, Spain
e-mail: jpiera@catsalut.cat; gerard.carot@catsalut.cat

A. Scheuer, J. Studzinski (eds.), *Digital Maturity in Hospitals*,
https://doi.org/10.1007/978-3-031-80704-6_8

Keywords

openEHR · Health information technologies (HIT) · Catalan eHealth Platform · Shared Electronic Health Record (HC3) · Patient-centred care

Abbreviations

EHR	Electronic Health Record
FHIR	Fast Healthcare Interoperability Resources
HIT	Health Information Technologies
MPI	Master Patient Identifier

1 Introduction

Catalonia, a vibrant and industrious region in the northeast corner of Spain, is renowned for its rich history and cultural heritage as well as for its progressive and innovative approach to healthcare (Pérez Sust et al. 2020). Home to a diverse population of over 8 million people, the region operates within a complex healthcare system that strives to balance universal coverage with quality, efficiency, and sustainability. Amidst the evolving landscape of global healthcare, Catalonia has embarked on a transformative journey towards adopting an open platform paradigm; an innovative approach in health information technologies (HIT) aimed at revolutionising how healthcare is delivered, managed, and experienced.

The open platform paradigm in health is a forward-thinking approach that emphasises the importance of people-centred care, interoperability, open standards, and the strategic use of digital technologies to enhance health outcomes. Initially described by the Apperta Foundation in the United Kingdom, it represents a shift away from proprietary, siloed systems towards a more integrated, flexible, and user-friendly digital health ecosystem (Apperta Foundation 2018). This paradigm champions the idea that healthcare should be accessible, transparent, and seamlessly connected, enabling the free flow of health data to support patient care, research, and innovation.

This chapter explores Catalonia's journey towards this paradigm, detailing the strategic decisions, challenges, and achievements along the way. It will delve into the initial motivations for change, the policy framework and strategies deployed, the stakeholder engagement process, and the implementation phases of this ambitious transition. Furthermore, the chapter will highlight the impact and outcomes of adopting an open platform approach, offering insights into how this initiative has shaped the healthcare landscape in Catalonia and what the future may hold.

As we embark on this exploration, it is essential to understand the context in which Catalonia's healthcare system operates—the intricacies of its structure, the challenges it faces, and the innovative spirit that drives its evolution. While examining Catalonia's approach to embracing the open platform paradigm, this chapter sheds light on a pioneering journey that could inspire and inform similar endeavours across the globe, ultimately contributing to the ongoing transformation of healthcare for the betterment of all.

2 Background

2.1 The Spanish and Catalan Healthcare Systems: An Overview

2.1.1 The Spanish Healthcare System

The Spanish healthcare system is characterised by its universal coverage, which ensures health protection and healthcare for all citizens as established by the Spanish Constitution of 1978. The healthcare system is funded from general taxation, supplemented by payroll taxes, contributions from employers, and health insurance funds for civil servants. The system operates under the National Health System (SNS, from Spanish *Sistema Nacional de Salud*), which provides services largely free of charge at the point of delivery, with some co-payments tailored to the income required for pharmaceuticals for those under 65. Private voluntary insurance plays a minor role, mainly offering services that complement the statutory system, such as reduced waiting times for specialised care (Bernal-Delgado et al. 2018).

Healthcare governance in Spain is highly decentralised, with significant powers devolved to the 17 regional governments (Autonomous Communities, in Spanish *Comunidades Autónomas*), each responsible for the organisation and delivery of health services within its territory. This structure allows for a regional health ministry that oversees healthcare policies and a regional health service that executes these policies, providing primary and specialist care. Primary care is the gateway to specialist services, which require referral from a General Practitioner (GP) (Larizgoitia and Starfield 1997; Dedeu and Marti 2018). Public health expenditure in Spain is mainly allocated to specialist care (both inpatient and outpatient), primary healthcare, pharmaceuticals, and preventive measures.

2.1.2 The Catalan Healthcare System

Catalonia, as one of Spain's regions, has developed its own distinct healthcare model within the framework of the decentralised Spanish health system. The Catalan government, through the Department of Health and the Catalan Health Service (CatSalut), has full competencies in health services, promoting a model that integrates both public and private healthcare providers into a unified public network. CatSalut acts as the purchaser of healthcare services under the terms of harmonised agreements and public tenders, ensuring quality control, while a diverse array of public and private organisations deliver these services.

The Health Plan for Catalonia 2021–2025 outlines the strategic direction for the healthcare system, addressing challenges such as an ageing population, lifestyle-related health risks, and the rising prevalence of chronic diseases (Departament de Salut 2021). The plan emphasises four strategic lines: (1) equal opportunities in

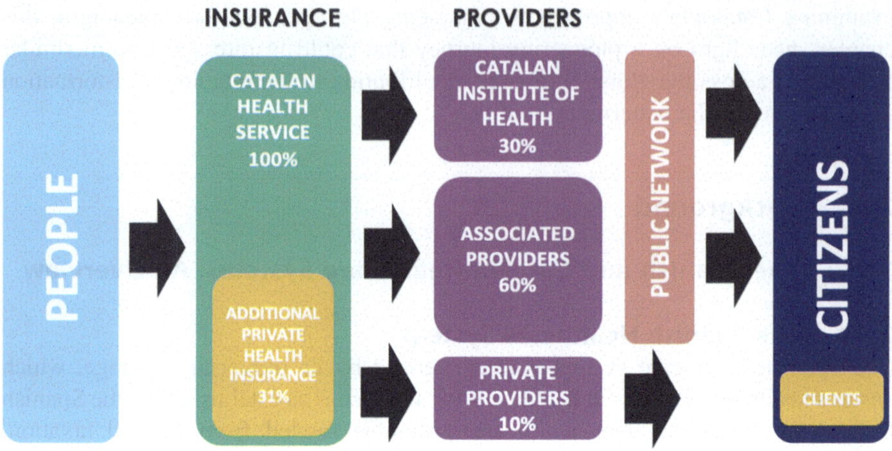

Fig. 1 Healthcare system organisation in Catalonia

health throughout life, (2) healthy environments, (3) integration of healthcare, and (4) transversal change levers (i.e. governance, health culture and citizen participation, leadership and participation of professionals, digital transformation, research and innovation towards health and system needs, and transparency).

Catalonia's healthcare system is marked by its provider diversity, with about 80% of specialised care and 20% of primary care delivered by providers not directly managed by the Department of Health. This has led to a separation of purchasing and providing functions, with CatSalut ensuring equitable access to healthcare services across the region (Fig. 1). The focus of the regional system on integrated care aims to improve coordination across different levels of care and between health and social services, facilitated by strategic planning and innovative use of information technologies (Aguilera et al. 2023; González-Colom et al. 2024; Baltaxe et al. 2019).

Despite its strengths, including high-quality care and broad access, the system faces several challenges, such as managing the needs of an ageing population, addressing health inequalities, and ensuring the sustainability of healthcare financing. Additionally, the fragmentation of IT systems deployed by the different service provider organisations, the rapid pace of technological advancements, and the increasing demand for personalised healthcare services have further complicated the landscape, necessitating innovative solutions to enhance system efficiency and patient outcomes.

2.2 Historical Perspective on the Adoption of Digital Health Technologies in Catalonia

Catalonia has long been at the forefront of integrating information technology into healthcare, with initiatives dating back several decades (Marimon-Suñol et al. 2010).

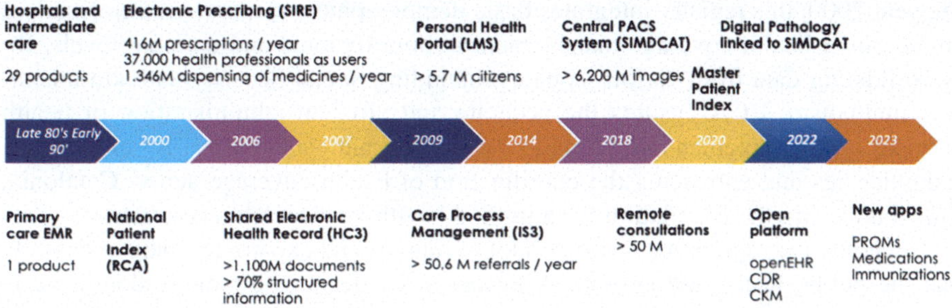

Fig. 2 Timeline of the main technologies and services conforming the Catalan eHealth Platform (1980s-2023)

Early efforts were focused on digitising patient records and administrative processes, laying the groundwork for more sophisticated digital health applications (Solans et al. 2021; Saigí-Rubió et al. 2021) Over time, the region has seen the implementation of various eHealth solutions and services, which nowadays conform the so-called Catalan eHealth Platform (Fig. 2).

These different initiatives carried out over time were driven by a vision to integrate service provider organisations and healthcare levels to achieve better health outcomes (Baltaxe et al. 2019; Cano et al. 2017; González-Colom et al. 2024). The idea was to improve healthcare delivery through information management and to facilitate more informed decision-making by healthcare professionals. However, despite the deployment of system-level technologies aimed at fostering a deeper integration across the Catalan ecosystem, the system remained hampered by fragmentation and difficulties in achieving semantic interoperability between different organisations and their respective IT systems.

In the following subsections, we describe the main technologies composing the Catalan eHealth Platform from a functional and technical perspective.

2.2.1 A Unique Patient Identifier: The Cornerstone of Data Integration

The Individual Health Card (TSI, from Catalan *Targeta Sanitària Individual*), issued by regional health authorities, embodies the commitment to accessible and interoperable healthcare services throughout Spain. With a unique personal identification code, the TSI facilitates the entitlement to and utilisation of National Health System services across different regions, ensuring continuity of care for citizens regardless of their location. The card is supported by regional databases that enable the efficient management and coordination of health services, underscoring the Spanish healthcare system's move towards greater integration and patient mobility (Spanish Government 2003). The interoperability of the TSI system, powered by a robust technical infrastructure, exemplifies Spanish leadership in adopting HIT to foster a people-centred healthcare ecosystem while avoiding patient identification errors (Riplinger et al. 2020; Parisi 2003; Sánchez 2011; Morris et al. 2014).

In Catalonia, the implementation of the TSI is linked to the Central Insured Registry (RCA), acting as a comprehensive database of insured individuals. Since the year 2000, this registry integrates basic demographic and insurance information, streamlining health coverage management at both regional and national levels. By consolidating data from various sources, including Social Security and administrative mutualism, RCA ensures the accurate and efficient administration of health services. Its implementation marks a significant step towards eliminating redundancies and enhancing the coordination of health coverage across Catalonia, thus reinforcing the foundation for a unified health service delivery system.

Catalonia is a region that receives a lot of tourists and expatriates, who eventually use the public healthcare system. A historical barrier to the patient identification model was the lack of a unique identifier for non-insured citizens. This situation was particularly complex during the COVID-19 pandemic, when many non-insured citizens had to be vaccinated (Vela et al. 2022a; Monterde et al. 2021). Because of that, since the year 2021, Catalonia has been promoting the introduction of a Master Patient Identifier (MPI) aiming to monitor every individual who accesses medical care within the Catalonian healthcare system, going beyond just tracking the insured population. The groundwork for Catalonia's MPI is established on the Fast Healthcare Interoperability Resources (FHIR) standard, an innovative protocol created by HL7 for the electronic exchange of healthcare data. Implementing MPI via FHIR offers a sturdy, expandable, and interoperable approach for the identification and monitoring of patients. Utilising the adaptable capabilities of FHIR, such as the Patient resource, allows to capture an array of patient identifiers and demographic details, thus guaranteeing precise patient alignment and recognition across varied healthcare providers and infrastructures.

2.2.2 ePrescription System: The First System-Level Service

The electronic prescribing system of Catalonia, designed to streamline medication management across the region, consists of two interconnected components: SIRE (from Catalan *Sistema Integrat de Recepta Electrònica*) and SIFARE (from Catalan *Sistema d'Informació Farmacèutic de Recepta Electrònica*) (Generalitat de Catalunya 2023; Government Spain 2010). Deployed in 2003, ePrescription was the first service deployed at scale across the Catalan health ecosystem and the first component of the Catalan eHealth Platform (Gilabert-Perramon et al. 2010).

SIRE, the integrated electronic prescription system, was specifically developed to be utilised for outpatient prescriptions across the Catalan public healthcare system. Its implementation led to a significant shift from traditional paper-based prescriptions to a digital format, improving the efficiency, accuracy, and safety of prescribing practices. The system facilitates the electronic generation, transmission, and tracking of prescription orders, allowing healthcare professionals to manage and review patients' medication plans digitally. This digital transformation not only reduces the administrative burden associated with paper prescriptions but also enhances the quality of care by minimising medication errors and improving prescription accuracy.

The SIFARE component complements SIRE by providing a platform for community pharmacies to manage the dispensation of medications prescribed electronically. Through SIFARE, pharmacies can access, review, and process electronic prescriptions, ensuring that patients receive the correct medications in a timely manner. The system supports data collection for invoicing purposes, enabling seamless financial transactions and reimbursement processes between community pharmacies and CatSalut.

The collaboration between SIRE and SIFARE exemplifies a holistic approach to medication management, from prescription by healthcare providers to dispensation by community pharmacies, fostering a more integrated healthcare ecosystem. The implementation of these systems has led to improved patient convenience, as individuals no longer need to physically visit healthcare providers for repeated prescriptions (e.g. chronic medication). Additionally, the systems enhance medication adherence by providing clear, accurate, and accessible medication plans, thereby contributing to better health outcomes for patients across Catalonia.

2.2.3 Shared Electronic Health Record of Catalonia: A Shift from Multiple EMRs to a Comprehensive EHR

Launched in 2008, the Shared Electronic Health Record of Catalonia (HC3) represents a landmark in the effort to consolidate patient health information into a single, accessible electronic health record (EHR) (Marimon-Suñol et al. 2010). This system pools data from various sources, including different EMRs and diagnostic tools, into a centralised repository. HC3 facilitates a longitudinal view of a patient's healthcare journey, enabling healthcare professionals across disciplines to access and contribute to a comprehensive patient record. The implementation of HC3 has significantly improved the quality of healthcare delivery, aiding in clinical decision-making and promoting a more collaborative approach to patient care (Aguilera et al. 2023; González-Colom et al. 2024; Cases et al. 2023; Ruiz-Ramos et al. 2024; Vela et al. 2022b; González-Colom et al. 2023; Valero-Bover et al. 2022a, b).

From a technical perspective, HC3 was conceived with the fragmentation of Catalonia's health information systems in mind. It aggregates information adhering to international standards for messaging and semantic interoperability, such as HL7-CDA, SNOMED-CT, and LOINC, ensuring that data can be seamlessly shared and understood across different healthcare providers. Security measures are rigorously applied, allowing access only through a virtual private network, safeguarded by multi-level authentication protocols to ensure patient privacy and data protection. This strategic approach to both integration and security has positioned HC3 as the core component of the Catalan eHealth Platform.

2.2.4 Personal Health Folder: In the Search for Patient Empowerment

The platform "My Health" (LMS, for Catalan *La Meva Salut*) represents a cornerstone in Catalonia's people-centric approach to healthcare digitalisation. First deployed in 2009, this personal health portal aims to empower patients by providing comprehensive access to their health information online. It reflects a broader shift towards engaging patients in their healthcare journey, moving beyond the traditional paternalistic patient-provider dynamic. LMS allows individuals to view medical records, test results, medications, vaccinations, digital imaging, and allergies, significantly enhancing transparency and fostering a more informed population. The platform is an example of Catalonia's commitment to leveraging digital solutions to bridge information gaps and improve healthcare experiences for its citizens.

LMS integrates information from HC3, ensuring that patients have a holistic and up-to-date view of their health status and ongoing treatments. Beyond this core feature, LMS is distinguished by its user-friendly interface and array of functionalities designed to facilitate active patient involvement in healthcare management. Patients can schedule, reschedule, or cancel appointments with healthcare professionals directly through the portal, thereby increasing accessibility to healthcare services and reducing administrative overhead for healthcare facilities. Additionally, LMS offers detailed insights into health plans, including service coverage, which aids patients in navigating the healthcare system more effectively. The platform also features a wealth of health resources and educational materials, providing guidance on preventive care and health maintenance.

Technologically, LMS feeds from information produced by other services and technologies of the Catalan eHealth Platform, such as HC3, SIRE, or synchronous and asynchronous remote consultations (Solans et al. 2021; Rodríguez-Molinero et al. 2023). The platform's architecture ensures secure and real-time access to health data, adhering to strict privacy and data protection standards to safeguard patient information. Security protocols are meticulously implemented, requiring robust authentication mechanisms for user access, thereby maintaining the integrity and confidentiality of health data.

The adoption of LMS among the population was significantly accelerated by the COVID-19 pandemic (moving from 600k users in March 2020 to roughly 6M in March 2024), which acted as a catalyst for increasing digital engagement with healthcare services. The platform expanded its functionalities to include digital certificates for COVID-19 tests and vaccinations, renovation of electronic prescriptions, and sick leave forms, leading to a substantial increase in registered users and daily access rates (from 2k to 170k unique daily accesses).

2.2.5 Care Process Management: In the Search for Normalising Care Pathways

The introduction of the Care Process Management system (IS3) in 2014 was a significant advancement towards integrating healthcare and social services. IS3 is designed to navigate the complexities of healthcare delivery that often spans multiple organisations and care settings. By fostering interoperability and data sharing, IS3 enables a seamless flow of information, supporting a continuum of care that is responsive to the evolving needs of citizens. This system illustrates Catalonia's commitment to a holistic approach to health and social care, leveraging digital technologies to enhance the coordination and efficiency of services.

The architecture of IS3 incorporates cutting-edge technologies to ensure robust interoperability across the health and social care spectrum. Features like a message broker and a rule manager facilitate centralised communication and the application of business logic, while monitoring tools and a terminological server ensure data consistency and meaningful use. This technical backbone supports the system's aim to improve healthcare experiences, reduce administrative burdens, and optimise resource utilisation across Catalonia's healthcare landscape. Through IS3, Catalonia has pioneered an integrated care model that embodies the future of health and social service delivery.

2.2.6 Federation of Technological Infrastructures and Services: Aiming to Optimise Resources and Increase Cybersecurity

The concept of federated infrastructures in Catalonia's healthcare system represents a transformative approach towards managing and sharing healthcare data across different services and providers. These platforms, which include initiatives like SIMDCAT (Digital Medical Imaging System of Catalonia) and DigiPATICS (Digital Pathology for the Catalan Institute of Health) (Temprana-Salvador et al. 2022), are core examples within the digital transformation strategy of the regional healthcare system. These federated infrastructures are centralised hubs that aggregate, store, and facilitate access to critical healthcare information, such as medical images and pathology reports, from multiple sources. From a clinical perspective, this approach significantly improves the efficiency of diagnostic processes, enhances the quality of patient care, and fosters collaboration among healthcare professionals. Furthermore, the deployment of such infrastructures allows reducing costs by taking advantage of the economies of scale while improving overall cybersecurity.

SIMDCAT was the first example to test this approach. It facilitates the management and access of medical imaging data across the healthcare system, eliminating redundancies and facilitating quicker, more accurate diagnoses. This platform allows healthcare providers to access a patient's imaging history from any location, supporting informed clinical decision-making and improved patient outcomes. Similarly, DigiPATICS integrates digital pathology data into a centralised model (leveraging the SIMDCAT infrastructure), enhancing the accessibility and efficiency of pathology services. These platforms not only improve clinical workflows but also pave the way for advanced analytical techniques, such as artificial intelligence solutions for diagnostics and treatment planning.

The transition to these federated infrastructures marks a shift from a fragmented model, where each healthcare provider managed its own infrastructure, to a more unified and efficient system. This centralised approach addresses challenges related to data sharing, interoperability, and cybersecurity, enabling a seamless exchange of information across Catalonia's healthcare ecosystem. As these platforms continue to evolve, they will play a crucial role in facilitating innovation, improving healthcare delivery, and ensuring the sustainability of the public healthcare system.

2.3 The Need for Change: Towards Patient-Centric Care

The limitations of Catalonia's digital health infrastructure started to emerge more clearly as the need for services that were both more integrated and focused on the patient increased. The healthcare data, trapped in proprietary systems and the associated efforts to make it shareable across various healthcare environments, resulted in inefficiencies, redundant efforts, higher costs, and obstacles to expanding innovative solutions (Fig. 3).

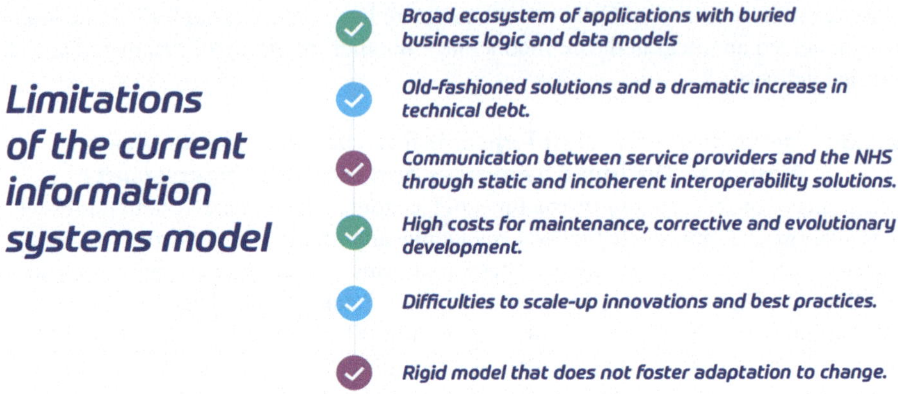

Fig. 3 Limitations of the Catalan information systems model in healthcare

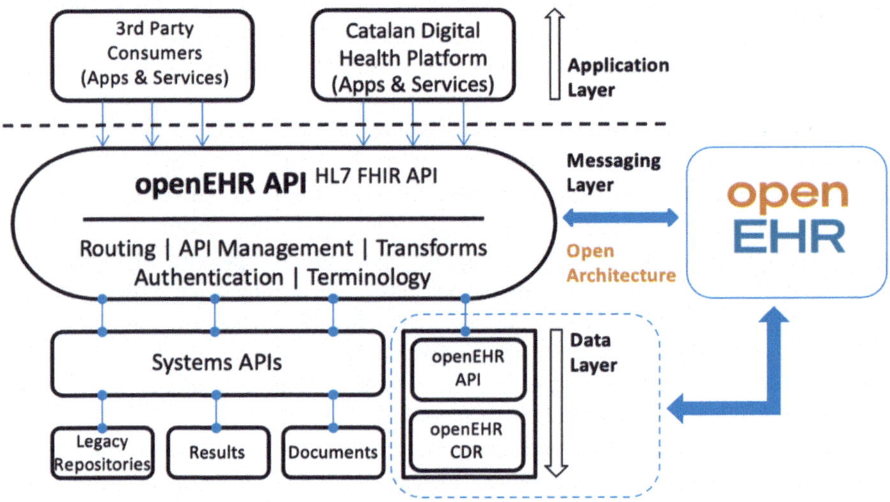

Fig. 4 High-level architecture for the open platform (adapted from the Apperta Foundation 2018)

The fragmentation of HIT was probably among the limitations that contributed the most to the onset of a strategic reflection process. The realisation that improving healthcare delivery required a foundational change in how health information was managed, shared, and utilised became apparent. In that context, the open platform paradigm emerged as the only solution, advocating for patient-centricity, data liquidity, and security in digital health solutions (Apperta Foundation 2018). It represented a significant departure from the status quo, emphasising the need for open standards, shared data platforms, and a collaborative approach to healthcare innovation (Fig. 4).

This new approach was driven by a clear vision: to transcend traditional barriers in health information exchange and unlock new possibilities for care coordination, patient empowerment, and health system efficiency (Pedrera-Jiménez et al. 2023). When aligning with the open platform paradigm, Catalonia set forth on a path to revolutionise its healthcare system, aiming to foster a more connected, accessible, and resilient health ecosystem. It was an acknowledgment that the future of healthcare would be increasingly digital and that embracing openness, collaboration, and innovation was essential to meeting the needs of both patients and healthcare providers in the twenty-first century.

3 The Journey Towards the Open Platform Begins

3.1 Laying the Groundwork: Policy Framework and Strategic Vision

Catalonia's transition to the open platform paradigm in health did not happen overnight. It has been the culmination of strategic planning, policy formulation, and a clear vision for the future of healthcare in the region. Central to this effort was the development of a comprehensive policy framework in 2017 that outlined the principles, goals, and actions required to achieve a seamless, interoperable health information system (Piera-Jiménez et al. 2020). A bottom-up approach, facilitated by IT leaders across Catalonia and participated by more than 300 individuals across the health sector, culminated with the first Digital Health strategy for Catalonia 2018–2022. Consisting of fifteen different strategic initiatives, grouped in 4 vectors, the plan was funded with 40 million euros for its implementation in January 2020 (Fig. 5).

The main objective of this strategy is integrated and people-centred care, supported by an information systems model that systematically uses health data to build value-added services, which encourages innovation, and is built on a collabo rative model based on open standards.

In order to respond to this main objective, a series of secondary objectives were identified:

1. Construction and deployment of the Catalan EHR.
2. Deployment of an information systems model based on the paradigm of open platforms.
3. Deployment of innovation through the systematic use of health data.
4. Establishment of a robust IT governance model with the participation of all stakeholders in the Catalan health ecosystem.

This policy groundwork was supported by a robust strategic vision from the Catalan Health Plan 2016–2020 that highlighted digital health as a key factor to transform the healthcare sector (Catalan Department of Health 2016). The goal was

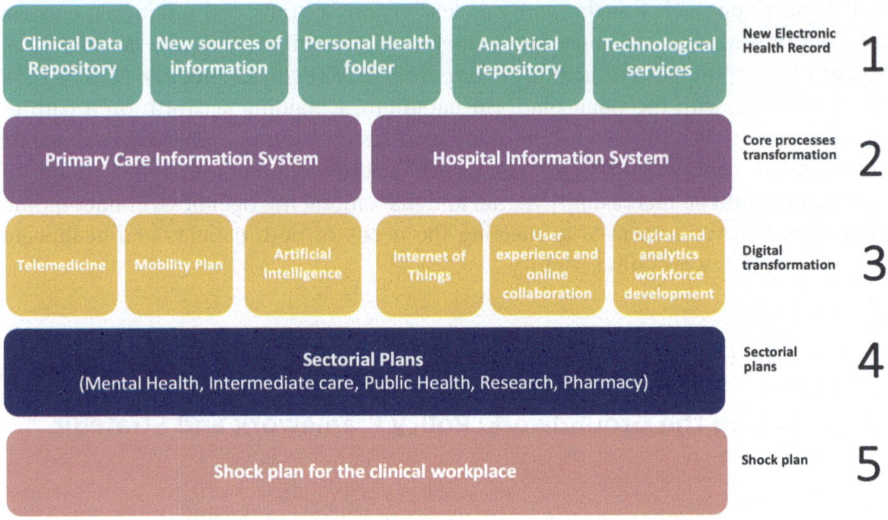

Fig. 5 Strategic initiatives and vectors in the Digital Health Strategy for Catalonia 2018–2022

clear: to create a unified health ecosystem that leverages digital technologies to improve care delivery, enhance patient outcomes, and optimise system efficiency by the systematic use of HIT.

The Catalan EHR will be the main outcome of the Digital Health Strategy for Catalonia becoming the backbone of a new information systems model in healthcare. This EHR seeks to offer a more horizontal vision in care delivery, overcoming the barriers of levels of care to move towards collaborative models between organisations and health providers, and to promote the sustainability of the health system through economies of scale.

3.2 Engaging Stakeholders: Building a Collaborative Ecosystem

Acknowledging that the success of the open platform paradigm hinged on widespread acceptance and cooperation, Catalonia has embarked on an extensive stakeholder engagement process. Workshops, seminars, and forums are being organised to gather input, address concerns, and foster a sense of ownership among all participants. Healthcare providers have been briefed on the benefits of semantic interoperability and the role of open standards in facilitating better care coordination. Furthermore, patients and patient advocacy groups have been engaged to ensure that the new system's design and functionality reflects their needs and preferences.

This collaborative approach is reinforced by a strong governance model led by CatSalut and participated by the service provider organisations (public and private). This is helping to build consensus on the importance of moving towards an open platform paradigm and has laid the foundation for its successful implementation.

Overall, this process aims to highlight the critical role of change management in navigating the transition from legacy systems, addressing resistance, and cultivating a culture of innovation within the healthcare sector.

3.3 Investing in Infrastructure: Technology, Training, and Resources

A key component of Catalonia's journey has been the investment in the necessary infrastructure to support the open platform paradigm. This includes upgrading existing IT systems, deploying new software solutions that adhere to open standards, and ensuring robust data security measures. Significant resources have been allocated to implement and maintain a centralised clinical data repository, which is core to the infrastructure and will facilitate the seamless exchange and analysis of health information across different care settings.

In addition to technological investments, Catalonia recognises the importance of human capital in realising its vision. Therefore, training programmes have been launched to equip healthcare professionals with the skills needed to navigate the new digital landscape effectively. These efforts will be complemented by public awareness campaigns aimed at educating patients on the benefits of the open platform approach and how data liquidity empowers them in their healthcare journey.

In the period 2024–2030, Catalonia will be investing additional 140 million euros in finalising the main components of the open platform, and in developing and implementing the Catalan EHR.

3.4 Implementation Plan: Current Situation and Next Steps

The implementation process of Catalonia's Digital Health Strategy and its main components (open platform, Catalan EHR, and IT governance model) is not a simple process and requires meticulous planning. This section summarises the planning proposal that should help achieve the goals set by the strategy for the year 2030 (Fig. 6).

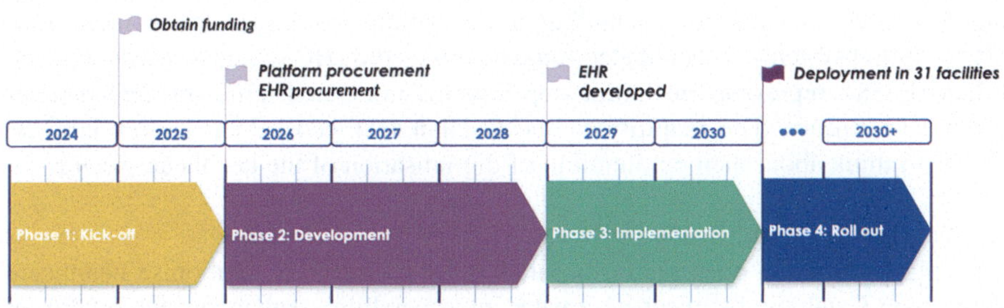

Fig. 6 Digital Health Strategy 2024–2030 implementation roadmap

Considering the complexity of the overall implementation strategy, a roadmap with four differentiated phases was developed:

Phase 1: Kick-off (2024–2025)—It corresponds to all the activities that must help to define the characteristics of the open platform and its associated services, as well as the functional definition of the Catalan EHR. In this phase, the IT governance model is deployed and the necessary funding is sought to address the next phases of the plan. In addition, the recruitment process is launched, which will allow progress to the next phase. The planning for this start-up phase is extremely detailed in order to ensure the success of the project.

Phase 2: Development (2026–2028)—In the second phase, the platform services and the Catalan EHR are developed. At the end of this phase, a minimum viable product of the EHR will be available.

Phase 3: Implementation (2029–2030)—In this phase, the Catalan EHR is deployed in the context of the Catalan Institute of Health and at the Hospital Clínic de Barcelona. These two processes of migration and implementation, extremely complex and with a great deal of change management, will be used to industrialise the deployment process for the forthcoming implementations.

Phase 4: Roll out (2030+)—In the last phase, the Catalan EHR is successively deployed in the different service provider organisations across the region.

4 Conclusion

The transformative journey of Catalonia towards the open platform paradigm in healthcare aims to mark a step forward in redefining how healthcare is delivered, managed, and experienced not only within the region but also as a model for global healthcare systems. This ambitious transition underscores the region's commitment to innovation, people-centred care, and the strategic employment of digital technologies to enhance health outcomes.

Key Achievements and Learnings

This initiative showcases the importance of strategic planning, stakeholder engagement, and a phased implementation approach. The establishment of a comprehensive digital health infrastructure, including the adoption of unique patient identifiers, ePrescription systems, a shared electronic health record (HC3), and a personal health folder (LMS), represents an initial step towards integrated, efficient, and patient-centred healthcare. The maturity of the Catalan HIT ecosystem has brought the region to think about the next iteration of digitalisation of the healthcare sector.

Implications for Healthcare Delivery

The shift towards an open platform paradigm is poised to revolutionise healthcare delivery in Catalonia by ensuring greater interoperability, enhancing the quality of care, and facilitating seamless access to health data. This approach should not only improve patient outcomes but also foster innovation, support research, and enhance the sustainability of the healthcare system.

Challenges and Opportunities
While the journey has been marked by notable achievements, it has also encountered challenges, including the fragmentation of HIT, data security, and the need for continuous stakeholder engagement. These challenges present opportunities for learning and improvement, particularly in enhancing system flexibility and resilience against evolving healthcare demands.

Future Directions
Looking ahead, Catalonia's journey towards an open platform paradigm in healthcare is far from complete. The continuous evolution of digital health technologies, coupled with the growing emphasis on personalised medicine and patient empowerment, will necessitate ongoing adaptation and innovation. Further investments in infrastructure, training, and research are essential to fully release the potential of this transformative approach.

Global Relevance
Catalonia's experience serves as a valuable blueprint for other regions and countries aiming to harness the power of digital health technologies to improve healthcare delivery. The lessons learned from Catalonia's journey underscore the critical role of collaboration, open standards, and a patient-centric approach in the successful implementation of digital health strategies. As this journey continues, it holds the promise of not only transforming healthcare in Catalonia but also inspiring and guiding similar endeavours worldwide, ultimately contributing to the betterment of global healthcare.

References

Aguilera CC, Vallespín EV, Escayola MC et al (2023) Evaluación poblacional del impacto del nivel socioeconómico en los resultados clínicos en pacientes con insuficiencia cardiaca en entornos de atención integrada. Rev Esp Cardiol 76(10):803–812. https://doi.org/10.1016/j.rec.2023.03.009

Apperta Foundation (2018) Defining an Open Platform. https://apperta.org/assets/Apperta_Defining_an_Open_Platform.pdf. Accessed 25 July 2024

Baltaxe E, Cano I, Herranz C et al (2019) Evaluation of integrated care services in Catalonia: population-based and service-based real-life deployment protocols. BMC Health Serv Res 19, 370:1–11. https://doi.org/10.1186/s12913-019-4174-2

Bernal-Delgado E, Garcia-Armesto S, Oliva J et al (2018) Spain health system review. Health Syst Transit. Copenhagen: WHO Regional Office for Europe 26(3):1–179

Cano I, Dueñas-Espín I, Hernandez C et al (2017) Protocol for regional implementation of community-based collaborative management of complex chronic patients. NPJ Prim Care Respir Med 27(1):44. https://doi.org/10.1038/s41533-017-0043-9

Cases L, Vela E, Santaeugènia Gonzàlez SJ et al (2023) Excess mortality among older adults institutionalized in long-term care facilities during the COVID-19 pandemic: a population-based analysis in Catalonia. Front Public Health 11:1208184. https://doi.org/10.3389/fpubh.2023.1208184

Catalan Department of Health (2016) Health Plan for Catalonia 2016–2020. https://scientiasalut.gencat.cat/bitstream/handle/11351/4040/pla_salut_catalunya_2016_2020_2016_ang.pdf?sequence=5&isAllowed=y. Accessed 25 July 2024

Dedeu T, Marti T (2018) Multiprofile primary health care teams in Catalonia, Spain: a population-based effective model of services delivery: good practice brief. World Health Organization. Regional Office for Europe. https://iris.who.int/handle/10665/345705. Accessed 25 July 2024

Departament de Salut (2021) Pla de salut de Catalunya 2021-2025. https://scientiasalut.gencat.cat/bitstream/handle/11351/7948/pla_salut_catalunya_2021_2025_2021.pdf?sequence=1&isAllowed=y. Accessed 25 July 2024

Generalitat de Catalunya (2023) What is an electronic prescription and how does it work? https://catsalut.gencat.cat/ca/serveis-sanitaris/atencio-farmaceutica/recepta-electronica/que-es-i-com-funciona/index.html#googtrans(calen). Accessed 25 July 2024

Gilabert-Perramon A, López-Calahorra P, Escoda-Geli N et al (2010) Receta electrónica en Cataluña (Rec@t): una herramienta de salud. Med Clin (Barc) 134:49–55. https://doi.org/10.1016/S0025-7753(10)70010-5

González-Colom R, Herranz C, Vela E et al (2023) Prevention of unplanned hospital admissions in multimorbid patients using computational modeling: observational retrospective cohort study. J Med Internet Res 25:e40846. https://doi.org/10.2196/40846

González-Colom R, Carot-Sans G, Vela E et al (2024) Five years of hospital at home adoption in Catalonia: impact, challenges, and proposals for quality assurance. BMC Health Serv Res 24:154. https://doi.org/10.1186/s12913-024-10603-1

Government Spain (2010) Real Decreto 1718/2010, de 17 de diciembre, sobre receta médica y órdenes de dispensación. https://www.boe.es/eli/es/rd/2010/12/17/1718/con. Accessed 25 July 2024

Larizgoitia I, Starfield B (1997) Reform of primary health care: the case of Spain. Health Policy (New York) 41(2):121–137. https://doi.org/10.1016/s0168-8510(97)00017-1

Marimon-Suñol S, Rovira-Barberà M, Acedo-Anta M et al (2010) Historia Clínica Compartida en Cataluña. Med Clin (Barc) 134:45–48. https://doi.org/10.1016/s0025-7753(10)70009-9

Monterde D, Carot-Sans G, Cainzos-Achirica M et al (2021) Performance of three measures of comorbidity in predicting critical COVID-19: a retrospective analysis of 4607 hospitalized patients. Risk Manag Healthc Policy 14:4729–4737. https://doi.org/10.2147/RMHP.S326132. Accessed 25 July 2024

Morris G, Farnum G, Afzal S et al (2014) Patient identification and matching. https://www.healthit.gov/sites/default/files/patient_identification_matching_final_report.pdf. Accessed 25 July 2024

Parisi LL (2003) Patient identification: the foundation for a culture of patient safety. J Nurs Care Qual 18(1):73–79. https://doi.org/10.1097/00001786-200301000-00010

Pedrera-Jiménez M, García-Barrio N, Frid S et al (2023) Can OpenEHR, ISO 13606, and HL7 FHIR work together? An agnostic approach for the selection and application of electronic health record standards to the next-generation health data spaces. J Med Internet Res 25:e48702. https://doi.org/10.2196/48702

Pérez Sust P, Solans O, Fajardo JC et al (2020) Turning the crisis into an opportunity: digital health strategies deployed during the COVID-19 outbreak. JMIR Public Health Surveill 6(2). http://publichealth.jmir.org/2020/2/e19106

Piera-Jiménez J, Rodríguez JR, Pérez P (2020) New Catalonian digital health strategy: a presentation. Health Manag 20(2). https://healthmanagement.org/c/hospital/issuearticle/new-catalonian-digital-health-strategy-a-presentation. Accessed 25 July 2024

Riplinger L, Piera-Jiménez J, Dooling JP (2020) Patient identification techniques – approaches, implications, and findings. Yearb Med Inform 29(01):81–86. https://doi.org/10.1055/s-0040-1701984

Rodríguez-Molinero A, Carot-Sans G, Escrig R et al (2023) Study protocol of a randomized controlled trial to assess safety of teleconsultation compared with face-to-face consultation: the ECASeT study. Trials 24:797. https://doi.org/10.1186/s13063-023-07679-1

Ruiz-Ramos J, Vela E, Monterde D et al (2024) Healthcare risk stratification model for emergency departments based on drugs, income and comorbidities: the DICER-score. BMC Emerg Med 24(1):23. https://doi.org/10.1186/s12873-024-00946-7

Saigí-Rubió F, Vidal-Alaball J, Torrent-Sellens J et al (2021) Determinants of Catalan public primary care professionals' intention to use digital clinical consultations (eConsulta) in the post–COVID-19 context: mixed methods study. J Med Internet Res 23(6):e28944. https://www.jmir.org/2021/6/e28944

Sánchez XM (2011) Errores de identificación de pacientes [Patient identification errors]. Enferm Clin 21(5):295–296. https://doi.org/10.1016/j.enfcli.2011.07.006

Solans O, Vidal-Alaball J, Roig Cabo P et al (2021) Characteristics of citizens and their use of teleconsultations in primary care in the Catalan public health system before and during the COVID-19 pandemic: retrospective descriptive cross-sectional study. J Med Internet Res 23(5): e28629. https://doi.org/10.2196/28629

Spanish Government (2003) Ley 16/2003, de 28 de mayo, de cohesión y calidad del Sistema Nacional de Salud. https://www.boe.es/buscar/pdf/2003/BOE-A-2003-10715-consolidado.pdf. Accessed 25 July 2024

Temprana-Salvador J, López-García P, Castellví Vives J et al (2022) DigiPatICS: digital pathology transformation of the Catalan health institute network of 8 hospitals—planification, implementation, and preliminary results. Diagnostics (Basel) 12(4):852. https://doi.org/10.3390/diagnostics12040852

Valero-Bover D, Fradera M, Carot-Sans G et al (2022a) Impact of the COVID-19 pandemic on the incidence of suicidal behaviors: a retrospective analysis of integrated electronic health records in a population of 7.5 million. Int J Environ Res Public Health 19(21):14364. https://doi.org/10.3390/ijerph192114364

Valero-Bover D, González P, Carot-Sans G et al (2022b) Reducing non-attendance in outpatient appointments: predictive model development, validation, and clinical assessment. BMC Health Serv Res 22:451. https://doi.org/10.1186/s12913-022-07865-y

Vela E, Carot-Sans G, Clèries M et al (2022a) Development and validation of a population-based risk stratification model for COVID-19 in the general population. Sci Rep 12:3277. https://www.nature.com/articles/s41598-022-07138-y. Accessed 25 July 2024

Vela E, Plaza A, Carot-Sans G et al (2022b) Data and care integration for post-acute intensive care program of stroke patients: effectiveness assessment using a disease-matched comparator cohort. J Integr Care 30(4):434–455. https://doi.org/10.1108/JICA-12-2021-0062

Dutch Hospital Reference Architecture ZiRA: A Tool for Collaboration and Cohesion

Antje Derksen and Mikis van Dijk

Abstract

An initiative to set up hospital reference architecture in The Netherlands started over 20 years ago and led to the publication of the ZiRA, the Dutch hospital reference architecture, in 2018. The vision was that with well-functioning hospital architecture, patients and healthcare professionals would benefit from improved IT quality and efficiency and that administrators would be able to manage IT more effectively and IT professionals could introduce innovative applications more quickly.

In this chapter, the writers provide an overview of the organization and key artifacts of the ZiRA, give examples of the usage and show the evidence of the added value of the hospital reference architecture. By analyzing whether the goals set 20 years ago have been achieved, this chapter gives an insight into how a reference architecture can contribute to the digital maturity of hospitals and also sets out important attention points and conditions for achieving and maintaining this value.

Does this mean that the ZiRA contributes to the digital maturity of hospitals in the Netherlands? As it encourages collaboration and cohesion, the ZiRA has proven to be relevant and valuable and is therefore an interesting instrument to take note of.

A. Derksen (✉)
Enterprise Architect, Ministry of the Interior and Kingdom Relations, The Hague, The Netherlands

Former ZiRA Governance Board, The Hague, The Netherlands
e-mail: Antje.Derksen@rijksoverheid.nl

M. van Dijk
Enterprise Architect, Nictiz, The Hague, The Netherlands

ZiRA Governance Board, The Hague, The Netherlands
e-mail: mikis.vandijk@nictiz.nl

A. Scheuer, J. Studzinski (eds.), *Digital Maturity in Hospitals*,
https://doi.org/10.1007/978-3-031-80704-6_9

Keywords

ZiRA · Reference architecture · Hospital IT · Interoperability · Digital maturity

Abbreviations

BCP	Business Continuity Plan
CIA	Confidentiality, Integrity, and Availability
DTC	Diagnostic Therapeutic Cycle
HCIM	Healthcare Information Model
ZiRA	Ziekenhuis Referentie Architectuur (Dutch Hospital Reference Architecture)

1 Introduction

At the end of the first decade of this century, a large number of hospitals in the Netherlands were confronted with similar IT developments and questions in and around their own hospital. This situation motivated the Dutch Association of Hospitals (NVZ) and Nictiz, the Dutch competence center for digital information management in healthcare, to launch a partnership under the name iHospital. The iHospital program was the organizational form in which collaboration between hospital information professionals was achieved with the care processes and guidelines being leading. Within this program hospitals expressed the need for a clearer and joint structure for IT, with which hospitals could learn from each other in IT and bring supply and demand together.

The vision was that with well-functioning hospital architecture (1) patients and healthcare professionals benefit from an improvement in quality and efficiency when using IT, (2) administrators will be able to manage IT more effectively, and (3) IT professionals can introduce innovative applications more quickly. The iHospital program published the first version of a hospital reference architecture in 2011 based on practical experiences from various hospitals and consisting of domains, business activities, and information objects. It was embraced by Dutch hospitals and the useful application created a call for expansion with architectural principles, a process model, and application functions.

This is where the development of the Dutch hospital reference architecture "Ziekenhuis Referentie Architectuur (ZiRA)" started with the support of the Dutch Ministry of Health, Welfare and Sport. In 2018, the ZiRA artifacts were published on www.ziraonline.nl. Since then, the ZiRA has been further expanded with practical examples and in 2023 the ZiRA translation to English in cooperation with The Open Group was published. A core group of iHospital members is still responsible for the management and further development of the reference architecture under the leadership of Nictiz. To stay current and to discuss developments, working conferences for iHospital members are held on a regular basis with presentations on themes such as network care and multidisciplinary consultation and the connection with national policy.

Dutch hospitals still recognize and embrace the ZiRA and the iHospital collaboration. Does this mean that the ZiRA contributes to the digital maturity of hospitals in the Netherlands? Did (and does) it contribute to (1) an improvement in quality and efficiency of healthcare in the Netherlands, (2) in managing IT more effectively, and (3) introducing innovative applications more quickly as was visioned two decades ago?

2 A Common Ground for Collaboration

Architecture is a means to help organizations maintain an overview, manage the complexity of information provision, and implement changes coherently. Architecture describes the design of organizations in frameworks and models. The starting point for ZiRA developers was to create a common framework tailored to their type of organization that could serve as a tool for complex organizational and IT issues and as a starting point for reuse so that not every hospital had to reinvent the wheel for itself. A common language and conceptual framework could contribute to the interchangeability and comparability of local architectural artifacts.

The ZiRA has been developed primarily for people who focus on the design of information provision in a hospital, not only enterprise and information architects but also information managers, solution architects, functional and technical designers, and functional administrators. The ZiRA is also applied to organizational issues, regardless of information provision. This means that the ZiRA is (in part) also intended for policy makers, advisors, and others involved in organizational and process changes.

The ZiRA has a twofold purpose. To develop, acquire, and maintain a framework and tools to support the organization of information delivery in hospitals as well, and to provide hospitals with a partnership focused on safe, effective, and meaningful information exchange. A common ground for collaboration, cohesion, and continuous improvement.

2.1 The ZiRA Layered Framework and Coherent Artifacts

The ZiRA consists of a coherent set of architectural artifacts, including architectural principles, five key reference model templates, theme pages, a book of practical examples, and downloadable files. The ZiRA can be used in whole or in part. Each part of the ZiRA can be adapted to the needs of an organization.

The core part of the ZiRA is the reference models that describe the architecture of hospitals at different levels and principles to provide direction for its improvement. The ZiRA follows the Nictiz Layer Framework (Sprenger 2019) in the structure and

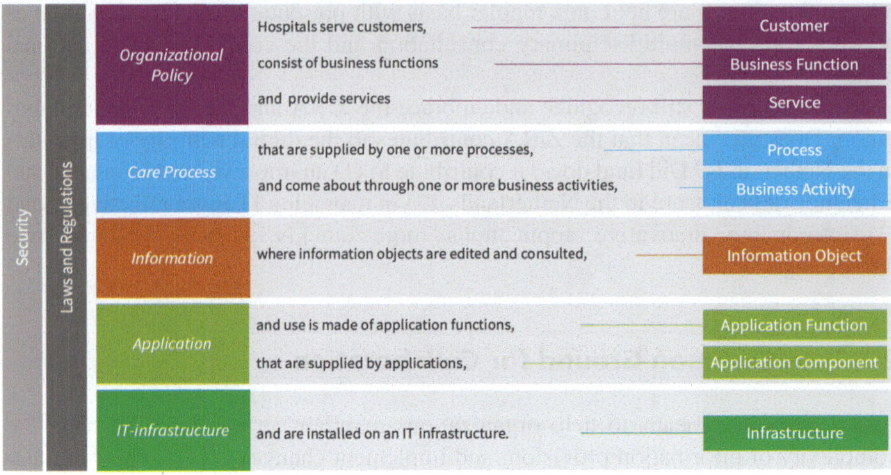

Fig. 1 The Nictiz Layer Framework with the ZiRA concepts

naming of the models and concepts that make up the reference architecture of hospitals (Fig. 1).

The organization layer of the ZiRA focuses on customers and a variety of services that hospitals provide through business functions. Business functions are the competencies or capabilities required for a hospital to accomplish its goals and objectives. The second layer is the care process layer, where business processes ensure services and consist of work processes that are provided through the realization of one or more business activities.

Business activities are the specific tasks or steps, associated with an organizational role, that constitute a business function or process. The information layer is concerned with the vast amount of data that is created, processed, and used during the execution of business activities in hospitals. The fourth application layer focuses on the wide range of software designed and written to facilitate the collection and use of information. Finally, the fifth layer points to the IT infrastructure like networks, databases, and storage facilities that support applications.

2.1.1 ZiRA Metamodel

The ZiRA metamodel indicates the relationships among the key ZiRA concepts and shows their coherence (Fig. 2). The color scheme matches the layered framework and is used in the models that constitute the ZiRA.

The top row of the metamodel corresponds to the organizational policy layer in the interoperability model. It identifies the core value proposition of a hospital: namely, providing services to customers. A Customer is often the patient but can also be a collaboration partner, health insurer, the government, or another actor that uses a hospital service. A Service is provided by a business process. The top row of the metamodel also includes the concept Business Function: a collection of

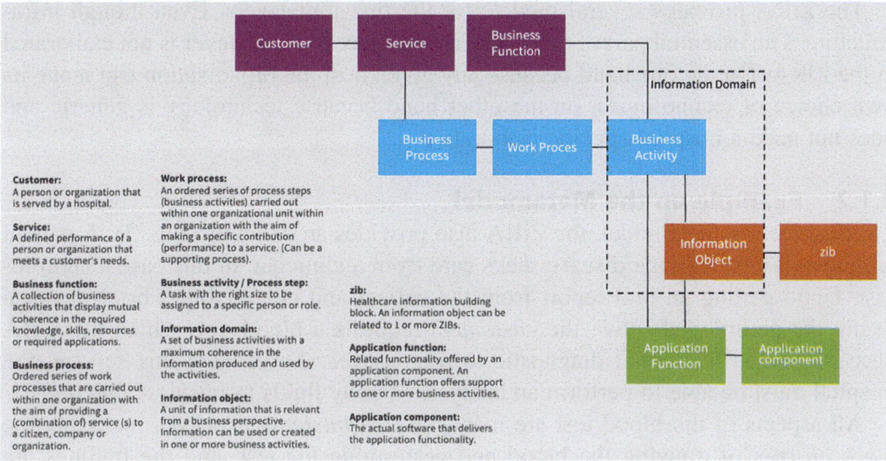

Fig. 2 The ZiRA metamodel

interrelated ("cohering") business activities that, together, produce the knowledge, skills, resources, and competencies necessary to create an organizational capability.

The second row of the metamodel represents hospital processes as starting with specific tasks carried out by people in specific roles that are combined into an ordered series of "process steps" that, all together, result in delivering services to customers. So, a Business Process is a sequence of work processes carried out within one organization with the aim of providing a combination of services to a person, company, or organization. A Work Process is a sequence of business activities carried out (typically) within a single organizational unit with the aim of making a specific contribution to a service. And a Business Activity or Process step is an activity that can be assigned to one person or one role.

The information layer of the metamodel identifies the concept Information Object and Healthcare Information Model (HCIM, or zibs in Dutch). An information object is defined here as a unit of information that is relevant from a business perspective. Information (objects) can be used or created in one or more business activities. A HCIM is a technology-independent, logical information model based on an ISO standard for Clinical Information Models (ISO 2022).

The fourth and final row of the ZiRA metamodel represents application functions and application components. An "Application Function" works alone or with other application functions and with components to support one or more business activities. An "Application Component" is the actual software that delivers the application functionality.

Besides this the ZiRA metamodel identifies an information domain expressed as the relationship between business activities and information objects. Information domains are clusters of highly interrelated business activities and information objects, and thus form a logical grouping for information provision.

The ZiRA provides several models for the first four layers. Even though infrastructure is an essential part of enterprise architecture this fifth layer is not elaborated in the ZiRA. On the one hand because any given hospital organization can make its own choice of technologies, on the other hand because technology is generic and does not need a hospital-specific elaboration.

2.1.2 Example on the Metamodel

To illustrate the metamodel, the ZiRA also provides an example (Fig. 3). A patient with symptoms of Lyme disease seeks care from a clinician, in this case a neurologist. Upon getting an oral report from the patient and conducting a brief physical exam, the neurologist (now, the *customer*) requests a blood test, which the ZiRA classifies as an additional diagnostic testing *service*. To provide this service, the hospital must be able to perform an analysis of body fluids (a *business function*).

All aspects of the blood test are managed (the *business process*), including the *work process* of drawing the blood and performing the test, and the findings are summarized (the *business activity*).

The process of confirming the (likely) presence or absence of Lyme disease in the patient generates a lab report (an *information object*), that is specified with the *HCIM* for a laboratory test result, with the assistance of the lab information management system that creates the lab file (*application component* and *application function*, respectively). The business activity and the information object are in the additional diagnostic testing *information domain*.

2.1.3 Architecture Principles

Besides the reference models on the 4 upper layers of the Nictiz Layer Framework (Sprenger 2019) the ZiRA also provides 12 architecture principles: rules that provide

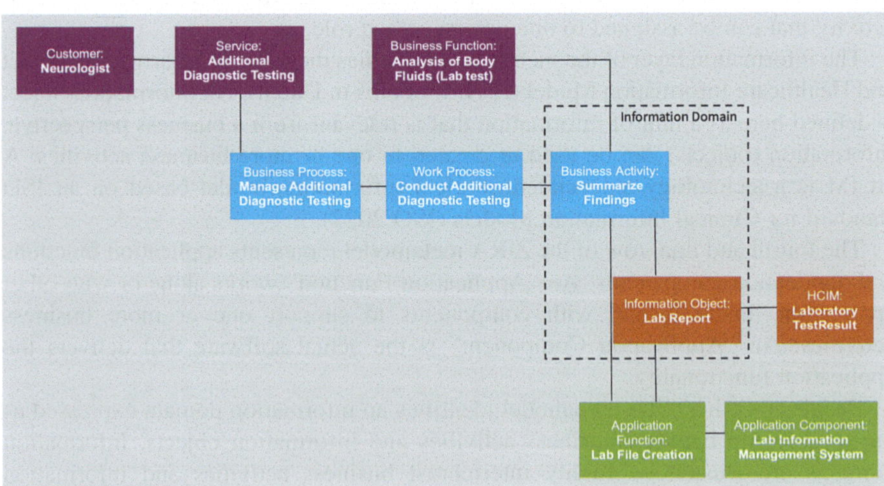

Fig. 3 ZiRA metamodel illustrated with an example

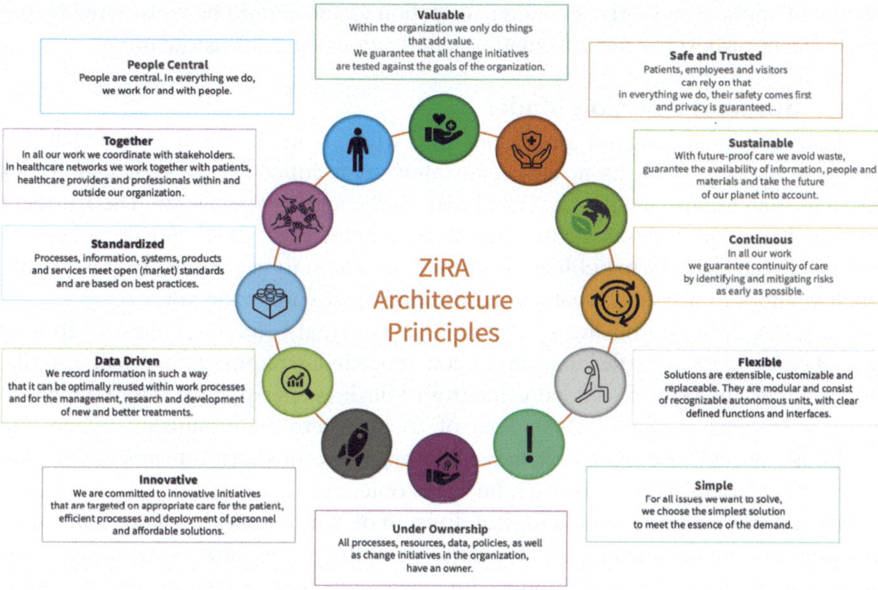

Fig. 4 ZiRA architecture principles

Nr.3	Layers					Sustainable
Description						With future-proof care we avoid waste, ensure the availability of information, people and materials and take the future of our planet into account.
Rationale						With sufficient attention to sustainability, we can maintain healthy business operations in the future. We keep healthcare good, accessible and affordable by focusing more on prevention [Source: IZA]. We also focus on reducing waste of materials and energy and ensuring the sustainable employability of our people and resources.
Implicatons	O	P	I	A	T	• *We take the impact on the 3 Ps into account in all choices: people, planet and profit.*
	O					• *As a hospital, we have a sustainability policy that is continuously tested and adjusted.*
			I	A	T	• *When making choices, we use: reuse comes before buying and buying comes before making it yourself.*
			I	A	T	• *We take the supplier's sustainability policy into account when selecting and purchasing new products and services.*
			I	A	T	• *Our information provision is future-proof and environmentally conscious.*
			I	A	T	• *We store our data in a future-proof manner so that it survives the lifespan of applications.*
			I	A	T	• *Data is not stored or duplicated without purpose.*
					T	• *We avoid unnecessary redundancy*

Fig. 5 Elaboration of the ZiRA architecture principle "Sustainable"

direction and structure to the design of organization, process, information, application, and technology. The ZiRA architecture principles are visualized and provided with a brief statement (Fig. 4).

In addition, a separate elaboration is available with a description, rationale, and implications for each ZiRA architectural principle. This is where implications are related to one or more of the layers and so provide a more concrete interpretation of a principle. An implication is a more concrete interpretation of the abstract architectural principle. Implications limit the design space by setting concrete frameworks or requirements.

Figure 5 shows the elaboration of the principle Sustainable. It shows, for example, that the implication "We store our data in a future proof manner so it survives the lifetime of applications" gives concrete direction to and should be considered by the development of designs for information, applications, and infrastructure.

2.1.4 Business Function Model

The ZiRA Business function model identifies what an organization does, regardless of how it does it. It looks at an organization as a collective of logical units, or functions, that require similar or related knowledge and competencies. The Business function model provides a neutral frame of reference in which no organization-specific choices (e.g., what clinical information system to use, what devices to buy, which vendors to use, and what services to outsource) are made yet.

The ZiRA defines a business function as an internal behavioral element formed by a set of business activities that shows coherence in the required knowledge, skills, and resources and provides added value from a business perspective. More popularly stated, a business function is an area of focus where an organization wants to structurally assign resources to achieve its objectives. In short, a business function is a capability required to achieve a business objective.

Business functions provide a logical division of activities within an organization. Functions can be decoupled, which creates flexibility in organizational design. It is important that the responsibilities for each business function are assigned unambiguously.

The business Function Model has multiple applications. It serves as a foundation on which other organizational aspects, such as processes and applications, can be plotted (Fig. 6).

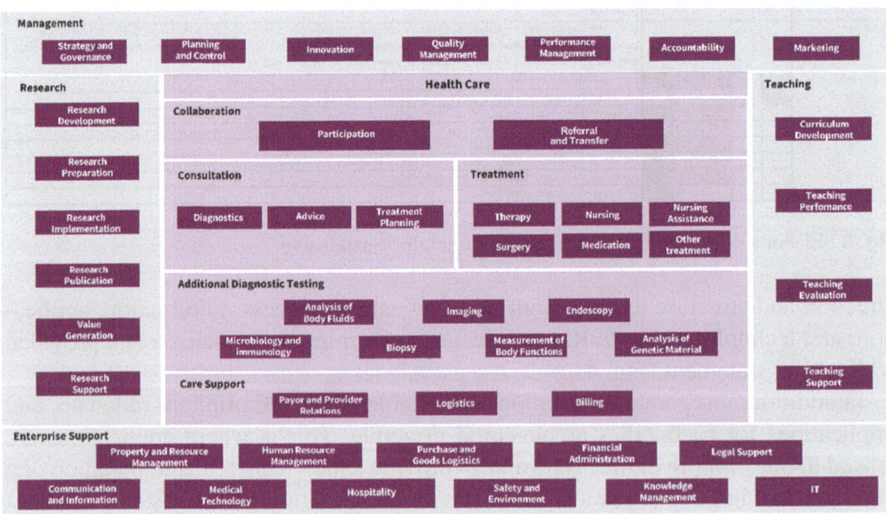

Fig. 6 The ZiRA Business function model

2.1.5 Process Model

As the ZiRA Metamodel (Fig. 2) describes, services are provided to customers via processes. A process consists of a sequence of activities required to achieve a certain result. ZiRA offers a reference architecture for the core care processes provided in a hospital. The ZiRA Process model describes processes on three levels: business processes that deliver healthcare, the work processes within each of these business processes, and the work processes consisting of process steps (or business activities).

Seven business processes are key to deliver healthcare in hospitals and make up the Diagnostic Therapeutic Cycle (DTC) (Fig. 7). The processes required to determine care needs precede and enable those involved in making a diagnosis. In turn, although the process of making a diagnosis may invoke the need to conduct additional diagnostic testing, the aim and purpose of making a diagnosis is to establish a treatment plan. Establishing a treatment plan may also invoke the need to conduct additional diagnostic testing. But the aim of establishing a treatment plan is either to treat the patient or to give the patient (or the patient's caregivers) advice. Treatment of the patient may also involve additional diagnostic testing (for example, when a surgeon finds a mass and a biopsy is required to determine whether it is benign or malignant). Finally, whether the patient receives advice or treatment, the final step in the DTC is transferring the patient out of the hospital. This cycle applies to inpatients and outpatients in the same way.

The ZiRA process model provides a basis for more detailed process descriptions and ensures clear coherence, a similar level of abstraction, and unity of language. Other applications of the ZiRA process model include gaining insight into the impact of changes or dependencies and transfer points between people and departments and facilitating discussions about ownership of business processes and the definition of projects.

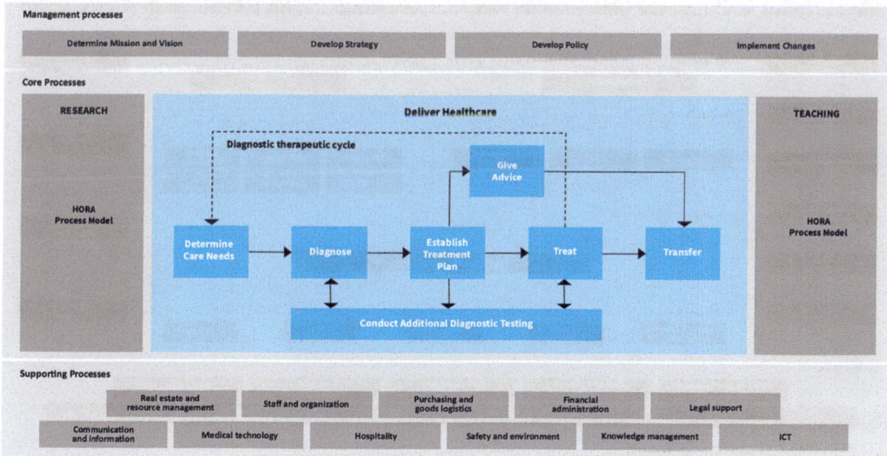

Fig. 7 The ZiRA process model for Healthcare delivery and Diagnostic Therapeutic Cycle

2.1.6 Information Model

The information model of the ZiRA (Fig. 8) describes the data that hospitals manage in a logical interrelationship by means of various views. It identifies which information objects exist within the Healthcare business domain. Information objects are units of information that are relevant from a business perspective. Examples of information objects are anamnesis, report, order, etc.

An information model is also called a business object model or a conceptual data model.

ZiRA describes the larger units of data (information objects) in a language that is broadly relevant and recognizable within hospital organizations. Just like the business function model, the information model is independent of the organization's design and IT; it therefore offers a stable frame of reference.

The information model promotes clarity of language regarding information. The information model is used to support organization-wide discussions about data management and data governance responsibilities. It can, for instance, be used as a tool for examining and analyzing information to support the execution and improvement of business processes or supporting organization-wide discussions about data management responsibilities. Other applications are classifying data for information security purposes, complying with varied and changing privacy regulations, or developing and maintaining a data dictionary to manage data assets.

Information objects are units of information that are relevant from a business perspective. Information can be used or created in one or more business activities. A view of information objects in the healthcare processes provides insight into the use of the information objects in the processes developed within the ZiRA. The work processes include which business objects are used in the process.

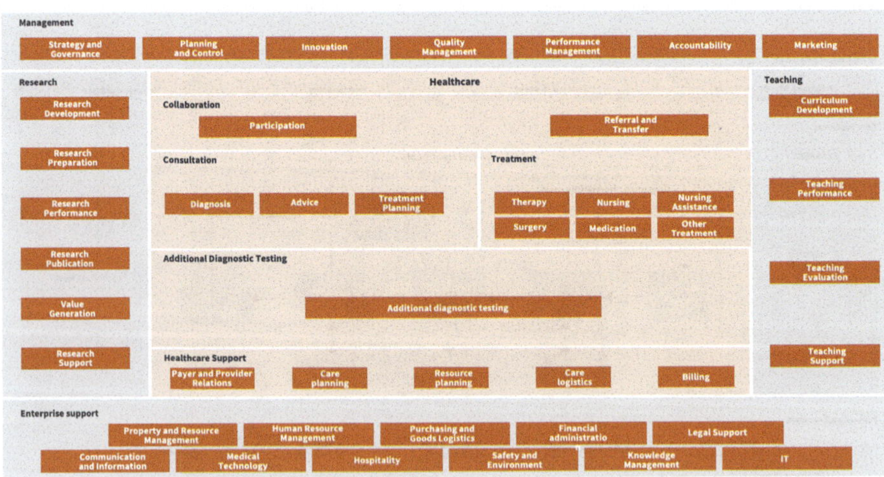

Fig. 8 The ZiRA Information domain model

2.1.7 Application Function Model

The Application Function Model (Fig. 9) consists of subgroups of software functionality that a hospital needs to support its business processes. An application function refers to a specific capability or type of functionality provided by a software application. An application function supports one or more business activities. Application functions are designed to support the overall goals and objectives of the organization. They are primarily defined by the functionality they offer and the data they manage. The model describes application functions at a logical level, independent of specific product choices.

Application functions are provided by application components, the actual software that provides the functionality. The ZiRA makes no statements about which components provide the functionality. The choice for "best of breed" solutions or integrated suites and therefore which products, services and solutions to adopt is left to the individual institutions.

The Application function model reflects the view of which application functions support business functions and business activities in the five major ZiRA business domains (management, research, healthcare, education, and business support). In this model, a distinction is made between application functions that directly support

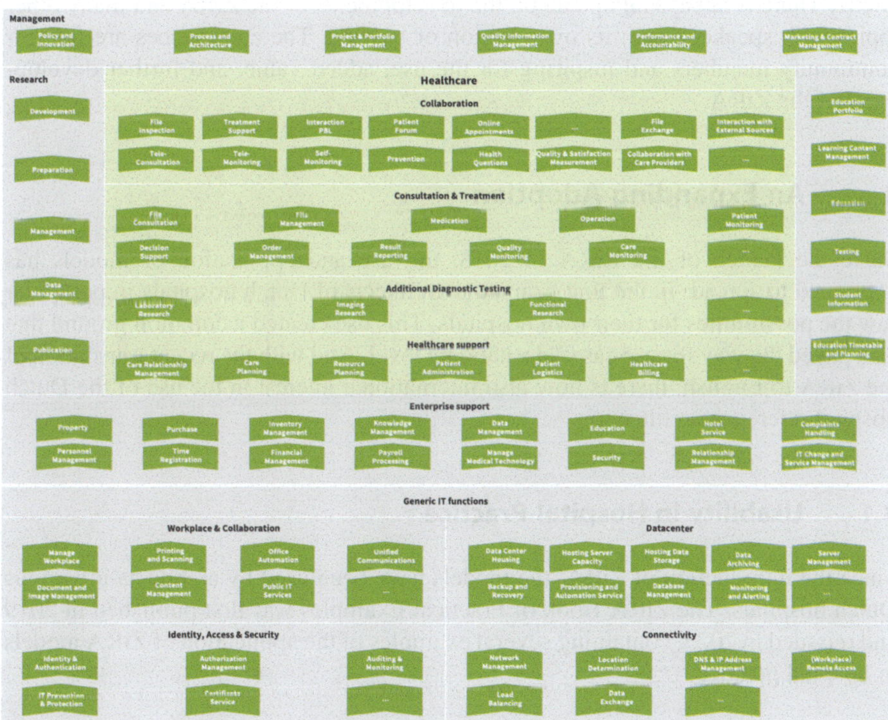

Fig. 9 The ZiRA Application function model

the business functions (specific applications) and applications that are used in many different business functions in all business domains (generic IT functions).

2.2 iHospital Working Conferences

In 2009, at the initiative of the Dutch Association of Hospitals (NVZ) and Nictiz, the partnership under the name iHospital was launched in The Netherlands. The target group of iHospital is mainly (information) architects from hospitals.

Hospitals often must deal with similar questions and developments regarding the design of information provision. Nictiz makes the connection and facilitates a platform on which hospitals can ask questions and share their knowledge and experiences. Not only with a hospital reference architecture but also by organizing work conferences.

At these conferences hospital (information) architects meet each other and share knowledge and experiences. It is a platform to discuss developments on themes such as network care and multidisciplinary consultation and the connection with national policy. Since 2009, more than 40 working conferences have been held. These conferences take place in the center of the Netherlands about two to four times a year. The agenda is determined by the participants themselves and facilitated by Nictiz. There is always an update on the developments of the ZiRA and many other topics with speakers from its own section or beyond. The conferences are free for community members and inspiring for the use, added value, and further development of the ZiRA.

3 An Expanding Adoption

Since the launch of the ZiRA in 2018, the use and application of models has continued to spread. In the first years, the architects of Dutch hospitals in particular saw the possibilities for their own hospitals. This use created a common ground that also paved the way for use cases at a national level. And with the recent translation of the ZiRA to English, there is now also international interest in the use of the Dutch hospital reference architecture.

3.1 Usability in Hospital Practice

Since the introduction of ZiRA, the models have been used by architects in various Dutch hospitals. The ZiRA Book of Practical Examples was first published in 2019 and reissued in 2023, containing several examples of the application of ZiRA models in hospital practice.

3.1.1 The Patient Care Path

One of the examples in the ZiRA Book of Practical Examples is the use of the Business function model (Fig. 6). This example illustrates the use of ZiRA to gain insight into which departments a patient visits during a specific treatment, by means of plotting this path on the ZiRA business function model (Fig. 10). This creates a shared and recognizable image, facilitates communication between different departments, and provides guidance for digital transfer agreements (Nictiz 2023).

It can also provide insight into how to improve the efficiency of the care pathway. In addition, it offers a starting point for optimizing the patient experience and patient care.

3.1.2 CIA Classification

The value of data or applications is established through classification in Confidentiality, Integrity, and Availability (CIA). A CIA classification (or BIV classification in Dutch) is a classification used within information security, which indicates the availability (continuity), integrity (reliability), and confidentiality (exclusivity) of information and systems.

Information systems and data are generally classified according to the CIA classification in order to determine the intended level of security. A highly classified system has for example, a CIA class of 333, a low classified system will be classified as 111. Appropriate measures can be taken based on this classification.

Figure 11 shows an example from the ZiRA Book of Practical Examples of how a Dutch hospital used the ZiRA Information domain model for the CIA classification within its organization (Nictiz 2023). Other Dutch hospitals saw this example as an

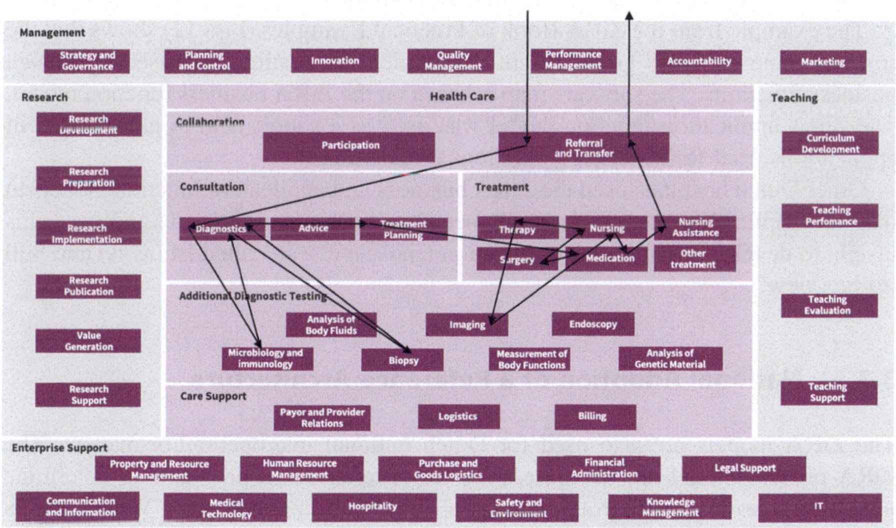

Fig. 10 The patient care path. A translation of an example from the ZiRA Book of Practical Examples

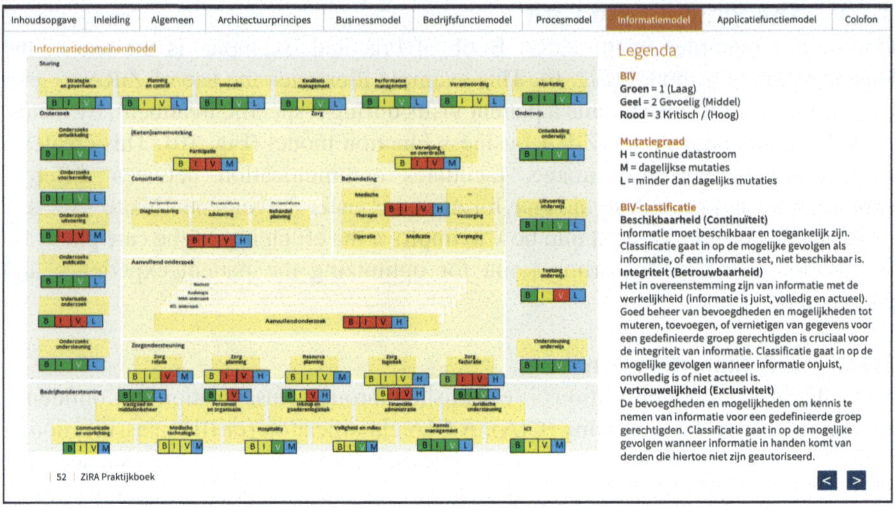

Fig. 11 CIA classification. An example from the ZiRA Book of Practical Examples

encouragement to carry out such a classification for their hospital with the use of the ZiRA Information domain model and to learn from the example given and its lessons learned.

3.1.3 Hospitals Complex Application Landscape

A frequently used application of the ZiRA models is to gain insight and overview of the complex application landscape within a hospital.

The example from the ZiRA Book of Practical Examples (Fig. 12) shows that the software map provides good insight into which applications are used for which business functions. The software map is based on the ZiRA business function model. The ZiRA application function model was used as a supplement to gain insight of applications used for generic IT functions (Nictiz 2023).

Other Dutch hospitals used the ZiRA business and application function models to map the applications in use during a merger between two hospitals and used this insight to develop a joint future application landscape, a so-called Ist (as-is) and Soll (to-be) view.

3.2 National Adoption of a Reference Architecture

The ZiRA models are also used for Dutch national interoperability matters. The ZiRA process model has been the basis for several Dutch programs to determine medically relevant data surrounding a specific condition or care path. With ZiRA as a nationally accepted common ground, these programs can quickly start making specific agreements and standards for structured data exchange.

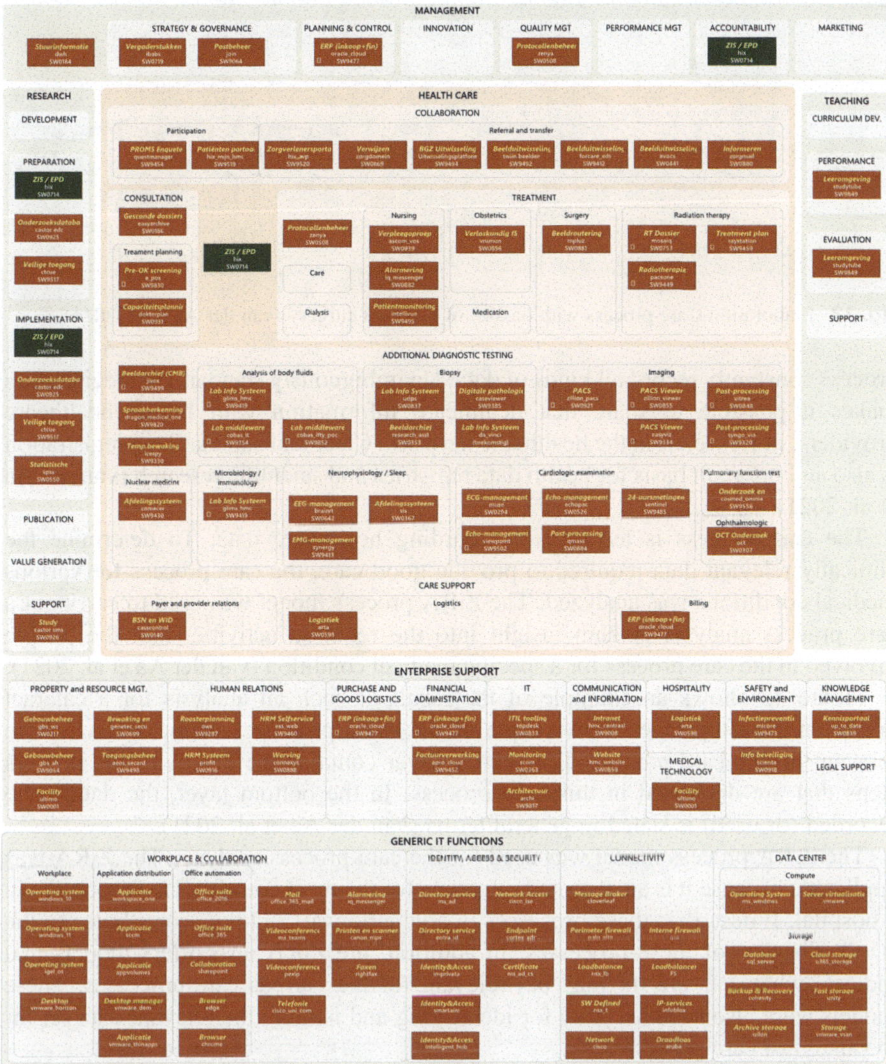

Fig. 12 Software map. A translated example from the ZiRA Book of Practical Examples

3.2.1 Registration at the Source

A national program in which Dutch hospitals and associations of medical specialists have worked intensively together over the past few years is "Registration at the source." The mission of this program was to achieve better care through the unambiguous registration of healthcare information for reuse under the motto "Registration at the source: that makes healthcare better!" (Registratie-aan-de-bron 2022)

Healthcare information is the common thread in the healthcare process. Doctors, nurses, the patients themselves, and other healthcare providers involved in the care

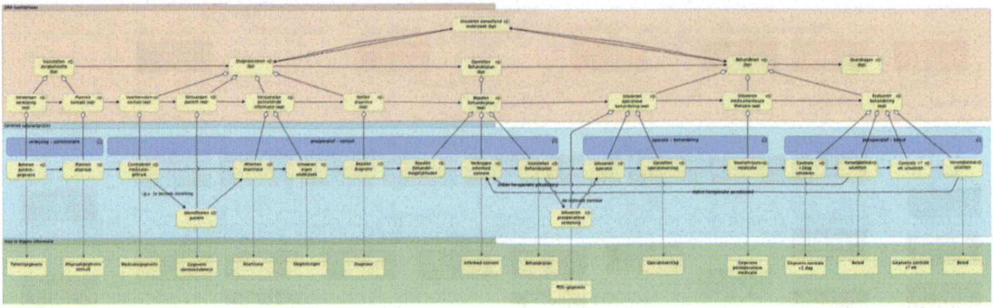

Fig. 13 Elaboration Care process with dataset of Cataract surgery (van der Aa et al. 2021)

process constantly record all kinds of data. Unambiguously standardized registration makes it possible to share that healthcare information with fellow healthcare providers, in-house, or in the healthcare network. Clear and standardized registration is also an important basis for health data registries and scientific research (van der Aa et al. 2021).

The care process is leading for recording healthcare data. To determine the clinically relevant data required to provide good care, the care process for various medical conditions was analyzed. The ZiRA process model was used to carry out a care process analysis to gain insight into the required activities and the people involved in the care process for a specific medical condition (van der Aa et al. 2021).

Figure 13 shows an example of the result of such an analysis for a cataract operation. The top layer shows some of the main processes of the ZiRA, where relevant in this care process. The second layer contains the details of the process steps that are discussed in this care process. In the bottom layer, the data to be recorded (or used) is stated in general terms (van der Aa et al. 2021).

The ZiRA process model was used for the care process analysis. The ZiRA was applicable because it is an accepted reference for the most common care processes in a hospital. It describes these processes in a generic way and thus contributes to the standardization of these processes. In addition, the ZiRA also offers models that indicate what data are (or may be) relevant for the various care processes. These models were also used as a tool for identifying and naming the relevant data set for the analyzed care processes.

This care process analysis looked at the parts of ZiRA that were important to the process in question, both in terms of activities and data to be captured. The results of the care process analysis provided input for agreements on reuse and information exchange. To enable reuse, the care process data were recorded unambiguously and standardized as much as possible. This was possible by using the same reference models as a starting point for each analysis.

3.3 International Knowledge Sharing

Since October 2023 the translation of the ZiRA in collaboration with The Open Group[1] is available as a Guide. All ZiRA reference models are now published in English and available for download from the ZiRA website.[2] With this translation, this Hospital Reference Architecture has become accessible and usable for an international audience.

Effective standards are essential to the establishment of information exchange in healthcare, which is necessary to improve healthcare delivery and outcomes. When each hospital adopts its own preferred terminology and proprietary approach to describing the systems that support clinical care, it creates a significant barrier to efforts to effectively exchange health information. Adopting a hospital reference architecture such as ZiRA provides a foundation to help hospital organizations overcome the communication challenges between disparate internal and external systems (Griffith et al. 2022).

3.3.1 Business Continuity Plans

The Healthcare Forum of The Open Group also utilized the English translation of the Dutch hospital reference architecture ZiRA to analyze and map essential services in a large UK NHS Hospital.

Hospitals worldwide must create Business Continuity Plans (BCPs) to ensure seamless operation and essential service delivery during and after disruptions such as natural disasters, technology failures, cyber incidents, utility failures, facility or infrastructure damage, or human errors. The output of the analyses and mapping on the ZiRA models derived for this UK NHS Hospital vastly improved the hospital staff's understanding of the relationships and dependencies upon critical system elements, enhancing the organization's readiness to address potential losses in operating capability resulting from key information technology (IT) application outages (Gordon 2023).

3.3.2 International Attention

Since the translation of the first part of the ZiRA was published, the ZiRA website has attracted significantly more international visitors. The ZiRA website has visitors from a lot of different countries, as shown by the website analytics (Fig. 14) of Global visitors to the ZiRA website. In the period from April 2023 to March 2024, the total number of measured visitors to the website is 4,300. Of these, 2,800 website users came from the Netherlands; 1500 visitors were from outside the Netherlands, a percentage of 35% and spread across all continents.

[1] The Open Group Hospital Reference Architecture Guide: The Complete and Expanded English Translation of the Dutch ZiRA: www.opengroup.org/library.

[2] Ziekenhuis Referentie Architectuur: https://sites.google.com/site/zirawiki/home.

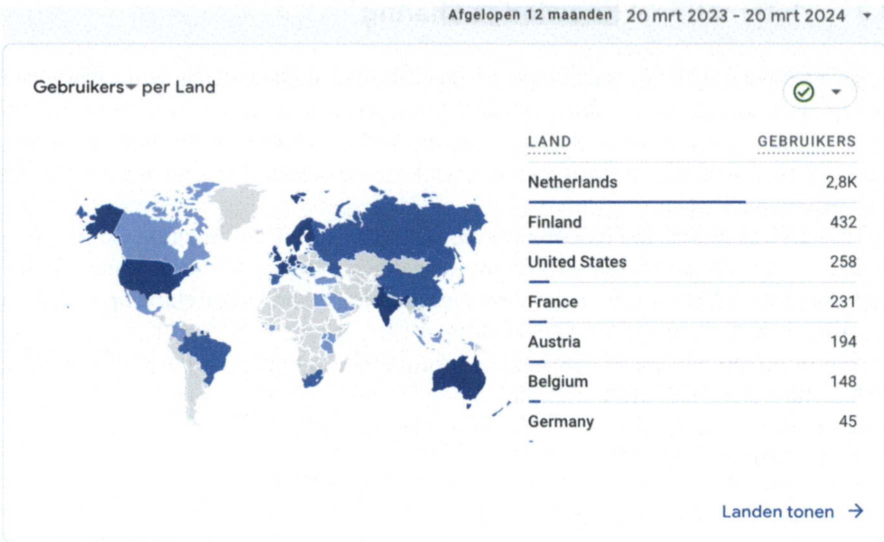

Fig. 14 Global visitors of the ZiRA website

4 A Relevant Tool for Structure, Management, and Optimization

To assess whether the ZiRA as a hospital reference architecture has contributed to digital maturity in hospitals, the generally accepted principle that architecture is one of the cornerstones on which digital maturity is built was taken as a starting point. With the hypothesis that the ZiRA promotes the use of architecture in hospitals, it would contribute to digital maturity. And while ZiRA's impact has never been formally evaluated, there is ample evidence that this is the case.

4.1 Practical Relevance

One way to measure the relevance of ZiRA is to assess whether it is used and attracts interest. First, let us look at the audience for whom and with whom ZiRA was originally developed: the Dutch community of information architects working in and for hospitals. It is safe to say that, almost 15 years after the first parts of ZiRA were published, there is continued interest. The development of the ZiRA and use cases for the reference architecture are a recurring agenda item at the iHospital conferences and often spark lively discussion. The fact that the iHospital community still feels very engaged can also be seen in the number of requests that are made for additions to the model or for keeping it up to date with current developments.

Secondly, the attention paid to the ZiRA Book of Practical Examples can be seen as an important indication of the practical relevance of the ZiRA. Based on these practical examples, the ZiRA administrator regularly receives requests for information from people who want to implement a similar application.

4.2 National and International Adoption

Apart from architects in Dutch hospitals the ZiRA attracts a broader audience. Within the Netherlands it is not only used in and by hospitals, but it has had a wider spread to many more organizations involved in healthcare-IT and policy. ZiRA is also receiving increasing attention from abroad.

One way to see this is by looking at the visitor statistics for the ZiRA online website (Fig. 14). Another indication is that it has been used in other countries: the British National Health Service (Gordon 2023) and the Swedish Karolinska University Hospital (Nictiz 2023) have used the ZiRA as a foundation for some of their hospital IT improvement projects.

A recent milestone is the adoption of ZiRA by the Open Group, who considers it a good practice example of the use of ArchiMate®,[3] and a valuable addition to their more abstract O-HERA™[4] healthcare reference architecture. The Open Group now offers the translated version of ZiRA to a worldwide audience as a Guide on The Open Group Library.

4.3 Working with Architecture

Another way to indicate that ZiRA has actually contributed to digital maturity is to argue that it has effectively led to a better and broader adoption of architecture in the Dutch hospital community. As a practical, proven, and broadly accepted toolset, it enables hospitals to make a quick start with architecture and helps those that have already started to improve their architectural capabilities. By using ZiRA's principles, models, and tools, hospitals have a baseline and can avoid starting from scratch, thereby increasing efficiency and reducing cycle times.

The common use of the ZiRA also leads to mutually recognizable architectural artifacts among hospitals, thus facilitating reuse. As working with architecture is one of the key prerequisites for digital maturity, the ZiRA can be seen as a significant contribution to the current level of maturity in Dutch hospitals.

The ZiRA models are also used for Dutch national interoperability matters. National programs encourage the uniformity of language and technology, thus

---

[3] ArchiMate® is a registered trademark and ArchiMate® Modelling Language, a standard of the Open Group.

[4] The O-HERA™ is a registered trademark, a Health Enterprise Reference Architecture of the Open Group.

contributing to standardization. This is a prerequisite for improving the interoperability of health information and enabling high quality care by supporting healthcare delivery and patient experience.

4.4 Supportive to Digital Transformations and Quality of Care

While the promotion of architecture is a relevant but not too specific merit of ZiRA, some of its applications contribute to digital maturity in even more demonstrable ways, such as the architecture principles when they are adopted. They emphasize that IT decisions should be made based on long-term benefits and impact for the hospital and its stakeholders rather than short-term or departmental goals. Another widely used and valued tool is the application-function model. Both internal hospital information architects and external consultancy firms regularly use it to visualize the application landscape in current state and the desired state, for instance in the case of a merger or after the implementation of a new hospital information system. The model makes it easy to detect functions that are served by multiple applications. Once such duplicates are detected, measures can be taken to phase them out, a relevant step in achieving a viable, effective, and mature IT landscape.

The ZiRA makes a broader connection between various important aspects that impact and add value to an organization. It supports digital transformations and helps to structure, organize, and optimize them.

5 A Shared Architecture Encourages Digital Maturity

Dutch hospitals recognize and embrace the ZiRA and the iHospital collaboration. But can we also conclude that the ZiRA contributes to the digital maturity of hospitals in the Netherlands? Did it contribute to (1) improve the quality and efficiency of healthcare in the Netherlands, (2) manage IT more effectively, and (3) deploy innovative applications more quickly, as envisioned two decades ago?

5.1 ZiRA's Contribution to Dutch Hospitals

A direct link between improving the quality and efficiency of healthcare in the Netherlands is not easy to demonstrate. What can be deduced from the findings is that ZiRA has made a valuable contribution to improving the digital information provision in hospitals. This in turn has contributed to the quality and efficiency of healthcare.

ZiRA has certainly contributed to more effective IT management, both in hospitals and for national interoperability. It provides a basis from which changes can start without reinventing and rebuilding the wheel.

Finally, the hospital reference architecture has ensured that digitalization transformations in healthcare have started faster by using the ZiRA models as an accepted common ground. These projects also had a more careful approach: they had more overview and structure, which provided insight into how to approach change properly. As a result, change goes better and requires less repair and recovery afterwards.

5.2 Maintaining Relevance and Coherence

The examples of use of ZiRA in the Netherlands show that a hospital reference architecture can contribute to the digital development of hospitals. Maintaining the ZiRA reference models in relation to national and international developments is important to keep a reference architecture relevant and up to date. Efforts are needed to keep the target group involved and to ensure sufficient knowledge exchange so that the reference architecture can continue to feed on best practices and maintain its coherence and topicality.

5.3 Continuous Improvement Needs a Common Ground

ZiRA is an open-source reference architecture based on international standards and best practices. It has been extensively applied in the Netherlands. Its usefulness is not limited to hospitals in the Netherlands and since the translation of the ZiRA in collaboration with The Open Group, it is available for use by an international audience. There are early signs that ZiRA is internationally applicable, or at least an inspiration for the further development of digital maturity in health care worldwide. A modest contribution, but not an insignificant one.

References

Gordon D (2023) Improving business continuity planning in hospitals. https://www.linkedin.com/pulse/improving-business-continuity-planning-hospitals-david-gordon-mbci-imnze/. Accessed 14 Mar 2024

Griffith JL, Lee J, Meijboom G et al (2022) ZiRA, The Dutch Hospital Reference Architecture, a tool to address a worldwide need. https://blog.opengroup.org/2022/06/23/zira-the-dutch-hospital-reference-architecture-a-tool-to-address-a-worldwide-need/. Accessed 14 Mar 2024

ISO (2022) ISO-13972:2022. https://www.iso.org/standard/79498.html. Accessed 14 Mar 2024

Nictiz (2023) Praktijkboek ZiRA. https://nictiz.nl/app/uploads/2023/09/Zira_Praktijkboek.pdf. Accessed 14 Mar 2024

Registratie-aan-de-bron (2022) Betere zorg door eenduidig registreren. https://www.registratieaandebron.nl/over-het-programma. Accessed 14 Mar 2024

Sprenger M (2019) Electronic information for health and care services: are we getting better? https://nictiz.nl/publicaties/electronic-infromation-for-health-and-care-services/. Accessed 14 Mar 2024

van der Aa C, Schwantje H, Aziz L (2021) Registratie aan de bron. Handleiding analyse zorgproces. https://amigo.nictiz.nl/uploads/a973ba20-e9ce-4821-9ace-c7c400e169c3/Handleiding-analyse-zorgproces-v1.0-3.pdf. Accessed 1 Feb 2024

The Nordic Way: Denmark's EHR Approach and Regional Integration

Bogi Elisasen, Claus Duadal Pedersen, and Aron Szpisjak

Abstract

The Nordic countries—Norway, Finland, Sweden, Denmark, and Iceland—are uniquely positioned to lead digital health innovation due to their shared characteristics, including high public trust, robust digital infrastructures, and the use of unique personal identification systems. This chapter examines the commonalities between the Nordic health systems and the evolution of digital health systems in Denmark, focusing on the development and digital maturity of Electronic Health Records (EHRs) within Danish hospitals. Denmark's success is driven by public trust in data management, a cohesive healthcare system funded through public means, and the early adoption of a unique electronic ID system (CPR number). The country's approach includes consensus-driven governance, the implementation of national communication standards, and an open market strategy that encourages innovation and competition.

This chapter delves into the assessment of digital maturity in Danish hospitals, which demonstrated a high level of digital maturity in a system characterized by local autonomy and an open market for EHR solutions. The Nordic Health 2030 initiative underscores the region's commitment to creating digitally enabled, sustainable health systems, with Denmark serving as a model. The potential establishment of a Nordic Health Data Space, aligned with the broader

B. Elisasen (✉)
Copenhagen Institute for Futures Studies, Risskov, Denmark
e-mail: be@cifs.dk

C. D. Pedersen
Sundhed.dk Fonden, Odense C, Denmark

A. Szpisjak
Copenhagen Institute for Futures Studies, Copenhagen, Denmark
e-mail: as@cifs.dk

A. Scheuer, J. Studzinski (eds.), *Digital Maturity in Hospitals*,
https://doi.org/10.1007/978-3-031-80704-6_10

European Health Data Space (EHDS), further highlights the region's leadership in digital health and digital maturity.

Keywords

Nordic health 2030 initiative · Denmark electronic health records · Humanome framework · Nordic interoperability standards · Sustainable health models · Digital maturity

Abbreviations

CC	Consensus and Agreement-Driven
CPR	Central Person Registry (Denmark)
EHR	Electronic Health Record
EHDS	European Health Data Space
HCP	Healthcare Professional
Kanta	National Health Information System (Finland)
KaPA	National Service Architecture (Finland)
MedCom	Danish Health Data Communication Standards
NSP	National Service Platform (Denmark)
SDSD	Health Informatics Organization in Denmark
SHS	Sustainable Health System

1 Introduction

The Nordic countries—Norway, Finland, Sweden, Denmark, and Iceland—share several foundational characteristics that make their healthcare systems uniquely positioned to lead in digital health innovation. These commonalities provide a strong basis for their collaborative efforts in transforming healthcare delivery and serve as a model for other regions.

1. **High Trust in Public Institutions:** Across the Nordic region, there is a profound level of trust in public institutions. This trust extends to the management and use of personal data, which is critical in the healthcare sector. Citizens in these countries are generally confident that their health data will be handled responsibly and securely, which facilitates the widespread adoption of digital health solutions. For instance, Norway's health network and Finland's Kanta Services benefit from this trust, enabling secure and efficient data exchanges across the entire healthcare system.
2. **Robust Digital Infrastructure:** The Nordic countries have invested heavily in building robust digital infrastructures that support comprehensive health

information systems. Norway's *Norsk helsenett*,[1] Finland's National Service Architecture (KaPA), and Sweden and Denmark's EHR systems are all underpinned by advanced digital frameworks that ensure interoperability, secure data exchange, and the seamless integration of healthcare services. These infrastructures are designed to be user-centered, technologically advanced, and secure, ensuring that both healthcare professionals and patients can rely on these systems for accurate and timely information.

3. **Unique Personal Identification Systems:** A critical enabler of digital health in the Nordic region is the widespread use of unique personal identification numbers. These IDs, such as Denmark's CPR number and Sweden's personal identity number, are used across all public and private sectors, including healthcare. This universal identifier allows for the seamless integration of health records and other personal data, making it easier to provide consistent and personalized care. It also simplifies access to health information for both patients and healthcare providers, as seen in Finland's MyKanta service[2] and Sweden's Journalen portal.
4. **Focus on Patient Empowerment and Transparency:** The Nordic countries emphasize transparency and patient empowerment through digital health platforms. Norway's Helsenorge, Finland's MyKanta (Eriksson-Backa et al. 2021), and Sweden's Journalen (Hägglund and Scandurra 2021) all provide patients with direct access to their health data, enabling them to take an active role in their healthcare. These platforms allow patients to view their medical records, manage prescriptions, and access other health services online. This level of transparency not only improves patient engagement but also enhances the overall efficiency and effectiveness of healthcare delivery.

Building on these commonalities, the Nordic Health 2030[3] initiative is a unifying effort to further integrate the healthcare systems across the region. This initiative encapsulates the collective vision of the Nordic countries to create the most digitally enabled, sustainable health systems in the world. Central to this vision are the guiding principles of the Humanome and the Sustainable Health Model. The Humanome represents a forward-thinking digital framework designed to enhance the usability of health data, giving individuals greater access and control over their personal health information. Meanwhile, the Sustainable Health Model provides a comprehensive approach to health management, facilitating real-time insights, early interventions, and overall population health (The Copenhagen Institute for Futures Studies 2024).

These principles are not merely aspirational; they are actively being implemented across the Nordic region, driving the transformation of healthcare systems toward a future where prevention, precision, and sustainability are paramount.

[1] https://www.nhn.no/

[2] https://www.kanta.fi/en/mykanta

[3] http://nordichealth2030.org/

Denmark stands out as a particularly illustrative example of these efforts. The country's journey in developing Electronic Health Records (EHRs) offers a compelling case study of how digital health initiatives can be effectively deployed to enhance healthcare delivery. By exploring Denmark's experience, we can gain valuable insights into how these digital frameworks are revolutionizing patient care, improving outcomes, and setting new standards for healthcare not only in the Nordics but across the globe.

2 The Development of Electronic Health Records in Denmark

2.1 Historical Background

From an external perspective, the development of digital solutions in the Danish healthcare system appears to be a long success story. As early as the 1960s and 1970s, the first rudimentary IT solutions for healthcare began to be developed. However, the significant leap forward occurred in the 1990s and early 2000s, when technology had reached a level—both technically and economically—that made it possible to achieve daily operational benefits that could offset the costs of acquiring, implementing, and operating electronic health records (EHRs) within the Danish healthcare system (Bernstein et al. 2003).

Today, the Danish healthcare system is characterized by the implementation of electronic records in all sectors, used in daily operations. This applies to private practitioners, such as podiatrists, physiotherapists, and chiropractors, as well as general practitioners, specialists, municipalities, and hospitals. All systems exchange data in a fully integrated and automated manner through national standards. Furthermore, citizens have access to approximately 90% of their own data via Sundhed.dk, and healthcare professionals can likewise access data from other hospitals or the primary sector through the same portal (Løhde and Bjerre 2023).

Several elements have contributed to Denmark's success in reaching this level, but three key factors stand out. First, there is a general trust among Danish citizens in public institutions to act responsibly and securely manage their data. Citizens not only trust public authorities to collect data but also expect them to use it efficiently. This means there is an expectation that information only needs to be provided once and will then be accessible to all relevant healthcare professionals.

Second, the Danish healthcare system is publicly owned, operated, and funded through taxes (Tikkanen et al. 2020). Internationally, there is a clear trend that countries with publicly funded and operated healthcare systems are better positioned to deliver cohesive digital solutions across sectors and service providers. This trend is seen not only in the Nordic countries but also in countries like Spain and the UK.

Third, Denmark was one of the first countries to introduce a unique electronic ID for all citizens—the CPR number. This ID is used across all public and private registries as an identification key, whether it involves health information, taxes, social data, libraries, banks, or insurance companies. The Danish CPR number was introduced in the mid-1960s as part of a major tax reform. It can thus be said that effective tax collection not only secures the financial foundation for the Danish healthcare system but also lays the groundwork for its effective digitalization.

2.2 The Key Principles of Danish Development

1. **Consensus and Agreement-Driven**
   Unlike other Nordic countries, Denmark has a tradition of governance where the state sets the framework through legislation, while regions and municipalities handle the practical execution of tasks. The governance of legislative implementation is managed through binding agreements between the parties in the healthcare system. This approach provides significant flexibility, allowing for quicker and more flexible adjustments to direction and resources compared to a cumbersome and often time-consuming legislative process.

   The practical implementation of national strategies is regulated through annual financial agreements between the state on one side and the regions and municipalities on the other. These agreements are crucial in creating consensus among healthcare stakeholders on the focus, direction, and goals for the digitalization of the healthcare system.
2. **The Cohesive Healthcare System**
   Despite being a relatively small country, the structure of Denmark's healthcare system is quite complex (The Ministry of Health 2017). Regions are responsible for running hospitals, while municipalities handle elder care, home care, rehabilitation, and prevention. General practitioners operate as private businesses under agreements with the regions, and private specialists, chiropractors, and physiotherapists also play a significant role in delivering healthcare services. This distribution of tasks, funding, ownership, and responsibility necessitates a high degree of coordination and communication to present a cohesive and unified healthcare system to citizens.

   Since the structural reform in 2007, this coordination task has largely been addressed through agreements between regions and municipalities on the daily operation of the healthcare system (Strandberg-Larsen et al. 2007). This agreement system is built on three pillars:
   (a) The first pillar consists of political framework agreements that define the economy, goals, tasks, and responsibility distribution among the parties.
   (b) The second pillar involves professional agreements that outline the desired professional processes and quality, such as cross-sector treatment plans, discharge and referral processes, and interdisciplinary conferences.
   (c) The third pillar is the digital support of the political and professional goals, ensuring that various activities are supported by digital solutions, whether it involves new communication standards, telemedicine solutions, or the use of video conferencing.

3. **Use of National Standards as a Requirement**
   The implementation of national communication standards truly began with the MedCom II project (1997–1999),[4] where the standards developed in MedCom I were implemented across all systems in general practice and hospitals. The development of national communication standards not only facilitated information exchange but also resulted in all Danish electronic health records being built around the same fundamental data model. The impact on data quality, semantic and classification standardization, and the ability to use data for clinical, administrative, and research purposes cannot be overstated.
4. **Open Market: Multi-Vendor Strategy**
   A fundamental principle in the Danish healthcare system has been that the procurement of IT solutions is up to the responsible organization, emphasizing the importance of establishing an open and free market. Competition ensured that individual organizations could purchase products that matched their needs in terms of price and quality. The only requirement for participating in the market was the ability to communicate with other organizations using MedCom standards (Dixon et al. 2022). This approach led to a vibrant and highly active industry of Danish companies in the 1990s and 2000s, which competed—sometimes fiercely—to supply the relatively small market with solutions. In 1999, there were 14 system suppliers for general practice, 5 municipal systems, 6 EHR systems for hospitals, and so on.

2.3 National Strategies: Key Milestones

The digitalization of the Danish healthcare system has, since the 1990s, been driven by a unique interplay between central governance and local innovation. National strategies have primarily aimed to ensure the nationwide implementation of solutions developed and tested in local projects, where there is consensus on their value to the healthcare system. At the same time, most strategies have allocated funds to carry out various development and testing projects, often within the MedCom project programs. Over the years, the balance between implementation and development has varied, but this interplay has overall helped to ensure both room for innovation and successful large-scale adoption, thereby maximizing the return on development investments.

1. **"Digitizing Paper"—1995**
   The first national strategy was largely driven by the experiences from the initial MedCom project, which resulted from a consolidation of various local projects focused on electronic communication, such as those in Copenhagen, Viborg, and Funen counties. These projects were initiated under pressure from forward-thinking general practitioners who found it unacceptable that one could send an

[4] MedCom 2 på tryk—https://medcom.dk/wp-content/uploads/2023/03/medcom2-paa-tryk.pdf

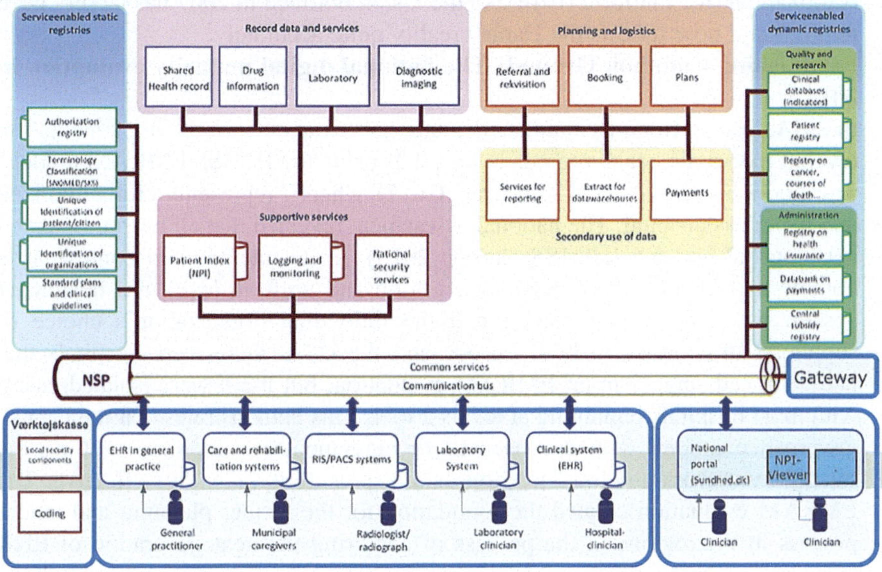

Fig. 1 The suggested National infrastructure by SDSD in 2007 (Larsen n.d.)

invoice electronically but not an electronic prescription between the doctor and the pharmacy.

2. **National Infrastructure 2008**

   In the strategy that was launched in 2008[5] the main focus moves from consolidating existing systems to a large-scale dissemination strategy to all hospitals and to new groups of health care professionals like physiotherapists, psychologists, etc.

   A key element of the 2008 strategy was the development of a proposal for a common public infrastructure, as described by SDSD (Fig. 1). The aim of the national infrastructure was to ensure the coherence between the systems and the seamless exchange of data and information.

3. **The 2011 Financial Agreement**

   The 2011 financial agreement established the principle that the party responsible for a task is also responsible for the IT aspect, a principle that remains in place today. This principle forms the basis for not introducing an all-encompassing IT system for the entire Danish healthcare system. At the same time, the agreement emphasized that regions and municipalities were obliged to actively participate in national solutions, such as the Shared Medication Record, Health Journal, and the Health Data Network. The agreement has since been updated to include the

[5]Sammenhængende Digital Sundhed i Danmark, National strategi for digitalisering af sundhedsvæsnet 2008 – 2012 – til fremme af befolknings sundhed samt forebyggelse og behandling.

National Service Platform (NSP), so the vision outlined in 2007 has largely been realized and now reflects the Danish reality quite accurately.

4. **Establishing Common Ground: The national digital maturity evaluation in 2014**
   In 2014, Danish Regions conducted a national assessment of the implementation level of electronic health records in hospitals using the HIMSS EMRAM model.[6] The model operates on an 8-step scale (0 to 7), where 7 represents a fully digitized and paperless hospital. The national assessment revealed that all hospitals scored between 5.2 and 5.6 (HIMSS Europe 2015). In an international context, it is notable that Denmark achieved such a high and uniform level in a healthcare system based on a free market and the individual organization's choice of suppliers. In contrast, in both Europe and the USA, there were hospitals that had achieved stage 7 in an EMRAM evaluation, but these were isolated cases, with most hospitals remaining at levels 2 to 4. This demonstrates that the Danish governance model, with a consensus or agreement-based process setting the overall framework and local freedom and responsibility, is highly effective. The EMRAM evaluation created the foundation for the further planning and tender process in the regions in the process of acquiring the next generation of EHR systems. Especially the common procurement process in the Capital and Selland regions was benefiting from information.

2.4 The Cohesive Healthcare System and Standardization

The implementation of EHRs in Danish hospitals cannot be understood without considering the broader context in which hospitals operate. The clear division of tasks among actors in the Danish healthcare system requires effective EHR systems and well-functioning electronic communication. This has meant that EHR procurement and implementation in hospitals have not only focused on internal use and functionality requirements. There has always been a specific task of ensuring that the systems support cross-sector task sharing and communication. This requirement has proven to be resource-intensive and challenging, particularly for new international suppliers entering the market. Especially for suppliers from the US market, adapting to local communication standards rather than relying on their usual standards has been a significant challenge.

National solutions, such as the Health Journal, where clinicians can access patient data from other hospitals and primary care records via Sundhed.dk through a direct link in their local EHR system, are highly effective in ensuring access to all relevant information without the need for everyone to use the same system nationwide. Communication standardization has ensured that data is available in the same format across all systems.

[6]HIMSS EMRAM model—https://www.himss.org/what-we-do-solutions/maturity-models-emram

2.5 Implementation Processes

The implementation of electronic health records (EHR) in Denmark has been a complex and multifaceted process, with several factors playing a central role. This section focuses on the various perspectives that have been crucial to the implementation processes.

1. **Management Perspective**
   The management perspective is vital for the successful implementation of EHR systems. In the Danish context, the early stages were driven by passionate individuals who, in some cases, advanced the development and implementation more in spite of, rather than because of, the top leadership in the counties. This changed significantly with the transition to regions and the introduction of national strategies from 2008 onward. The implementation of EHRs and their daily use became both a task and a priority for the regional leadership. This shift meant that the necessary financial resources were allocated, and procurement and implementation processes were professionalized and standardized. The focus on management ensured clear strategic communication about the direction of the organization, along with the necessary leadership to secure the required changes in workflows and procedures.
   An interesting aspect of the Danish EHR development is the tradition of closely following projects with an evaluation process. This applies at the national level through the EHR Observatory[7] and in several local projects. For instance, the SHIFT-EHR project produced five major evaluation reports documenting both results and the process. In many cases, evaluations were also used as a management tool for early warnings if any part of the implementation process was not functioning optimally. By maintaining a management focus on evaluations, quick interventions could be made if interviews or observations identified staff groups struggling with the systems or where there was strong professional resistance.
2. **Implementation and the Interaction Between Private Suppliers and the Public Sector**
   Over time, there has been a significant change in the interaction between public organizations acquiring EHR systems and the companies supplying these products. During the first wave of development and implementation (1996–2007), 13 different projects were initiated under the national HEP program. These projects were accompanied by a national evaluation that annually presented the results at the EHR Observatory's conferences. This was a crucial element in ensuring knowledge sharing across different projects and counties.
   All the projects were characterized by local enthusiasts who, with varying levels of support from the broader organization and top management, embarked on developing and implementing an EHR system in collaboration with a Danish

[7] Nøhr et al., Digitalisering af det danske Sundhedsvæsen, Syddansk Universitetsforlag, 2020.

supplier or the Danish division of an international IT company. The starting point for system development was often an EHR system for general practice or another clinical system. The development of these systems was heavily based on close interaction with clinical users, involving direct contact between developers and clinicians, and continuous adaptation of the systems as knowledge and experience evolved. This period was marked by a pioneering spirit and expectations of creating significant business opportunities both nationally and internationally. There was a strong sense of collaboration and an understanding that these were developmental projects with unclear goals, driven by ambitions to "create the best EHR system in Denmark" or "together forge a new business venture." In many of these projects, the formal procurement process was minimal, and the requirements specification could simply be an agreement to deliver "an EHR system." This led to highly dynamic development processes that produced solutions well-tailored to support clinical workflows and processes. The initial projects, centered around smaller hospitals or single departments, sometimes struggled to scale to the next hospital. Additionally, vague contracts between hospitals and suppliers sometimes resulted in disagreements over deliveries, timelines, and payments.

The second wave of implementation was a consequence of the structural reform that came into effect in January 2007.[8] This reform reduced the 14 counties and the Capital Region Hospital Community to 5 regions. At the onset of the regions, at least 15 different EHR systems had been implemented across hospitals in Denmark, some from the same supplier but with vastly different configurations and coverage across counties. The regions then initiated a consolidation process, either through regional tenders or by extending an existing system from one county to cover the entire region, ultimately reducing the number of EHR systems to 5. However, the consolidation was not solely driven by the structural reform; several counties had already begun work on expanding a comprehensive record system covering all hospitals in the county. For example, in 2003, Funen County launched a tender that resulted in the acquisition of COSMIC EHR from Cambio for implementation across all hospitals.

After the national EMRAM assessment in 2014, the final consolidation process began, with the two eastern Danish regions, the Capital Region and Region Zealand, conducting a joint tender won by EPIC. The subsequent implementation process faced significant resistance from a large portion of clinical staff, numerous media cases, and considerable political turbulence (Allen 2019). During the implementation, it became clear that the need for adjustments to the technical solution, integration with other systems and national standards, and especially the harmonization of workflows across departments and hospitals had been underestimated.

In Region Southern Denmark, a tender process was carried out in 2017-18, won by Systematic, which already provided EHR solutions to the North and

[8] https://www.ism.dk/Media/7/A/Strukturaftale.pdf

Central regions. This means that today, there are only two EHR systems in use across Danish hospitals: Sundhedsplatformen (EPIC) in Eastern Denmark and Systematic in Western Denmark (Hertzum et al. 2022).

2.6 Perspectives

1. **Better Utilization of Data and Cross-Sectoral Cohesion**
   With the significant standardization and widespread adoption of EHR systems in Denmark over more than 25 years, there is no doubt that the Danish healthcare system holds some of the most comprehensive data in the world (Jordt 2021). Additionally, due to shared standards and the requirement for reporting to national databases, this data is of very high quality. In 2024 a new structural reform of the Danish health care system was pass by the parliament. The reform have an increased focus on prevention and moving care and treatment from hospitals to primary care, the well-developed eHealth environment will be an important basis for the implementation of the structural changes.

   There has been, and continues to be, an ongoing discussion, particularly in recent years, about how the healthcare system can better utilize the knowledge embedded in this data. Until now, research and quality improvement have been the primary beneficiaries of this data. The national quality organization, RKKP[9] operates more than 80 national quality databases that receive automatic reporting from various EHR systems, primarily to monitor clinical quality and develop new clinical guidelines and procedures.

   In the research domain, it is particularly registry-based research that has significantly benefited from the data available in Denmark. However, there is also an understanding that the potential of Danish data is far from fully realized, and especially with the development of AI solutions, there are great opportunities to create better solutions that can support individual clinicians in their daily work through more effective use of data. To better and more effectively utilize data, the Danish healthcare system is working on creating a "single entry point" to data, where researchers and companies can access data. However, several technical, legal, and ethical challenges remain unresolved, making the exploitation of Danish data complex.
2. **International Trends**
   Two international trends are influencing the development of EHR systems in Denmark: the adoption of new standards and their implementation, particularly the transition to FHIR as the primary standard in the Danish healthcare system,[10] and the trend toward the use of the OpenEHR standard as the basis for new tenders. The second international influence comes from the EU and its decision to establish the European Health Data Space (EHDS), which could potentially have

[9] https://www.rkkp.dk/

[10] https://hl7.dk/FHIR-implementations-1.0.0.html

a significant impact on the Danish system, especially concerning the costs associated with its implementation.

(a) **New Standards—Market Disruption?**
The Danish infrastructure has always relied on the use of international standards, functionally and semantically adapted to the Danish context. The first standards implemented in the 1990s were UN-EDIFACT. The national Danish version of these standards has since been maintained and certified by MedCom.[11] However, the industry has moved on to new standards, with the HL7-FHIR standard now gaining widespread adoption. This means that in the coming years, the Danish healthcare system faces a significant and costly task of transitioning to new standards, requiring substantial development of all systems. The task is coordinated by MedCom, but the costs are borne by each healthcare organization.

In the short term, the implementation of the FHIR standard will not result in significant functional changes but will open up opportunities to more actively explore solutions based on OpenEHR[12] in Denmark. The developments in this area are also of interest in the Danish context, particularly in terms of potentially returning to a more open market where multiple vendors can offer solutions to the healthcare system. The developments in the EU with OpenEHR, particularly in Catalonia, are being closely monitored, with ongoing assessments of the advantages and costs that such a model could have for the Danish healthcare system.

(b) **EHDS—European Health Data Space**
The EHDS is an EU initiative aimed at creating a common digital infrastructure and rules for handling, exchanging, and accessing health data across member states. EHDS has two primary objectives: Primary use of health data: Improving healthcare and treatment by ensuring that patients and healthcare professionals have easy access to relevant health data regardless of the country within the EU. Secondary use of health data: Promoting research, innovation, and policy development by allowing secure access to anonymized health data for researchers, companies, and public authorities. EHDS aims to strengthen cooperation within EU healthcare systems, promote innovation, and ensure better patient outcomes.

EHDS will undoubtedly improve the integration between systems in many European countries where digital infrastructure is not as advanced as in Denmark. However, there will also be potential benefits for the Danish healthcare system, primarily concerning the secondary use of data. It is also clear that implementing the new EU requirements will have significant practical and economic consequences (Hägglund et al. 2024). Three levels of possible implementation are being considered based on the following models:

[11] https://medcom.dk/standarder/

[12] https://openehr.org/

(i) **Minimum Model**
The minimum model is limited to data already shared or planned to be shared nationally. It provides for the display of received data from abroad via a common portal solution. Additionally, for sent/shared data, it relies on national infrastructure and existing systems, continuing the project with patient summaries and ePrescriptions, and supplementing with MedCom discharge letters, the Laboratory Answer Database, etc. Furthermore, citizens' involvement is limited, offering the addition of data either not at all or only via existing solutions, e.g., My Health Overview (Min Sundhedsplatformen), the Shared Medication Record, etc.

(ii) **Intermediate Model**
The intermediate model builds on the above, implementing infrastructure to make more Danish data accessible for sharing abroad. This includes the ability to share prescriptions, discharge letters, and laboratory results from the Danish side, as well as making data from digital solutions for citizens available in the EHDS. Citizens' involvement is thus expanded to include citizen-reported data, and we also start to address some of the consequences related to research use.

(iii) **Maximum Model**
The maximum model is a complete implementation of EHDS, making all relevant Danish data accessible for both primary and secondary use. It involves implementing comprehensive infrastructure for data sharing, integration of all relevant systems, and ensuring full compliance with EHDS standards.

The cost levels of the three models naturally vary greatly. Current estimates predict that the minimum model will cost between 2 and 5 billion DKK for development and implementation, with an annual operating cost of approximately 500 million DKK.

3 Nordic Health 2030: A Vision for 2030

In 2019, the Copenhagen Institute for Futures Studies, under the auspices of the Nordic Health 2030 initiative, convened prominent health stakeholders from Denmark, Norway, Sweden, Finland, and Iceland. The purpose of these gatherings was to initiate a series of strategic workshops aimed at re-envisioning the future of health within the Nordic region. These workshops served as a platform for open discourse on the current challenges, identifying the requisite resources, partnerships, and leadership strategies necessary for transitioning to a health paradigm predicated on personalized prevention.

This initiative was in alignment with the Nordic Council of Ministers' overarching objective to render the Nordic Region the most integrated in the world by 2030, with a particular emphasis on health. A broad consensus emerged among stakeholders, both public and private, advocating for the adoption of shared values and the implementation of innovative, collaborative frameworks for the management, delivery, and regulation of healthcare. This consensus was encapsulated in a final report, published in October 2019, which outlined novel concepts for the future of Nordic health.

These concerted efforts engendered a sustained interest in regional healthcare collaboration, giving rise to the ongoing Nordic Health 2030 Movement. This movement seeks to integrate and operationalize the insights gleaned from the initial workshops to enhance population health, leadership, and cooperative action across the Nordic countries.

While the initial phase catalyzed a range of activities and highlighted the urgency of addressing health challenges within the region, it also underscored the necessity for stakeholders to prioritize specific areas of engagement and formulate robust action plans. Consequently, a second multi-stakeholder initiative, titled "Nordic Health 2030—From 'What' to 'How'," was launched, focusing on building consensus and devising action plans for systemic transformation. The prevailing ambition is to position the Nordic region as a global leader in health, utilizing strategic foresight tools to adeptly navigate and shape the future health landscape.

The vision for Nordic Health 2030 is articulated as follows:

- Through cross-border and cross-sector collaboration, the Nordic region aspires to be home to the healthiest and most interconnected populations and to develop and sustain the most digitally enabled, data-driven, integrated, and sustainable health systems globally.
- By 2030, the Nordic region has embarked on a transformative journey, spearheaded by the Inner Nordic Health Area initiative, with the goal of establishing the most interconnected, digitally empowered, and sustainable health systems in Europe. This evolution is propelled by robust cross-border and sectoral collaborations, with a focus on health sustainability and the integration of health considerations across all policy domains, particularly through the holistic application of data and digital technologies in healthcare management.
- A pivotal aspect of this transformation is the establishment of the Humanome, a digital health model designed to optimize data usability and enhance individuals' access to, utilization, and control over their health data. This initiative underpins the Nordic health systems' shift toward precision medicine and preventive care, with the objective of reducing the preventable disease burden.

3.1 The Humanome Model

Central to this paradigm shift is the Nordic Health 5/5 aspiration, which advocates for a strategic reallocation of health budgets: 5% of GDP is dedicated to late intervention and treatment, while the remaining 5% is allocated to early intervention and prevention (Fig. 2). This balanced approach is intended not only to maintain

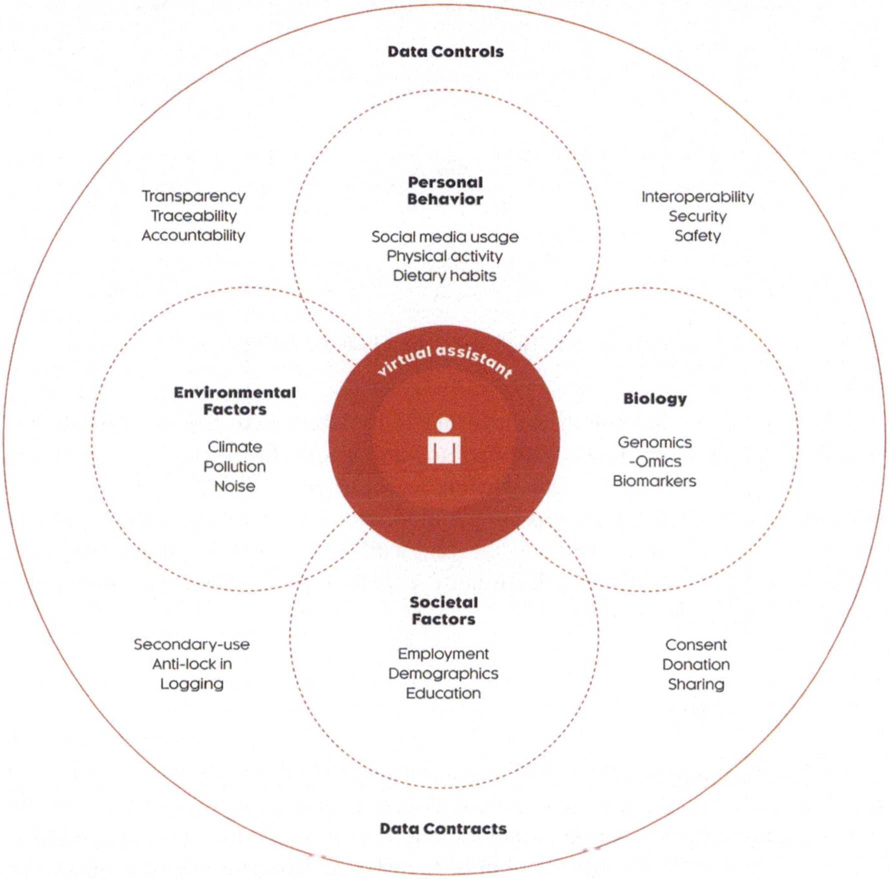

Fig. 2 The Humanome Model (The Copenhagen Institute for Futures Studies 2024)

health expenditure around 10% of GDP but also to foster a sustainable health model that prioritizes early detection and precise interventions.

3.2 Sustainable Health Model

Another significant concept is the Sustainable Health Model, which elucidates the dynamics of health and health-relevant data exchanges among individuals, organizations, and health systems (Fig. 3). This model facilitates the generation of real-time health insights, promotes sustained individual well-being, and supports the application of both individual and population-level data for timely interventions and research. Additionally, it enhances the efficiency of healthcare systems and the overall health of populations.

Fig. 3 The sustainable health model (The Copenhagen Institute for Futures Studies 2024)

By 2030, the Nordic countries have markedly intensified their collaboration at clinical and regulatory levels, striving toward a sustainable health model that integrates the principles of One Health and Planetary Health. This collective endeavor, coupled with increased cooperation with European and global partners, has positioned the Nordic region at the vanguard of a preventive health paradigm, emphasizing the widespread application of secondary prevention to maintain population health for as long as possible.

4 Discussion

For the Nordic countries, the EHDS aligns closely with their ongoing digital health initiatives. However, given the Nordic region's advanced digital infrastructure and strong tradition of cross-border collaboration, there is a growing desire among these countries to establish a Nordic version of the EHDS. This regional implementation would allow the Nordics to leverage their unique commonalities to create a more agile and efficient health data space that could be implemented more rapidly than the broader EU initiative. This is a high focus for the Nordic Health 2030 Movement.

The idea behind a Nordic Health Data Space is to capitalize on the region's existing strengths in digital health. The Nordics have already made significant progress in integrating health data systems within their borders and across borders through initiatives like the Nordic Interoperability Project and others. By creating a regional health data space, the Nordic countries could further enhance the interoperability of their systems, streamline data exchange, and ensure that the benefits of the EHDS reach their populations sooner.

Moreover, a Nordic Health Data Space could serve as a model for the broader European implementation, showcasing how a region with advanced digital capabilities and strong collaborative ties can effectively operationalize the goals of the EHDS. This approach would not only benefit the Nordic populations but could also provide valuable insights and best practices for other EU countries as they work toward their own EHDS implementations.

In summary, while the EHDS offers a promising vision for the future of health data in Europe, the Nordic countries are eager to accelerate this vision within their region. By establishing a Nordic version of the EHDS, they aim to quickly harness the potential of integrated health data systems, further solidifying their position as leaders in digital health innovation. This regional initiative could pave the way for a more connected, efficient, and patient-centered healthcare system across the Nordics, setting a high standard for the rest of Europe, especially built on the values and concept of the Nordic Health 2030 Movement.

References

Allen A (2019) Lost in translation: Epic goes to Denmark. https://www.politico.com/story/2019/06/06/epic-denmark-health-1510223. Accessed 1 Sep 2024

Bernstein K, Bruun-Rasmussen M, Vingtoft S et al (2003) Modelling and implementing electronic health records in Denmark. Stud Health Technol Inform. 95:245–250

Dixon BE, Schmidt T, Nøhr C (2022) Facilitating HIE in Denmark: the story of MedCom, a Danish health information organization. Health information exchange (2nd edn), pp 579–601. https://doi.org/10.1016/B978-0-323-90802-3.00031-9

Eriksson-Backa K, Hirvonen N, Enwald H, Huvila I (2021) Enablers for and barriers to using My Kanta – A focus group study of older adults' perceptions of the National Electronic Health Record in Finland. Informatics for Health and Social Care 46(4):399–411. https://doi.org/10.1080/17538157.2021.1902331

HIMSS Europe (2015) European hospitals EMRAM maturity overview: CIO summit 2015 session 1. https://na.eventscloud.com/file_uploads/0cf548ab2f4eaeafd0a6b4ce615f0399_Hoyt_Session_1_European_Hospitals_EMRAM_Maturity_Overview_CIOSummit.pdf. Accessed 1 Sep 2024

Hägglund M, Scandurra I (2021) User evaluation of the Swedish patient accessible electronic health record: system usability scale. JMIR Human Fact 8(3):e24927. https://doi.org/10.2196/24927

Hägglund M, Kharko A, Bärkås A et al (2024) A Nordic perspective on patient online record access and the European health data space. J Med Internet Res 26:e49084. https://www.jmir.org/2024/1/e49084

Hertzum M, Ellingsen G, Cajander Å (2022) Implementing large-scale electronic health records: experiences from implementations of epic in Denmark and Finland. Int J Med Inform 167:104868. https://doi.org/10.1016/j.ijmedinf.2022.104868

Jordt HH (2021) The Danish framework for health data. https://www.ehalsomyndigheten.se/globalassets/ehm/3_om-oss/sa-jobbar-vi-med-e-halsa/denmark_the-danish-framework-for-healthdata-6-1.pdf. Accessed 1 Sep 2024

Larsen O (n.d.) Health IT in Denmark: a cultural responsibility https://www.healthinnovationforum.org/article/health-it-in-denmark-a-culture-of-public-responsibility/. Accessed 1 Sep 2024

Løhde S, Bjerre M (2023) Digitalisation in Danish healthcare. In Healthcare Denmark. https://healthcaredenmark.dk/media/sghmh0in/digitalisation_2024_onlineversion.pdf. Accessed 1 Sep 2024

Strandberg-Larsen M, Nielsen MB, Vallgårda S et al (2007) Denmark: Health system review. Health Syst Transit 9(6):1–164

The Copenhagen Institute for Futures Studies (2024) Nordic health 2030 from what to how. https://cifs.dk/nordic-health-2030-download. Accessed 1 Sep 2024

The Ministry of Health (2017) Healthcare in Denmark: an overview. https://sum.dk/Media/C/A/Healthcare-in%20denmark%20an%20overview%20english-V16-dec.pdf. Accessed 1 Sep 2024

Tikkanen R, Osborn R, Mossialos E et al. (2020) International health care system profiles Denmark. https://www.commonwealthfund.org/international-health-policy-center/countries/denmark. Accessed 1 Sep 2024

The European Health Data Space and the Future of Cross-Border Healthcare

Lucas Deimel, Maike Hentges, and Rainer Thiel

Abstract

With the creation of the European Health Data Space (EHDS), the EU aims to foster the exchange of patients' health data across national borders for primary use in healthcare and secondary use for research and innovation. The draft Regulation builds on years of efforts to gradually establish and roll out secure cross-border exchange of health data, thereby boosting implementation of digital health products and services, also in hospitals. However, achieving interoperability for continuous healthcare within and across EU Member States remains challenging.

In this chapter, we discuss the current state of the EHDS and its potential impacts on hospitals' digital maturity. At this stage of the draft Regulation, these cannot be determined with certainty, but the EHDS promises to pool technological, ethical, and political advancements of over two decades and add a new component to inpatient care delivery and clinical research: the ability to access health data of patients from another EU Member State, with the patients maintaining control over such data.

A certain degree of digital maturity in hospitals, particularly in the domains of cybersecurity, data management, and exchange/interoperability, is required to fully utilise the potential of the EHDS, in addition to establishing cross-organisational relationships.

Keywords

European Health Data Space (EHDS) · Cross-border healthcare · Primary use · Secondary use · Health data exchange

L. Deimel (✉) · M. Hentges · R. Thiel
empirica Communication and Technology Research (empirica), Bonn, Germany
e-mail: Lucas.Deimel@empirica.com; maike.hentges@empirica.com; Rainer.Thiel@empirica.com

A. Scheuer, J. Studzinski (eds.), *Digital Maturity in Hospitals*,
https://doi.org/10.1007/978-3-031-80704-6_11

Abbreviations

EEHRxF	European Electronic Health Record Exchange Format
EHDS	European Health Data Space
EHR	Electronic Health Record
GDPR	General Data Protection Regulation
HDABs	Health Data Access Bodies
IPS	International Patient Summary
NCP	National Contact Points
NCPeH	National Contact Points for eHealth
TEHDAS	Joint Action Towards the European Health Data Space

1 Towards the European Health Data Space

The European Health Data Space (EHDS) offers significant benefits: enabling access to personal electronic health data anytime and anywhere, facilitating the sharing of this data with healthcare professionals across national borders, and supplying researchers and policymakers with essential data to enhance population health.

The COVID-19 pandemic has clearly demonstrated the importance of digital services in the health domain and has triggered an important acceleration in the uptake of digital health tools and services. The EHDS is meant to be a consistent framework for shared health data access and the use of such data for research, e.g., for providers of digital health services and products, thereby tackling persisting difficulties in the access to and exchange of electronic health data by patients, health professionals, healthcare providers and researchers. However several questions remain unanswered in the EHDS draft Regulation, not least for patients, healthcare providers and health professionals as the primary users (and owners) of electronic health data. Will the EHDS actually translate into value for them and what challenges will they face? Additionally, what are the implications for hospitals striving to achieve digital maturity, and how will this transformation impact patient care?

The EHDS is envisioned as the first of nine sector-specific European data environments to be established under the European Strategy for Data, potentially serving as a blueprint for future Common European Data Spaces (European Commission 2024a). In March 2024, the European Parliament and the Council reached a provisional agreement on the Commission's proposal for the EHDS Regulation (Council of the European Union 2024). The final legal text is still under development and expected during 2024, followed by implementing acts within 2 years of entry into force. At the earliest, the general application of the EHDS Regulation could be assumed in 2026/27, whereas data exchanges for primary and secondary use are scheduled to start in 2028.[1] At this stage of the legislative process, where changes are still possible, the precise impacts of the EHDS cannot be determined with certainty.

The draft Regulation (as of 18 March 2024) promises to pool technological, ethical, and political advancements of over two decades and add a new component to inpatient care delivery and clinical research: the ability to access health data of patients from another EU Member State, with the patients maintaining control over such data (European Commission 2022a). This is based on and complements citizens' rights regarding the processing of their personal data, as outlined in Regulation (EU) 2016/679, General Data Protection Regulation (GDPR).[2] It also connects with several other legal texts, proposals, and initiatives, including Directive 2011/24/EU,[3] the Medical Device Regulation,[4] the Artificial Intelligence Act,[5] and the EU cybersecurity framework.[6]

Ongoing projects and supporting actions, such as the Joint Action Towards the European Health Data Space (TEHDAS),[7] the HealthData@EU Pilot[8] for secondary use, and the Joint Action extended EHR@EU Data Space for Primary Use (Xt-EHR Joint Action),[9] are focused on the inception, dissemination, and scale-up of minimum specifications, guidelines, services, and infrastructures for cross-border data exchange for primary and secondary use, laying the foundations for the EHDS.

The ultimate goal is to create a Single Market for health data under the control of individuals, with specified rules to ensure the quality of health data and the interoperability of IT infrastructures for data exchange. By doing so, the European Commission aims to prevent further fragmentation in the digitalisation of health data among Member States. Over a period of 10 years, the EHDS could liberate more than EUR 10 billion: EUR 5.5 billion stemming from improving access to and exchange of health data in healthcare (primary use), and EUR 5.4 billion from better use of electronic health data for research, innovation, and policy making (European Commission 2022b).

---

[1] https://ec.europa.eu/commission/presscorner/detail/en/QANDA_24_2251

[2] Regulation (EU) 2016/679 of The European Parliament and of the Council of 27 April 2016 on the protection of natural persons with regard to the processing of personal data and on the free movement of such data, and repealing Directive 95/46/EC (General Data Protection Regulation). OJ L 119, pp 1–88. https://eur-lex.europa.eu/eli/reg/2016/679/oj

[3] Cross-border Healthcare Directive. Source: https://eur-lex.europa.eu/legal-content/EN/TXT/?uri=celex%3A32011L0024

[4] https://eur-lex.europa.eu/legal-content/EN/TXT/PDF/?uri=CELEX:32017R0745

[5] https://artificialintelligenceact.eu/ai-act-explorer/

[6] EU cybersecurity framework (NIS Directive). Source: https://www.enisa.europa.eu/topics/cybersecurity-policy/nis-directive-new

[7] Second Joint Action towards the European Health Data Space (TEHDAS2). Source: https://tehdas.eu/

[8] EHDS HealthData@EU Pilot. Piloting an infrastructure for the secondary use of health data. Source: https://ehds2pilot.eu/

[9] Xt-EHR Joint Action Project overview. Source: https://www.xt-ehr.eu/overview. Accessed 28 June 2024

In this chapter, we aim to shed light on the EHDS in the context of digital maturity of hospitals, discussing the potential impacts and challenges in terms of the (primary) use and (secondary) re-use of electronic health data. We will also consider potential spill-over effects of the EHDS beyond the European Union.

2 Primary Use of Health Data: MyHealth@EU

The European Parliament has been equipping Union citizens with an increasing range of tools and rights in cross-border healthcare settings. Examples include the European Health Insurance Card (EHIC), information access via National Contact Points for cross-border healthcare (NCPs), cross-border recognition of prescriptions, and remote access to own medical records.

In the context of the EHDS, the systematic, structured, and consensual sharing of citizens' electronic health data via the MyHealth@EU infrastructure will become a mandatory digital service for Member States. National Contact Points for eHealth (NCPeH) are expected to lead the legal and technical adaptations necessary for a seamless flow of electronic health data between two Member States. More importantly, the NCPeH informs citizens and healthcare providers about the opportunities presented by accessing electronic health data of citizens from another Member State. In 2023, healthcare providers in 11 Member States, including the Netherlands, Estonia, and Finland, were connected via MyHealth@EU, enabling the exchange of ePrescriptions, eDispensations, and Patient Summaries through operational NCPeH. In the first half of 2024, the NCPeH in two more Member States, Greece and Latvia, went live (Fig. 1), however, many countries still need to join (European Commission 2024b). The expansion of cross-border exchanges to include other electronic health data, such as laboratory results and hospital discharge reports, is planned for a later stage.

The EHDS sets common standards to overcome the fragmented landscape of commercial electronic health record (EHR) systems across Europe and the diverse national catalogues of standards, classifications, and terminologies specific to each Member State's individual requirements. These common standards involve two key components: the European Electronic Health Record Exchange Format (EEHRxF) and a European logging component for EHR systems (Council of the European Union 2024).

The EEHRxF specifies sets of data relevant for cross-border exchange and defines their format, while the logging component stipulates the detailed capture of information whenever healthcare providers access personal health data electronically. Providers of software for EHR systems will be obliged to undergo a self-assessment and provide a declaration of conformity with these specifications for EHR systems placed on the market. The draft Regulation also proposes the establishment of a public EU database for registering such interoperable EHR systems, as well as labelled wellness applications that connect to and supply data to EHR systems or national electronic health data access solutions. While the EEHRxF has been a voluntary instrument for Member States since 2019, it will be elevated to secondary

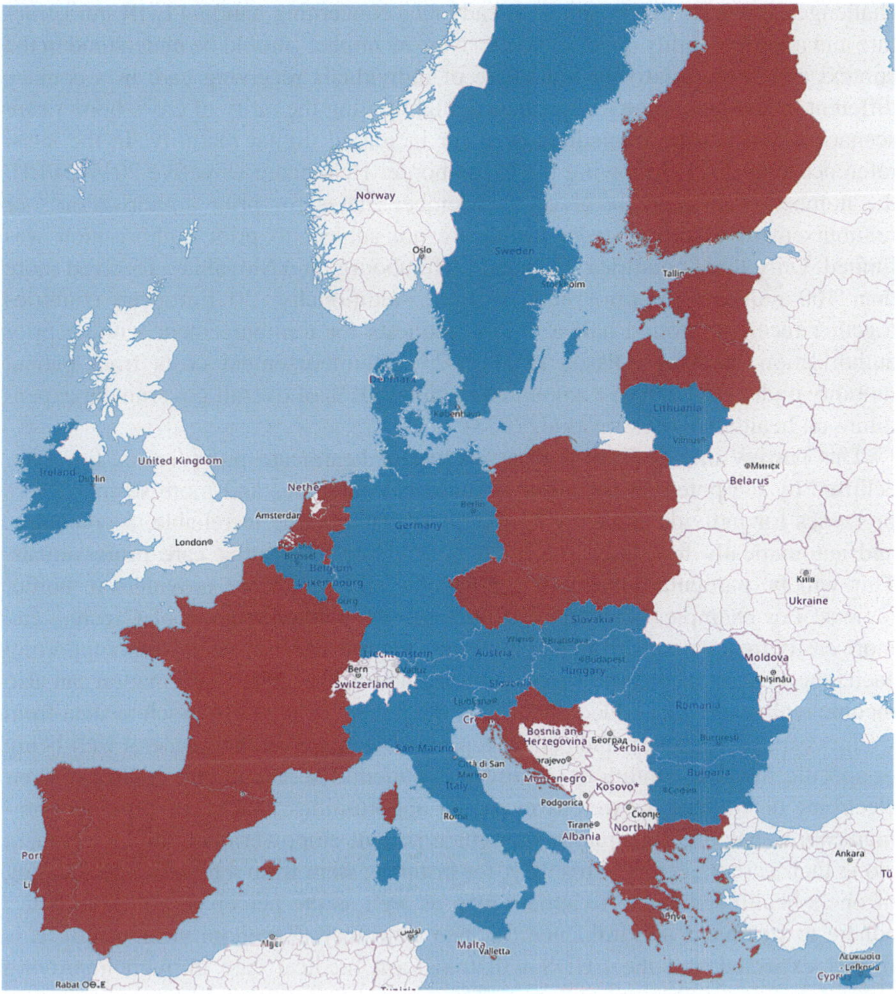

Fig. 1 Overview of Member States with an operational National Contact Point for eHealth (in red) as of Q1 2024 (eHDSI Monitoring Framework (KPIs) 2024a)

legislation once the EHDS Regulation enters into force. Compliance with the Regulation and its secondary legislation will be mandatory for any EHR system and software entering the market 3 years after the Regulation takes effect. Member States are free to impose conformance requirements and conditions for endorsement on any existing or future systems or architectures used by healthcare providers.

At this point, we expect the foreseeable and immediate impacts of the EHDS on hospitals and care providers to be relatively narrow and should not add to the challenges and tasks that hospitals already face concerning national EHR infrastructure and interoperability developments. 'Narrow impact' should be understood in the context of (1) the relatively low share of individuals receiving care in a country different from their country of residence, highlighting the rarity of cross-border care scenarios, and (2) the immediate need for increased digital maturity. In the latest reference year 2021, following the Cross-border Healthcare Directive 2011/24/EU, the number of requests received by Member States for prior authorisation[10] or reimbursement of cross-border healthcare not subject to prior authorisation was limited. Only three countries—France, Luxembourg, and Slovakia—received more than 100 requests for prior authorisation. Additionally, 20 European countries together received around half a million requests for reimbursement without prior authorisation in 2021 (Olsson et al. 2023). Reimbursement costs from patient mobility under the Directive amounted to only 0.01% of overall government expenditure on healthcare (Olsson et al. 2023).

This narrow impact from the perspective of healthcare providers is, however, eclipsed by the potential for improved medical outcomes and more seamless care processes for individuals receiving treatment. This is due to reliable, accountable, and automatically translated information available to treating care professionals. This will be particularly relevant for patients and healthcare providers in border regions. For example, hospitals in Bohemia, the southern region of Czechia, can more easily treat patients from neighbouring Austria and vice versa, thus supporting each other in case of bed shortages or emergencies. The shared information can also include real-world data added directly by the patient in the EHR, such as data from wellness apps, wearables, and patient-reported outcomes on their general well-being or specific medical conditions. Ultimately, patients' access to and control of their electronic health data, as well as increased opportunities for patient–provider communication, has the potential to strengthen patient empowerment and engagement along their care trajectory. This may, for instance, stem from a better understanding of one's health status and treatment plan as well as the perceived ability to better adhere to treatment and influence the care trajectory (Zanaboni et al. 2020). It is further expected that the EHDS will transform medical care through improving clinical decision-making, enabling more personalised care, and ultimately improving patient outcomes and safety.

More direct implications for hospitals involve the maturity of IT systems, as the successful implementation of the EHDS will largely depend on addressing interoperability issues. The EEHRxF was previously built upon Health Level Seven (HL7) standards (European Commission 2019), which could be replaced by the more recent HL7 Fast Health Interoperability Resources (FHIR). For example, current guidelines on the European Patient Summary are compatible with the ISO:27269 Health

[10] *Prior authorisation* means authorisation that patients need to receive from their health insurance provider before their travel abroad to be guaranteed reimbursement of their healthcare services in the country of treatment. Source: https://health.ec.europa.eu/system/files/2022-02/crossborder_enhancing-implementation_info-provision_en.pdf

Table 1 Number of hospitals in EU countries with an NCPeH that are operational with MyHealth@EU services (eHDSI Monitoring Framework (KPIs) 2024b)

Country	No. of hospitals
Croatia	130
Czech Republic	39
Estonia	20
Luxembourg	4
Malta	1
Netherlands	2
Portugal	64
Spain	258

informatics—International Patient Summary (IPS) standard, for which HL7 FHIR implementation guidelines exist (eHealth Network 2023). The IPS can be regarded as the international formalisation of the European Patient Summary guidelines. Both stipulate a range of code systems, including SNOMED CT Global Patient Set (GPS), ICD-10 (and ICD-11 once available), ATC, and LOINC, which should be used complementarily where possible. This complementary approach acknowledges that these systems overlap in their use cases and are employed to varying extents across the EU.

Under the EHDS, these standards could become the defining specifications for both cross-border and national data exchange. Member States will be obliged to adopt these standards nationally and could choose to link these to funding or reimbursement in the healthcare system. Therefore, it is in the best interest of hospitals to plan ahead and consider the EU Patient Summary and related specifications for future procurements.

As of early 2024, 518 hospitals in eight Member States are connected to MyHealth@EU and able to exchange Patient Summary data (Table 1). More countries and hospitals are expected to join this list in the coming years until participation in MyHealth@EU will become mandatory with the entry into force of the EHDS Regulation. Progress towards full coverage of hospitals will ultimately come down to many factors, such as NCPeH set-up in Member States, development and refinement of the related specifications (e.g., ePrescription) as well as the capacity of hospital IT-departments to comply with these specifications.

3 Secondary Use of Health Data: HealthData@EU

Focused primarily on research and innovation outside the clinical research domain, the new decentralised cross-border HealthData@EU infrastructure shall provide access to health data from EU Member States for both private and public sector research and development initiatives. In this context, the 2-year HealthData@EU pilot project kicked off in October 2022.

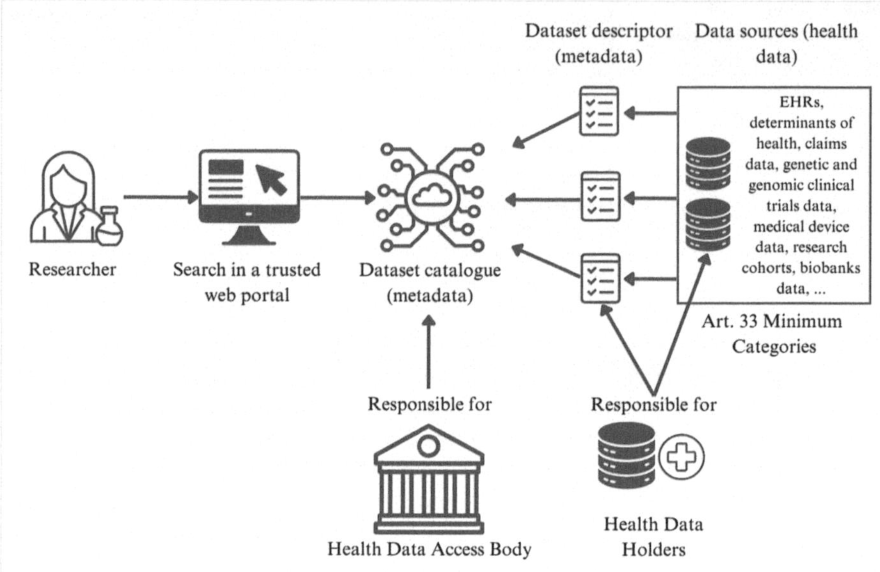

Fig. 2 Simplified process of data access for secondary use (Adapted from Langefeldt and European Commission Directorate-General for Health and Food Safety 2023)

HealthData@EU connects data platforms in a network infrastructure as a pilot version of the EHDS and publishes guidelines for data standards, quality, security, and transfer. The EHDS draft Regulation defines several purposes for secondary use, including supporting policymaking, regulatory activities, research, innovation, and the development of health products. This also extends to training artificial intelligence (AI) algorithms, such as those used in medical devices to provide more personalised care (Council of the European Union 2024).

Article 33 of the EHDS draft Regulation[11] outlines the minimum categories of electronic health data available for secondary use. These categories include data from EHRs, data impacting health (e.g., social, environmental, and behavioural determinants), genetic, genomic, and proteomic data, person-generated electronic health data, and data from registries and other databases (Fig. 2).

According to Article 36 of the proposal,[12] Member States will need to establish one or more Health Data Access Bodies (HDABs) for the secondary use of electronic health data. These bodies will ensure that such data is made available in pseudonymised or anonymised format to data users (e.g., researchers) by data holders (e.g., healthcare providers such as hospitals). HDABs will set up Secure

[11] https://www.european-health-data-space.com/European_Health_Data_Space_Article_33_(Proposal_3.5.2022).html

[12] https://www.european-health-data-space.com/European_Health_Data_Space_Article_36_(Proposal_3.5.2022).html

Processing Environments (SPEs) for data users, who will be provided with the requested data and tools for their research. The draft Regulation further stipulates that data users must inform the HDABs of the results of research conducted with datasets retrieved using HealthData@EU. HDABs will then communicate results on potential health benefits or impacts (e.g., on risk factors for disease development, new promising treatments) back to the data subjects.

By means of implementing acts, the European Commission will further specify the minimum characteristics data holders must provide. As depicted in Fig. 2, these will enter into a national dataset catalogue describing in machine-readable format the metadata of all datasets available in a given Member State (European Commission 2022b; EHDS 2022). It is expected that these specifications will build on the EEHRxF, which will become an EU-wide standard (ibid.).

The EHDS draft Regulation introduces a data quality and utility label indicating the characteristics of a dataset, thereby aiming to support data users in their process to identify relevant and suitable datasets (Council of the European Union 2024). Based on Article 56 of the proposal,[13] the label is to be applied by health data holders (e.g., hospitals). Initiatives funded under Horizon Europe[14] will advance the work on this new utility label starting 2024 in order to provide and pilot a common framework that shall limit the burden on data holders in the roll-out and application of the label (European Commission 2022a). Implementing acts determining the elements of the label, as well as minimum specifications for cross-border datasets to promote harmonisation, comparability, and cross-border research, will follow the adoption of the EHDS Regulation.

Regarding processing of personal (health) data, the EHDS provides an additional and specialised regulatory framework which applies in parallel to national data protection legislation. Existing additional layers of legislation will not be prioritised in the context of intra-Union cross-border secondary use of health data, as long as data users choose to follow the EHDS framework to gain access to data via HDABs.

Further, although existing research has demonstrated European citizens' positive attitudes towards health data sharing, there are reservations about the potential use of this data for commercial purposes (Skovgaard et al. 2019). These concerns are echoed by some patient advocacy groups, consumer organisations, digital rights advocates, and data protection entities, who fear that patients may not be the true focus of the EHDS Regulation proposal (European Digital Rights 2023). Addressing such concerns will be crucial for ensuring the success of the EHDS and maintaining public trust in the secondary use of health data. Developing voluntary health data training programmes for citizens and launching public communication campaigns, for instance, in hospitals, as done by the French Health Data Hub in collaboration with the Ministry of Health and the data protection authority (CNIL), could help to

[13] https://www.european-health-data-space.com/European_Health_Data_Space_Article_56_(Proposal_3.5.2022).html

[14] https://research-and-innovation.ec.europa.eu/funding/funding-opportunities/funding-programmes-and-open-calls/horizon-europe_en

raise awareness about the conditions for the secondary use of health data (TEHDAS 2023a).

The effects of the EHDS on strengthening the secondary use of health data in the EU are expected to be manifold. At the population level, the EHDS promises to provide regulators with easier access to crucial health data, enabling them to make timely and informed decisions, particularly in the event of future pandemics. By offering new potential for research and innovation, the EHDS could facilitate more efficient progress on EU initiatives such as the EU Cancer Plan, the Pharmaceutical Strategy for Europe, and activities by the European Commission Directorate-General for Health Emergency Preparedness and Response (HERA),[15] which was launched in 2021 in response to the COVID-19 pandemic.

3.1 Potential Impacts and Challenges of the EHDS for Hospitals

The EHDS requirements will affect various digital domains within hospitals, including cybersecurity, data management, and data governance, two-way enabled health information exchange (interoperability and international standardisation), as well as EHR systems. As hospitals will strive to meet these requirements, they will navigate complexities such as compliance, resource allocation, and technical integration. Historically, the implementation of health-related EU legislation has faced challenges and resulting delays, as observed with the Medical Device Regulation, the EU Clinical Trials Regulation, and the GDPR. Complexities arise from compliance requirements, the associated administrative burden, limited human and financial resources, idiosyncrasies and readiness of healthcare systems, a fragmented health data landscape, and differing interpretations of the legal texts, among other factors (Peseckyte 2023; Marelli et al. 2023). To this end, it is fair to assume that these historical challenges will likely persist with the implementation of the EHDS. The EHDS draft Regulation currently lacks detail and guidance for the technical implementation, and thus may not be able to overcome complexities resulting from varying solutions and national infrastructures (Raab et al. 2023). Notwithstanding these complexities, the EHDS and its requirements for Member States have the potential to enhance the digital maturity of healthcare systems and hospitals across the EU. By leveraging digital solutions for provider-to-provider communication and connectivity, the EHDS can promote collaboration between providers and healthcare professionals across specialties and sectors at both national and cross-border levels. Additionally, the EHDS can foster cooperation between healthcare providers and digital health authorities, which are to be designated in each Member State to monitor and report on the implementation of the EHDS.

- First and foremost, the translation and exchange of health data externally via the mandatory MyHealth@EU infrastructure necessitates that all hospital

[15] https://health.ec.europa.eu/health-emergency-preparedness-and-response-hera_en

information and IT systems can connect to national infrastructures via the NCPeH to enable two-way data exchange. This involves gradually implementing the priority categories for primary use, including Patient Summary, ePrescription, medical images, laboratory results, and discharge reports.
- Secondly, the mandatory self-certification scheme for EHR systems by their providers requires that EHR systems used by healthcare providers comply with common interoperability and security specifications. However, there is debate about whether the self-certification scheme is sufficient to ensure adherence to relevant data protection requirements (e.g., KBV 2022).

The detailed data quality requirements and technical specifications will be defined later through implementing acts. Hospitals will then need financial and technical support and guidance to upgrade their information systems where needed to facilitate high-quality data capture and exchange internally and externally, for instance through local, regional, or national procurement. Depending on the level of hospitals' maturity in data management processes, establishing and maintaining such infrastructures and functionalities will require significant investments in economic and human resources (TEHDAS 2023b). An early impact analysis of the EHDS by the Dutch Ministry of Health, Wellbeing, and Sports estimated data holders' initial costs for making high-quality data available for secondary use to HDABs to be between EUR 7 to 12 million, primarily due to necessary staff training (Ministry of Health, Wellbeing, and Sports 2022). In addition to investments in infrastructure and personnel, preparing for implementation will also require changes at healthcare providers' governance, legal, and organisational levels, such as conducting data quality and security audits.

To foster data quality, availability, and usability, strengthening the digital capacities of health professionals through training and upskilling will be essential. Incorporating digital (health) skills into the training of health professionals and micro-certification programmes will become increasingly important, especially for larger universities and teaching hospitals, as they deal with a large volume of data and are leading the healthcare research. However, some uncertainties still surround the obligations of healthcare professionals and providers, such as sending hospital data to a central EHDS facility. Questions remain regarding liability for health data in EHRs, additional bureaucratic and administrative workloads, and costs incurred for high-quality data registration, coding, updating, maintenance, exchange, and retrieval. To actively participate in the EHDS, hospitals and health professionals will need guidance on how and what data to make available through the EHDS and what the obligations of healthcare providers as data holders are (e.g., pseudonymisation for HealthData@EU, applying the data quality and utility label).

In any case, the increased requirements for healthcare providers must not compromise the quality of their healthcare service delivery (Marelli et al. 2023). At this stage, whether and how the responsibilities of healthcare providers and health professionals will be practically realised, incentivised, or remunerated is unclear. Further challenges may arise from the existence of different disease classification and remuneration systems across countries. Overall, this has led to concerns that the

EHDS might inadvertently hinder rather than support the work of healthcare professionals and researchers (Marelli et al. 2023).

At the same time, it is crucial to ensure that patients and citizens are well-informed about their rights regarding the primary and secondary use of their electronic health data. To facilitate this, health professionals need to receive clearly defined responsibilities with regard to their role in informing their patients about access to and use of the data in their EHR. Currently, health providers' responsibilities differ between countries in this regard, and responsibilities are not always clearly assigned.

The EHDS will play a pivotal role in standardisation and harmonisation of electronic health data and access services and overall healthcare data management within hospitals. Despite local or regional efforts to achieve standardised health data access, collection, and storage, most Member States are still on their path towards nationwide interoperability, which currently hinders seamless data sharing and the effective use of data (EIT Health 2024). In this regard, the EHDS presents a great opportunity for advancing national data exchange. Given the varying degrees of maturity, connecting all healthcare providers to the EHDS in the envisioned timeframe may seem ambitious. The mechanism by which stakeholders will adhere to established standards, cybersecurity protocols, and harmonised data and services can be regarded as an important indirect effect of the EHDS for hospitals. Furthermore, with the experiences from the COVID-19 pandemic, the role of electronic health data in population health has become increasingly important and can be further developed and defined within hospitals.

Overall, the EHDS initiative appears to receive more enthusiastic support from the academic research and innovation communities rather than healthcare organisations. Despite their key role as data holders, healthcare providers' awareness about the EHDS is reportedly still rather low, resulting in a lack of proactive measures by providers to plan for EHDS implementation (EIT Health 2024). Therefore, explicitly examining the benefits, pros, and cons from a hospital and patient perspective is imperative to foster the trust, support, and cooperation of patients and providers. This assessment should extend to primary, secondary, and tertiary uses, care aspects, research, and educational purposes. Showcasing specific use cases to healthcare providers and health professionals with demonstrated benefits will be crucial to gaining their buy-in as key stakeholders for EHDS implementation (EIT Health 2024).

Given the nature of EU legislation, any impact of the EHDS at the level of healthcare providers will be delayed by at least 3 years, likely more. The EHDS is expected to apply from 2027 alongside the first implementing acts. However, regarding the exchange of data categories for primary use (Patient Summary, ePrescription, eDispensation) and secondary use in general, the EHDS will most likely not apply before 2029.

Finally, it has become clear that the EHDS—despite its often-appraised potential for society and industry alike—will add a new regulatory layer to an already busy landscape. The increase in legislative density is likely to be accompanied by a more pronounced shift from on-premise to cloud-based IT and cybersecurity solutions at the level of healthcare provider organisations to reduce risks.

3.2 Ongoing EU Activities and Outlook

The EHDS has the potential to play a significant role in international electronic health data standardisation and harmonisation activities, extending beyond the EU. This influence can span various aspects, including data, registries, and process and outcome data. By aligning with international standards and promoting their use, the EHDS can contribute to a more cohesive and interoperable global health data ecosystem, possibly benefiting both EU and non-EU entities in their healthcare data management.

In the realm of international research and innovation activities, the EHDS can catalyse collaborative initiatives and partnerships, particularly in addressing rare diseases where the pooling of data globally would be highly beneficial to gain a larger basis for research on rare diseases. Furthermore, potential spill-over effects encompass critical areas such as pandemic preparedness through monitoring and surveillance. Specifically, the EHDS could serve as a model for international collaboration in tracking and responding to pandemics, leveraging its standardised and harmonised approach to health data for more effective global health emergency preparedness efforts.

In the compromise text of the EHDS Regulation proposal, the co-legislators provided further details on the conditions for harmonised data sharing outside the EU. These include, for instance, the data localisation requirement within the EU for secure data storage and processing for secondary use, unless adequacy decisions determine that the non-EU country offers an adequate level of data protection (European Commission 2022b).

In addition, the EHDS could facilitate international comparisons on data-related aspects of health and healthcare systems. The harmonised approach could allow for meaningful benchmarking and insights into health outcomes, system efficiencies and capacities, and various other parameters relating to the digitalisation of health systems and population health. This comparative analysis can contribute to a better understanding of trends and gaps, enabling countries to assess, plan, and enhance their capacities for the primary and secondary use of health data as well as overall health system planning based on shared insights and good practices. For instance, TEHDAS developed a benchmarking toolkit for comparing national situations in health data management against the EHDS Regulation proposal along the axes of data collection and quality, data infrastructure and governance, available human, technical, and financial resources, as well as training and capacity-building needs (TEHDAS 2023c).

Overall, analysing the concrete impacts of the EHDS appears challenging as the EHDS does not yet exist and work on the final text is ongoing. However, years of efforts to establish and roll out cross-border exchanges of Patient Summary, ePrescription, and eDispensation (MyHealth@EU) have impacted national implementation of digital health tools, standards, concepts, and applications. Some European countries have made significant advancements in national developments and are gradually adopting new legislative frameworks to enhance the primary and secondary use of electronic health data, possibly accelerated by the EHDS Regulation proposal. For instance, Spain has recently amended existing and adopted new legal acts to guarantee cross-border healthcare services (Patient Summary, ePrescription/eDispensation) and to specify responsibilities of the Ministry of Health and Regional Healthcare Services in providing these services (European Commission 2024c). In 2024, the Netherlands will introduce new legislation on the conditions under which hospitals can join the Dutch NCPeH (European Commission 2024c).

Further EU-wide initiatives aiming to enable and advance the secondary use of health data, the establishment of large research infrastructures and initiatives in bio-medical and clinical research have been beneficial in boosting cross-border collaboration in research and innovation activities (HealthData@EU). Moving forward, the development of country roadmaps with incremental steps, timelines, dedicated resources, and funding instruments, as well as detailed technical implementation guidelines for hospitals and other key stakeholders, will support the EHDS implementation. This will also entail an analysis of responsibilities and potential related burdens for hospitals as data holders to identify and allocate the necessary resources.

Acknowledgements We thank our experts at empirica for their insights and contributions to the development of this article, particularly Carola Schulz, Veli Stroetmann, and Nick Guldemond.

References

Council of the European Union (2024) Proposal for a Regulation on the European Health Data Space – analysis of the final compromise text with a view to agreement. https://www.consilium.europa.eu/media/70909/st07553-en24.pdf. Accessed 27 May 2024

eHDSI Monitoring Framework (2024a) KPI-1.1 - Number of Countries with Operational NCPeH. https://experience.arcgis.com/experience/77f459be23e545b48f46a79cfaf19423/page/1_1/. Accessed 27 May 2024

eHDSI Monitoring Framework (2024b) KPI-1.9.2 - Hospitals operational with MyHealth@EU services. https://experience.arcgis.com/experience/77f459be23e545b48f46a79cfaf19423/page/1_9_2/. Accessed 27 May 2024

eHealth Network (2023) Guideline on the electronic exchange of health data under Cross-Border Directive 2011/24/EU: Patient Summary. https://health.ec.europa.eu/system/files/2023-10/ehn_guidelines_patientsummary_en.pdf. Accessed 26 April 2024

EIT Health (2024) Implementing the European Health Data Space across Europe. https://eithealth.eu/wp-content/uploads/2024/04/EIT_Health_ThinkTank_Implementing_the_EHDS_across_Europe_23.04.24.pdf. Accessed 28 June 2024

European Commission (2019) Annex to the Commission recommendation on a European electronic health record exchange format

European Commission (2022a) Developing a data quality and utility label for the European Health Data Space. https://cordis.europa.eu/programme/id/HORIZON_HORIZON-HLTH-2023-TOOL-05-09. Accessed 24 April 2024

European Commission (2022b) Proposal for a Regulation on the European Health Data Space COM (2022). Provisional agreement as of 18 March 2024. https://www.consilium.europa.eu/media/70909/st07553-en24.pdf. Accessed 01 July 2024

European Commission (2024a) A European strategy for data. https://digital-strategy.ec.europa.eu/en/policies/strategy-data. Accessed 20 June 2024

European Commission (2024b) MyHealth@EU monitoring framework (KPIs): KPI-1.1 – number of countries with operational NCPeH. https://experience.arcgis.com/experience/77f459be23e545b48f46a79cfaf19423/page/1_1/. Accessed 23 July 2024

European Commission (2024c) MyHealth@EU monitoring framework (KPIs): KPI-7.1 – Laws and regulations affected and amended to enable cross-border services interoperability. https://experience.arcgis.com/experience/77f459be23e545b48f46a79cfaf19423/page/1_1/. Accessed 23 July 2024

European Digital Rights (2023) Joint public letter to EU lawmakers on patients' rights in the European Health Data Space. https://edri.org/wp-content/uploads/2023/04/Joint-public-letter-on-consent-in-EHDS-2.pdf. Accessed 28 June 2024

European Health Data Space (2022) Articles of the European Health Data Space (EHDS), Article 55, Dataset description. https://www.european-health-data-space.com/European_Health_Data_Space_Article_55_(Proposal_3.5.2022).html. Accessed 28 June 2024

Kassenärztliche Bundesvereinigung (2022) Europäischer Gesundheitsdatenraum. Stellungnahme der KBV zum Verordnungsentwurf der EU-Kommission über den Europäischen Raum für Gesundheitsdaten COM(2022)197 vom 3. Mai 2022. https://www.kbv.de/media/sp/2022-12-16_Stellungnahme_EHDS_KBV_final.pdf. Accessed 23 July 2024

Langefeldt O, European Commission Directorate-General for Health and Food Safety (2023) European Health Data Space: harnessing the power of health data for patients, research and innovation. AEPD-ENISA 2023 Conference. https://www.enisa.europa.eu/events/aepd-enisa-2023-conference/owe-langfeldt.pdf. Accessed 23 July 2024

Marelli L, Stevens M, Sharon T et al (2023) The European health data space: too big to succeed? Health Policy 135:104861. https://doi.org/10.1016/j.healthpol.2023.104861

Ministry of Health, Wellbeing and Sports (2022) Financiële impactanalyse European Health Data Space [Financial impact analysis European Health Data Space]. Available via https://open.overheid.nl/documenten/ronl-188e974c295399237f91d9fa6053ed8b4850a371/pdf. Accessed 1 July 2024

Olsson J, De Smedt L, De Wispelaere F (2023) Data on cross-border patient healthcare following Directive 2011/24/EU – Reference year 2021. European Commission Directorate-General for Health and Food Safety. https://health.ec.europa.eu/document/download/8410fcfe-ec52-4cf7-9e3c-1535e55788e8_en?filename=crossborder_2021_patient-healthcare_data_en.pdf. Accessed 27 May 2024

Peseckyte G (2023) Delayed implementation of medical device regulations pose shortage risks. https://www.euractiv.com/section/health-consumers/news/delayed-implementation-of-medical-device-regulations-pose-shortage-risks/. Accessed 20 June 2024

Raab R, Küderle A, Zakreuskaya A et al (2023) Federated electronic health records for the European Health Data Space. Lancet Digit Health 5(11):e840–e847. https://doi.org/10.1016/S2589-7500(23)00156-5

Skovgaard LL, Wadmann S, Hoeyer K (2019) A review of attitudes towards the reuse of health data among people in the European Union: the primacy of purpose and the common good. Health Policy 123(6):564–571. https://doi.org/10.1016/j.healthpol.2019.03.012

TEHDAS (2023a) Qualitative study to assess citizens' perception of sharing health data for secondary use and recommendations on how to engage citizens in the EHDS. https://tehdas.eu/tehdas1/app/uploads/2023/03/tehdas-study-to-assess-citizens-perception-of-sharing-health-data-for-secondary-use.pdf. Accessed 23 July 2024

TEHDAS (2023b) Sustainability plan for secondary use of health data in the European Health Data Space. https://tehdas.eu/tehdas1/app/uploads/2023/09/tehdas-sustainability-plan-for-ehds.pdf. Accessed 1 July 2024

TEHDAS (2023c) Country factsheets – Mapping health data management systems through country visits: development, needs and expectations of the EHDS. Available via https://tehdas.eu/tehdas1/app/uploads/2023/04/tehdas-mapping-health-data-management-systems-through-country-visits.pdf. Accessed 28 June 2024

Zanaboni P, Kummervold PE, Sørensen T, Johansen MA (2020) Patient use and experience with online access to electronic health records in Norway: results from an online survey. J Med Internet Res 22(2):e16144. https://doi.org/10.2196/16144

Part III

Key Enablers for Future Digital Maturity Strategies

Digital Patient Engagement and Mature Hospitals

Lorren Pettit

Abstract

Patient participation, often termed "patient engagement," is recognized as crucial in achieving quality care and positive health outcomes. Patient engagement facilitates collaboration between healthcare professionals and patients, addressing individual needs and leading to effective and personalized healthcare interventions. Digital health solutions, such as access to personalized healthcare information, remote monitoring, and telemedicine, enhance patient engagement by promoting improved patient–provider communication and empowering individuals in managing their health. The rise of patient-centric care (PCC), where specific health needs and desired outcomes drive healthcare decisions, underscores the importance of digital health tools in facilitating patient empowerment. This shift necessitates evaluating healthcare systems based on their digital patient engagement offerings, as seen in various digital maturity assessments. Although traditionally overlooked, patient engagement is increasingly recognized as a key indicator of a digitally mature healthcare organization. This chapter also distinguishes "patient engagement" from "patient empowerment" and introduces the concept of the "digital front door," encompassing various digital tools used to enhance patient experience. Finally, it presents a digital patient empowerment framework, emphasizing the need for a cohesive strategy in deploying digital health tools to meet patient engagement goals.

Keywords

Patient engagement · Digital front doors · Patient-centric care (PCC) · Digital maturity assessments · Patient participation

L. Pettit (✉)
Independent consultant; formerly Digital Health Analytics at CHIME—College of Health Information Management Executives, Nolensville, TN, USA

A. Scheuer, J. Studzinski (eds.), *Digital Maturity in Hospitals*,
https://doi.org/10.1007/978-3-031-80704-6_12

Abbreviations

ADA	Americans with Disabilities Act
CTA	Calls-to-Action
PCC	Patient-Centric Care
PHR	Personal Health Record
EMR	Electronic Medical Record
EHR	Electronic Health Record
PGHD	Patient-Generated Health Data
TTS	Text-to-Speech
WCAG	Web Content Accessibility Guidelines

1 Foreword

Ask any care manager to identify what's arguably the most important predictor of patient outcomes and you're likely to get a simple answer: the patient.

Patient participation has long been understood to play a pivotal role in ensuring quality care and positive health outcomes. Frequently referred to as "patient engagement," efforts by providers to involve patients in their care create an environment where healthcare professionals are poised to more readily collaborate with their patients on care decisions, address individual patient needs, and ultimately contribute to more effective and personalized healthcare interventions.

Efforts to encourage patient engagement have been enabled through a wide array of digital health solutions. Technologies such as those providing patients access to personalized healthcare information, remote monitoring, telemedicine, etc. work to promote conditions for enhanced patient–provider communications thereby empowering individuals to assume a more active role in managing their health.

The importance of digital health as a patient engagement enabler has been elevated by the market's embracement of patient-centric care (PCC). In PCC, an individual's specific health needs and desired health outcomes drive health care decisions and quality measurements. Partnering with patients, providers pursuing PCC goals aim to treat their patients not only from a clinical perspective but also from an emotional, mental, spiritual, social, and financial perspective as well. To accomplish these ends, providers have increasingly turned to digital health solutions to help facilitate the empowerment of their patients.

Given the progressive nature of PCC in the delivery of care, it is essential to consider a provider organization's digital patient engagement offerings when evaluating the maturation of a healthcare system. Indeed, several digital maturity assessments like CHIME's (College of Health Information Management Executives) annual Digital Health Most Wired survey and the DigitalRadar Hospital assessment in Germany among others do include patient engagement issues into their criteria lists. Though patient engagement has not always been included in digital maturity assessments (Duncan et al. 2022), the PCC movement underscores the criticality of such considerations. Arguably, digitally mature healthcare organizations should be defined in part by the degree to which they offer and support digital patient engagement tools.

2 Patient Engagement vs. Patient Empowerment

The term "patient engagement" is widely used throughout the healthcare industry. Credited with contributing to improved outcomes and experiences of care, patient engagement has long been cited by healthcare insiders as a valued component of any care delivery system.

Yet, while many individuals may have a basic understanding of what patient engagement entails, the lack of a universally shared definition and the interchangeability of terms such as "patient participation" and "patient empowerment" (Hickmann et al. 2022; Longtin et al. 2010) introduce ambiguity into patient engagement conversations. As a result, it is essential to begin this chapter with a clear definition of what is meant by the term "patient engagement." Or more pointedly, what it does not entail.

Staudt (2007) has conceptualized patient engagement as a construct comprising two distinctive yet overlapping parts: attitudinal engagement and behavioral engagement. Informally referred to as patient "buy in" (Yatchmenoff 2005), **attitudinal engagement** refers to the emotional investment and commitment to treatment that patients express, as manifested in their attitudes, beliefs, and treatment expectations. **Behavioral engagement**, on the other hand, centers around behavioral indicators of the patient's involvement in their care, such as attending physician visits, arriving to appointments on-time, and adhering to treatment recommendations.

Distinguishing attitudinal engagement from behavioral engagement is a significant exercise for providers to undertake. By doing so, different barriers to care, and therefore solutions, can be uncovered and addressed. Barriers to attitudinal engagement reflect cognitive, psychological, and/or emotional factors internal to the patient such as the lack of motivation, unclear expectations, and poor provider experiences (Becker et al. 2015; King et al. 2014). Barriers to behavioral engagement involve social/environmental factors such as access to transportation, availability of childcare, or scheduling difficulties (Harvey and Gumport 2015; Kreyenbuhl et al. 2009; Yu et al. 2019).

While both constructs are important for providers to address, it is the position of this author that reducing behavioral engagement barriers (or stated more positively, empowering patients) presents a more favorable starting point. The objective, tangible nature of behavioral barriers arguably makes it easier for healthcare providers to isolate and identify potential "empowerment" solutions for patients, compared to the subjective nature of attitudinal engagement barriers.

Patient empowerment is a multi-dimensional process that helps individuals gain control over their lives and increases their capacity to act on issues that they themselves define as important. Digital patient empowerment is realized when patients have the digital health resources needed to achieve the control desired. By directing attention to factors such as clear communication, accessible health information and streamlined processes within their control, providers create an environment conducive to digital patient participation.

3 The Digital Front Door

Another term requiring attention at the outset of this chapter is "digital front door."

Like "patient engagement," "digital front door" is an ill-defined term with no universally agreed upon explanation. Generally used to refer to digital health technologies used by healthcare providers to manage and improve a patient's experience during their healthcare journey, the term has grown in popularity spurred on in large part by the expanding array of technologies falling under this broad heading (e.g., patient portals; telehealth; symptom checkers; etc.). Name aside, the digital front door goes beyond pre-service activities. It's a holistic concept involving a wide array of stakeholders within a provider organization (e.g., Marketing; IT; Clinicians; Patient Experience; Finance; etc.) that delivers an omnichannel experience at every touchpoint leveraging seamless, frictionless, and value-added technology allowing consumers to interact with healthcare providers in the way they prefer.

4 A Digital Patient Empowerment Framework

Pulling the concepts of "patient empowerment" and "digital front door" together highlights the need for providers to have a cohesive strategy in the deployment of digital patient empowerment tools. With the digital front door encompassing varied stakeholders at every touchpoint along the patient's journey, it's possible for providers to offer patients an array of disparate digital tools which address select parts of the journey while ignoring others. To avoid this problem, providers need a framework which not only categorizes the varied types of digital patient empowerment tools, but also serves as a roadmap to advance consumers and patients toward the organization's patient engagement goals.

In this chapter we offer a framework for a digital patient empowerment strategy. One that considers both the digital tool's function and form.

5 Form Follows Function

The concept of "form follows function," first articulated by American architect Louis Sullivan (1856–1924) in his essay, "The Tall Office Building Artistically Considered," argues that a building's design should be dictated by its intended use. Essentially, a structure's shape should emerge naturally from its functional requirements. This principle has since been embraced in many other industries as a foundational guideline for effective design.

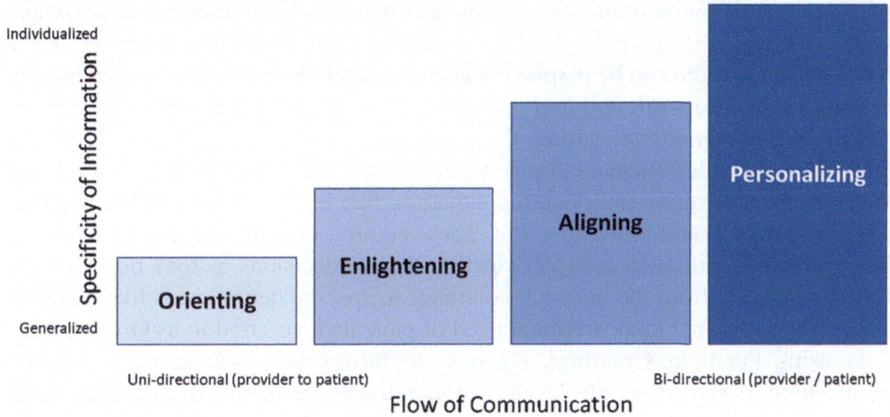

Fig. 1 Functional patient empowerment needs

For the digital health community, "form follows function" means the purpose of a digital health tool (its function) informs how the tool is to be designed (its form). As the primary function of a healthcare provider's digital health solution is to foster patient engagement, we need to start there: what are the varied empowerment needs of patients?

We propose a framework (Fig. 1) which categorizes patient empowerment needs into four progressive functional domains; Orienting; Enlightening; Aligning; Personalizing. Once these empowerment needs are understood, we can then consider the design (form) of digital tools.

6 Function

Understanding the function of a digital health solution is a critical starting point. It lays the groundwork for informed analysis, allowing individuals to navigate challenges, optimize performance, and make well-informed decisions. Without a solid understanding of function, efforts may be misguided, leading to inefficiencies or misinterpretations. Therefore, starting with a clear comprehension of function not only fosters a deeper understanding but also ensures a more strategic and purposeful approach in addressing various tasks or challenges.

6.1 Orienting

Healthcare professionals often make numerous assumptions about what patients want and how capable they are in navigating the healthcare system to access these needs. It is commonly suggested that patients desire access to their health data, wish to have a role in care decisions, or seek empowerment and engagement. While these beliefs may hold some truth, they assume a certain level of familiarity and competence in managing the complex healthcare environment. The reality, however, is that the healthcare system can be perplexing and overwhelming, not only for patients but even for those who work within it—let alone individuals facing health challenges.

The difficulties in navigating the healthcare system are substantial, posing challenges for both patients and providers. Patients often feel as if they are being placed in an unfamiliar and daunting situation, like hunters forced into a complex expedition they would prefer to avoid. They require as much guidance in accessing their medical information and participating in care decisions as they do in simply finding their way from the hospital's entrance to specific departments like imaging. Accordingly, the first empowerment need of patients is referred to as Orienting.

In using the term Orienting, we refer to information offered by a provider organization designed to introduce and familiarize its targeted audiences (e.g., consumers and patients) to foundational information about the provider organization.

Orienting arguably represents some of the most important and certainly most frequently accessed information during the initial part of a patient's health journey. Orienting is a uniquely unidirectional activity with almost all communiques flowing from the provider organization to the intended consumer/patient audiences.

The information individuals typically look for when familiarizing themselves with a topic, organization, or event often adheres to a basic framework called the "Five Ws." Originally linked to journalism, this method uses fundamental questions—*Who, What, Where, When, Why, and How*—to guide effective information gathering and problem-solving. Healthcare organizations can leverage the Five W formula as a guide in empowering a patient's orientation needs (Table 1):

The **Identification** section ensures that patients can trust the digital tools representing the provider organization by clearly associating the organization's legal entity with the content. This verification process involves confirming that all digital information has been reviewed and endorsed by the organization, establishing a foundation of trust for patients.

In the **Leadership** component, providing information about the primary leaders, such as their personal messages or professional biographies, goes beyond a mere listing of names. It humanizes the organization and encourages a sense of accessibility and transparency, thereby helping to reduce patient anxiety, especially in environments that might trigger a fear response.

For the **Organizational Story**, an increasing number of organizations are moving away from the dry, corporate style of storytelling and opting for a more engaging, narrative approach. This shift helps patients connect on a personal level, enhancing recall and emotional engagement. The story should highlight not only what the organization does, but also why it exists and how its values influence its operations.

Regarding **Performance**, while performance metrics can often be scattered across various sources, coalescing this data into a cohesive narrative is crucial for patient understanding. By doing so, the organization allows patients to make informed judgments on service quality and organizational efforts to maintain or elevate performance standards.

Table 1 The Five Ws of patient engagement (Petit 2020)

Category	Question addressed	Description
Access	**Where/ When**	Details when and where patients can access services. Includes contact points, hours of operation, and time frames for specific activities (e.g., visitation hours).
Identification	**Who**	Clarifies the entity behind the digital tool, ensuring patients recognize it as an authoritative source of information representing the healthcare organization. Establishing this link builds trust in the content provided.
Leadership	**Who**	Introduces key organizational leaders (e.g., CEO, Board Chair) through names, messages, or biographies, humanizing the organization and enhancing transparency and trust.
Organizational Story	**Why**	Communicates the organization's purpose, mission, and values. Emphasizes storytelling to make the message memorable and relatable, avoiding a sterile corporate tone.
Payment	**How**	Explains how patients can manage their financial responsibilities, including payment methods, insurance coverage, and available financial support services.
Performance	**How**	Provides information on how well the organization performs its services, including performance metrics, quality ratings, and improvement initiatives to aid patient decision-making.
Real-Time Informing	**When**	Offers real-time updates on service availability, such as Emergency Department wait times, to set accurate expectations for patients.
Services/ Providers	**What**	Outlines the range of clinical services offered and profiles the clinicians affiliated with the organization, ensuring clarity on the organization's primary offerings and expertise.

The **Real-Time Informing** component is essential to set realistic expectations for patients considering the use of a service at a specific moment. Providing real-time updates such as current emergency room wait times not only improves transparency but also helps patients make decisions that align with their immediate needs.

Finally, in the **Services/Providers** category, it is important to detail not just the clinical services offered but also provide relevant information about the credentials and affiliations of the clinicians. This enables patients to understand the expertise available and make more informed choices regarding their care providers.

6.2 Enlightening

Once a patient's interest is piqued, there are numerous ways they might seek to engage with the provider organization as a resource. One of the most effective paths to deepen patient engagement involves addressing their personal health concerns. Patients often turn to reputable health institutions—be it a physician or a healthcare organization's website—seeking reassurance that others share their experiences and that solutions are available to address their concerns. In essence, these patients are

Table 2 Categories of enlightening health content (Pettit 2020)

Category	Description
Care Environment	The **Care Environment** refers to the physical and social characteristics of a healthcare setting that can influence the way services are accessed or delivered. Understanding the care environment is crucial for effective patient engagement, as it helps contextualize the processes involved in care delivery. Addressing potential inefficiencies, such as redundant intake forms, can improve patient compliance and experience.
Clinical	The **Clinical** category covers educational content on medical topics, which is central to a provider organization's role. Patients often seek to deepen their knowledge about clinical matters, particularly from institutions that provide treatment for their conditions. This information is essential for patients who wish to understand their health concerns and explore appropriate care options.
Experiential	**Experiential** content focuses on helping patients navigate the emotional and psychological impact of healthcare interventions. By offering resources such as patient experience videos, advocacy services, or peer testimonials, provider organizations can alleviate fear and uncertainty. This category aims to prepare patients emotionally for their care journey, addressing their self-preservation concerns and making them feel supported and informed.

looking to place their individual health experiences into a broader context by gaining a more comprehensive understanding of the topic at hand. To meet this need for validation, healthcare organizations should offer digital tools designed to educate and encourage these information-seeking individuals.

The **Enlightening** domain, the second category in the patient empowerment framework, focuses on assisting patients and consumers by linking their personal health concerns with established knowledge and resources. This approach aims to foster a deeper understanding of specific health topics, helping patients make sense of their individual health needs in relation to trusted external sources. While the communication in this domain primarily flows from the organization to the patient, certain aspects involve dynamic engagement that may require active participation from the patient (Pettit 2020).

Educating patients plays a crucial role in the Enlightening domain, as it is a key step toward enhancing patient engagement. It is vital because it empowers patients to make more informed decisions about their care. One of the core objectives of patient-centered care is to treat patients as active partners in the care team, and this partnership is only effective if patient preferences are acknowledged. Informed, shared decision-making is heavily dependent on patient education. When individuals are well-informed about their health conditions and potential treatment options, they are better positioned to make choices that align with their preferences and desired outcomes.

The type of information an individual seeks in educating themselves on a health topic of course varies by the specific issue. That said the type of digital Enlightening health content provider organizations offers patients arguably partitions into three categories (Table 2):

6.3 Aligning

The components of the digital empowerment framework discussed up to this point have functioned to attract, inform, and retain patients. With most of these digital empowerment resources easily delivered via the healthcare organization's website or even a patient portal, this information is highly static in nature. Patients largely assume a passive role being the primary receptors of information the provider organization shares. Unless the website purposes to be nothing more than an "electronic brochure" for the organization, websites should offer patients the opportunity to take some type of action to support the provider organization's efforts.

The ability of a website to influence audience behavior is most noticeable when it supports product purchases. The success of converting site visitors into customers can be evaluated through metrics like "click-through rates" (the proportion of users who click on a specific link compared to the total number of viewers of a page, email, or advertisement) and the number of completed transactions. While these metrics work well for retail-oriented websites, their applicability to healthcare provider websites is less straightforward. This is because healthcare services present unique marketing challenges that complicate the use of traditional online measures for tracking patient engagement.

The primary issue stems from the nature of what is being marketed. Unlike retail, where tangible goods are the focus, healthcare organizations primarily offer services. Consequently, the website's content should emphasize these services. However, marketing a service differs significantly from marketing a product. There are four key distinctions between promoting healthcare services and marketing consumer goods that provider organizations need to consider in their digital strategies (Table 3):

The second major challenge is the inherent irrationality of the healthcare marketplace. Unlike conventional markets, healthcare operates under unique dynamics that complicate typical marketing strategies. Several characteristics make it difficult to

Table 3 Considerations in promotion health services (Pettit 2020)

Characteristic	Description
Intangibility	Services cannot be seen, touched, or physically experienced before being purchased, making it difficult for patients or consumers to assess the value of the service in advance.
Perishability	Services are produced and consumed simultaneously, meaning they cannot be stored or inventoried for future use. Once a service is provided (e.g., a patient consultation), it exists only in that moment and cannot be reused.
Inseparability	The delivery and consumption of services happen at the same time. In healthcare, this means that the service experience is created only when a provider (e.g., a nurse or physician) performs a task (e.g., administering medication or conducting a procedure) directly in the presence of the patient.
Variability	Due to the human element involved in service delivery, the quality and outcome can vary widely depending on who provides the service, the time, and the specific circumstances. This variability can occur between different providers or even within the same organization.

Table 4 Primary "irrational" aspects of the healthcare market (Pettit 2020)

Irrational component	Description
Limited Consumer Knowledge	Most patients lack a clear understanding of the costs and specific details of the healthcare services they receive, making it challenging for them to make informed decisions.
Disconnected Supply and Demand	There is little correlation between the availability of healthcare providers and the pricing of services. Typically, service prices are determined by third parties (e.g., insurance companies or government bodies) rather than by direct market forces.
Non-Economic Consumption	Healthcare decisions are driven more by immediate health needs than by pricing considerations. Economic principles such as discounts or promotions have minimal impact on patient behavior when a person requires urgent care.
Emergency Service Obligations	In most countries, healthcare providers cannot refuse emergency care based on a patient's ability to pay. This mandate means that financial constraints do not prevent access to emergency services, unlike most other industries.
Need to Offer Unprofitable Services	Unlike businesses that can focus solely on profitable offerings, healthcare providers must deliver a broad range of services, even those that are not financially viable, in order to provide comprehensive care.
Historical Data Gaps	The healthcare industry has historically underinvested in consumer data collection and analysis, making it challenging for marketers to develop informed strategies. While data practices are improving, significant gaps in patient insight remain.
Scarcity of Substitutes	In healthcare, there are few alternatives for patients needing specialized or acute care. This scarcity limits patient options and makes it difficult to apply typical consumer choice models found in other markets.

apply standard market principles to healthcare, requiring digital media leaders to recognize and navigate these complexities (Table 4).

Provider organizations frequently find it challenging to establish effective metrics for measuring website success, largely due to the inherent complexities of the healthcare industry. Despite this, incorporating Calls-to-Action (CTAs) is critical for engaging patients digitally and aligning with the organization's objectives—like how other industries use CTA Click-Through Rates as a standard performance measure. CTAs serve as prompts that motivate visitors to take specific actions, such as "download now," "view more," or "add to cart." They play a pivotal role in converting passive users into active participants, whether by influencing purchasing decisions or encouraging desired health-related behaviors.

6.4 Aligning

Aligning is defined as digital content designed to activate and channel patients to pursue some type of behavior supportive of the provider organization's efforts.

Table 5 Types of patient alignment (Pettit 2020)

Resource Alignment	Intimacy
Financial Support Resource Alignment involves using financial resources, such as cash or other assets, to show support for the organization. Although the perceived value of money varies, it is a more standardized form of support because cash is replaceable.	**Staff Commitment** Intimacy involves personal expressions of support that go beyond monetary contributions, such as sharing experiences on social media or applying for a job at the organization. These actions signify a greater level of personal commitment.
Less Personal As a result, Resource Alignment is less personal compared to Intimacy. It lacks the uniqueness and emotional investment seen in more intimate expressions of support.	**Deeply Personal** Intimacy-based support is highly individualized and irreplaceable. Since it involves non-financial resources, it is not as susceptible to commoditization and holds a deeper personal value.

Under this rubric, the measure of success is for patients to "align" their behaviors with those expounded by the provider organization. As such, the flow of communication under the Aligning domain is necessarily bi-directional, with opportunities for patient-initiated communications more extensive than one would see under the Orienting or Enlightening domains.

The **Aligning** function helps identify specific market segments within the broader population. Although Aligning content might be displayed to a wide audience on a provider organization's website, its primary aim is to motivate defined groups to take specific actions. For instance, providing details about community Smoking Cessation programs targets a particular audience (smokers seeking to quit) who meet the criteria to respond to that call to action. Since the goal of Aligning is to prompt patients to act, this type of content is naturally more personalized compared to the information provided in Orienting or Enlightening stages.

To engage specific patient groups, organizations must leverage both rational and emotional triggers that resonate with the patient's concerns and values. By channeling this interest through digital platforms, Aligning becomes a crucial element in building the organization's digital culture. The use of electronic tools to encourage patients to align their behavior with the organization sets a standard for how patients are expected to engage with the organization digitally.

While success in the Aligning stage is measured by the actions patients take, not all actions hold the same value. Some require more substantial personal commitment than others. Aligning acknowledges these differences by categorizing patient engagement into two distinct types of alignment (Table 5):

6.5 Personalizing

The rise of the "Web Economy" has significantly transformed the way businesses operate. Companies like Amazon, Uber, and Netflix have built their success by leveraging technology and information to create smooth, customer-centric digital experiences. This shift has set new standards for what consumers expect across various industries, including healthcare, where digital tools now play a crucial role in service delivery.

The healthcare sector has embraced these changes, incorporating digital health tools such as personal health records (PHRs), mobile health applications (mHealth apps), and patient portals to facilitate patient participation in their own care. The rapid expansion of low-cost wearable devices that track patient-generated health data (PGHD)—such as physical activity, weight, blood pressure, heart rate, and even blood glucose—has further enabled this transformation. Healthcare organizations are increasingly utilizing these data sources to improve patient engagement by integrating personal health information into their systems, ensuring easy patient access, and offering personalized care plans. This approach empowers patients to achieve their health and wellness objectives and fosters a digital culture where being "digitally known" by the provider is seen as a positive aspect of care.

The notion of being "digitally known" is central to the **Personalizing** domain, the fourth component of the digital empowerment framework. Unlike the other domains (Orienting, Enlightening, and Aligning), which are primarily designed for public-facing websites, Personalizing deals directly with patient-specific data. These three earlier domains help establish the organization's digital capabilities and signal to potential patients the type of digital culture they can expect. Advanced offerings in these areas can attract technologically savvy consumers.

Personalizing, however, goes a step further by focusing on individuals who are already recognized patients. In this domain, all tools and content are tailored specifically for known patients, made available through secure digital environments such as patient portals. Because of the sensitive nature of this information, only individuals with an existing relationship with the organization can access these tools after registering. The goal of Personalizing is to enable patients to use their health information to take informed actions toward achieving their wellness goals, making this domain the most precise example of how digital resources can create a personalized healthcare experience (Pettit 2020).

Three main features define this concept:

1. **Patient-Specific Data:** Personalizing revolves around using data and information exclusive to each patient. It is only accessible to patients or consumers with a known relationship to the organization.
2. **Secure Digital Environment:** Given the confidential nature of the content, Personalizing tools must be provided within a secure environment (e.g., patient portals), ensuring data privacy and compliance.
3. **Action-Oriented Goals:** The aim is to empower patients to leverage their personal health data to take specific actions that support their individual health and wellness targets.

Of all the domains, Personalizing offers the broadest range of digital tools and content. To bring structure to this variety, the components within this domain are categorized into three key elements:

Empowering Access

The first component of the **Personalizing** framework is **Empowering Access**, which focuses on equipping patients with the necessary resources to take informed actions toward achieving their health goals. To move forward in their health journey, individuals need access to tools and information that guide them in the right direction. Personalizing accomplishes this through two key resources (Pettit 2020):

- **Information:** Providing patients with accurate and relevant information is essential for enabling them to make well-informed choices about their health. Various types of secure, patient-specific information can be shared by provider organizations to support this goal, ensuring that patients have a solid foundation of knowledge regarding their care.
- **Tools:** While information is the starting point, having the appropriate tools allows patients to translate that knowledge into action. This step bridges the gap between being informed and actively managing one's health. To facilitate this process, healthcare organizations can offer a range of digital tools, such as appointment reminders, health trackers, and educational resources. These tools can be applied at different stages of the patient's healthcare journey, both before and after clinical encounters, empowering patients to follow through on their health decisions effectively:
  - *Planning Tools*: The first grouping of tools is designed to help facilitate the patient's initiation of healthcare services. Tools supporting a patient's health planning demands such as self-scheduling appointments and appointment reminders fall into this category.
  - *Assessing/Tracking:* The second category of tools focuses on supporting patients in self-managing their health. Effective self-management includes handling the medical, emotional, and day-to-day aspects of their condition with confidence. To facilitate this, healthcare organizations can provide digital tools that help patients track and evaluate their health status regularly. These resources, such as self-management questionnaires, patient-reported experience measurement (PREM) surveys, and patient-reported outcome measurement (PROM) tools, empower patients to actively monitor their health and make informed adjustments to their care plans.
  - *Convenience Tools*: These tools support patients and their caregivers while care is being delivered and include such things as patient entertainment systems, meal selection and ordering, and feedback tools for improvement suggestions.
  - *Education Tools*: The fourth and final grouping of tools supports the informational needs patients have about their health, treatment, and recovery. Digital tools under this rubric include videos, treatment plans, etc.

Empowering Control

The second aspect of **Personalizing** focuses on giving patients control over their health data management. This component highlights an evolved digital culture where patients are empowered to manage the accuracy and use of their health-related information. To support this, healthcare organizations can provide patients with administrative access to their medical records and pan-organizational personal health records (PHRs), enhancing transparency and engagement.

A key distinction in this component lies between electronic medical/health records (EMRs/EHRs) and personal health records (PHRs). While EHRs and EMRs are often used interchangeably because they store clinical, prescription, and patient communication data, EHRs enable data sharing across multiple, unrelated practices, whereas EMRs are restricted to a single location. PHRs differ significantly, as they are managed by patients or their authorized representatives, rather than by healthcare providers. Though PHRs include medical information, they may not capture every detail from various providers. When described as "pan-organizational," the PHR encompasses all of a patient's health records, irrespective of which organization holds the data, creating a comprehensive view of the patient's health.

Medical Records

Healthcare organizations can enable patients to take greater responsibility for managing their medical records and offer control in the following ways (Table 6):

Table 6 Patients' responsibility in managing medical records (Pettit 2020)

Medical record management feature	Impact on patients	Additional information
Record Correction Requests	Allows patients to review their medical records and identify inaccuracies or incomplete entries.	Patients can then request changes to correct errors, ensuring data accuracy.
	Empowers patients by giving them a way to maintain the integrity of their medical information.	Healthcare providers have the discretion to accept or reject these amendments based on legal and clinical guidelines.
Set Privacy Controls	Enables patients to restrict what portions of their health data are accessible to non-care team members.	Patients can define who can view their information and control the extent of the data shared.
	Helps safeguard sensitive health information by allowing patients to set personalized privacy preferences.	This feature supports compliance with privacy regulations and ensures patient autonomy over their data.
Electronic Record Transmission	Allows patients to electronically share their medical records with recipients of their choice.	For example, patients can send their data to a new provider, family member, or research institution.
	Promotes data portability, giving patients control over the distribution of their health information.	This capability is particularly useful for speeding up the transfer of medical data between organizations or for data donation purposes.

Pan-Organizational Personal Health Record

Empowering patient data management control responsibilities can extend to the patient's cross-provider Personal Health Record (PHR), if the provider organization is involved in a pan-organizational personal health record effort. By granting PHR access permissions to pan-organizations, patients can designate who (beyond the various care teams) can view their personal health records. Like the administrative privilege associated with a singular EHR noted earlier, this component of personalizing involves the patient allowing access to their complete pan-organizational personal health record.

Optimizing Empowerment

The third component of **Personalizing** centers on **curation**. This represents the most sophisticated level of digital engagement, as it involves carefully selecting, organizing, and delivering digital health tools and resources tailored to meet each patient's unique needs and preferences. Personalization becomes especially powerful when patients receive customized and dynamically adjusted health targets based on their specific risk factors. Through curation, healthcare organizations empower patients by offering highly individualized guidance and support that aligns closely with their health goals and conditions. This approach enables organizations to support patients in the following ways (Pettit 2020):

e-Consultations

e-Consultation is a method of delivering healthcare services remotely through digital channels. It allows patients to connect with medical professionals and access specialized expertise that might not be available locally. By offering virtual consultations, patients can receive timely medical advice regardless of their geographical location or the time of day, providing flexibility and convenience that traditional in-person visits may lack.

Patient-Generated Health Data (PGHD) Services

PGHD refers to health-related information that is created, recorded, or collected directly by patients. It includes data such as health history, treatment history, biometric measurements, symptoms, and lifestyle choices. PGHD differs from traditional clinical data in two ways: First, it is primarily captured by patients rather than healthcare providers. Second, patients have control over how this data is shared or distributed. PGHD is typically collected through various devices, including mobile health (mHealth) applications, wearable technology, and remote patient monitoring systems.

6.6 Form

Having established the varied patient empowerment needs digital tools function to address, we can now turn our attention to the design considerations of a digital tool, referred to here as Form.

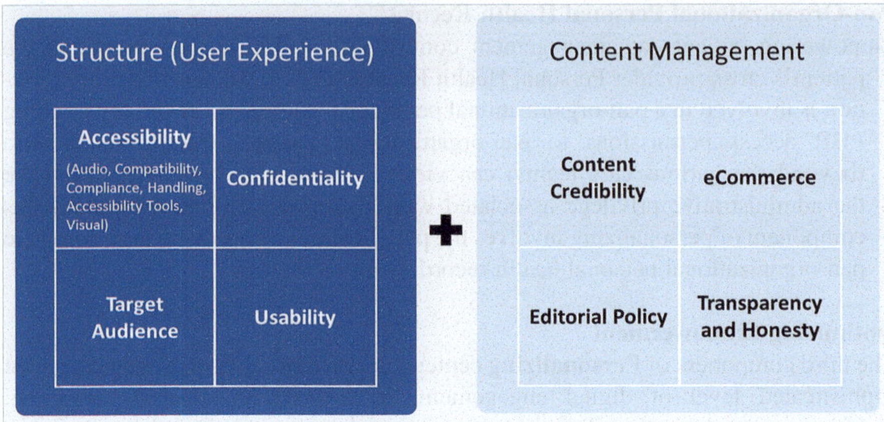

Fig. 2 Patient empowerment Form characteristics of digital health tools (Tool Design)

It should be noted upfront that in our digital empowerment framework, Form is not related to the creative production of websites and digital tools. While arguably a part of the "design" rubric, our framework cares not that a digital health tool uses the "Zig-Zag" or "Full Screen Photo" screen layout. This is not to diminish the value of layouts or the creative components of digital tool design (Fig. 2). On the contrary, a visually appealing digital tool layout is key to keeping patients activated.

Rather, Form in our framework focuses on the technical design factors providers should consider when offering digital empowerment tools. Here we limit our discussion to two main categories of issues:

1. Structural Issues Impacting Digital Tools
   The first set of issues providers should consider centers on user experience. The subjects under this heading are divided into four sub-categories (Pettit 2020):
   (a) Accessibility
   The Web is inherently structured to be inclusive for everyone, regardless of their hardware, software, language, location, or abilities. This principle is particularly crucial for web-based tools provided by healthcare organizations, as they often serve individuals with varying levels of physical abilities. **Web accessibility** ensures that digital tools are designed and developed to be usable by people with disabilities. The most effective way to achieve this is by adhering closely to the guidelines set by the **Web Accessibility Initiative (WAI),**[1] which provides standards to enhance usability for all individuals—including structural/semantic elements in HTML code, separating design features into Cascading Style Sheets (CSS) files and adding alternate text to all graphical elements.

[1] www.w3.org/wai

Accessibility concerns involving digital tools can be collapsed into six topic areas:

Audio

A common method of sharing content on websites or apps is through embedding short videos. These time-based media files often rely on audio to convey information to users. To ensure accessibility for individuals with hearing impairments, web or app designers should implement the following features:

Audio Controls: Users should have the ability to control audio playback independently of their device's settings. This includes options to pause, stop, or adjust the volume directly within the website or app interface.

Auditory Descriptions: For audio-only content, an alternative format should be provided, such as a text document that conveys the same message, including dialogue and descriptions of any relevant background sounds. This ensures that users with hearing impairments receive the full context of the content being shared.

Compatibility

Healthcare providers should make sure that their patient-facing content can be presented on as many browsers and end user devices as possible. Only if websites, apps, or other digital tools are compatible with most (if not all) operating systems, browsers, and other potentially limiting software or hardware related factors, there is a good chance that the targeted audience is able to access the organization's information properly. While there's no "magic bullet" that allows a website or app to be run or look great on all browsers or devices, developers should test their digital tools on a wide range of available products.

Compliance

Laws protecting the rights of individuals with disabilities have steadily expanded, aiming to eliminate barriers in all aspects of daily life. While physical accessibility is often the first concern that comes to mind, digital accessibility is equally crucial. Organizations with public-facing websites, apps, or digital tools are increasingly being scrutinized for compliance with accessibility standards. Digital content is considered accessible when proactive measures are taken to remove barriers, ensuring that individuals with disabilities can fully interact with websites, apps, or other digital platforms.

The specific laws governing digital accessibility differ by region. In the United States, the primary legislation is the Americans with Disabilities Act (ADA). Although the ADA does not explicitly mention digital platforms, US courts have interpreted Title III of the ADA to apply to websites. For a website to be ADA-compliant, it must ensure that people with disabilities can access and navigate the content seamlessly. The most widely recognized standard for achieving digital accessibility is the Web Content Accessibility Guidelines (WCAG) 2.1, which outline technical specifications for making digital content accessible across various devices, including desktops, laptops, tablets, and mobile devices. WCAG addresses the needs of individuals with diverse disabilities, such as auditory, cognitive, neurological, physical, speech, and visual impairments.

Handling

Many websites and apps leverage automated functionalities to enhance the user's experience. Yet some of these inclusions are not appropriate for all patients and should allow for manual control.

Animation Control: Animated interfaces can cause headaches, nausea, dizziness, and other health problems for many people, especially those with vestibular disorders, epilepsy, and migraine sensitivities. However, anyone can experience similar issues if subjected to excessive motion on the screen for a longer period. Providers should ensure their digital tools always provide more than one way for users to manage the tempo (e.g., stop; adjust) of animated web content.

Interactive Content: If the content of a digital tool involves user interaction, there should be no time restrictions on users or users should be allowed to postpone or suppress the time parameters, except when these imply an emergency. When an authenticated session expires, the user should be able to continue their activity without losing their data after being re-identified.

Keyboard Control: A foundational factor for a digital health tool to be considered disability compliant is that any interactions and information from a web browser that can be accessed with a mouse must also be accessible and controlled using a keyboard.

Keyboard Navigation: Digital health users should have the ability to navigate the entire tool using a keyboard, not just a mouse. Keyboard navigation operates differently, as actions are directed by the website itself rather than by manual cursor movements. When using a keyboard, the **TAB** key is used to move forward through elements, while **SHIFT + TAB** is used to move backward. These keys allow users to jump between focusable elements such as links, input fields, and buttons. As users progress through the website, they move between these elements in the order defined by the source code, which may not always align with the visual layout of the page.

Accessibility Tools

Website and mobile app accessibility software enables businesses to modify their digital platforms, making them accessible to individuals with disabilities or unique needs. These tools often work in tandem with physical assistive devices to deliver an inclusive digital experience for users. Accessibility solutions enhance traditional web and app development environments by providing developers with additional resources to design content that meets accessibility standards, ensuring that all users can effectively engage with the information provided.

Table 7 Meeting accessibility standards (Pettit 2020)

Feature	Description
Data Entry	Digital health users may need to input data into web or app-based forms. To support accurate data entry, the tool should offer clear instructions and guidelines. For forms involving legal or financial commitments, users should be able to review and edit their input before submission. Additionally, users should be informed about how to correct errors in their entries.
Data Entry Errors	When users make mistakes during data entry, the tool should alert them immediately, clearly identify the specific error, and provide instructions to help rectify it. This ensures that users can accurately complete forms without confusion or frustration.
Definition Tool	Healthcare often involves complex terminology that can be unfamiliar to patients. Websites or apps should provide a definition tool to clarify specific words, phrases, acronyms, or abbreviations that are industry-specific or otherwise challenging. This helps patients better understand the content presented.
Discoverable Language	Providing content in the user's native language is crucial for engagement, particularly in multicultural communities. Tools should support multiple languages and offer a seamless way to switch between them. The ideal approach is to host each language on separate URLs and interlink them for easy navigation, ensuring users find the appropriate version quickly.
Pronunciation Tool	Medical terminology often includes words derived from Greek or Latin that are difficult to pronounce. The tool should include a pronunciation guide to help users correctly articulate complex terms, making the content more approachable and reducing language barriers for non-experts.

Visual

Developers tend to design and assess the digital tools they create, using their own equipment's specifications. Using a fast hardware device and a state-of-the-art browser, they can turn their digital web-based efforts into a fabulous work of art. But not every person using a digital resource is viewing it with the latest device or browser. Moreover, many sites or apps neglect to address the challenges of visually impaired site visitors or app users. When a website or app is built without regard to proper web/app design, they become inaccessible to people with vision loss who use access technology (Tables 7 and 8).

(b) Confidentiality (Privacy and Data Protection)

Confidentiality pertains to the privacy and protection of information, governing who is authorized to view, share, and utilize it. Data with low confidentiality levels may be classified as "public" and poses minimal risk if accessed by unintended individuals. Conversely, information with high confidentiality is deemed sensitive and must be safeguarded to prevent identity theft, account breaches, system compromises, legal liabilities, reputational harm, and other serious outcomes.

(c) Target Audience and Usability

Table 8 Visual consideration in website development (Pettit 2020)

Feature	Description
Contrast Ratio Color	The contrast ratio measures the difference in luminance between the brightest (white) and darkest (black) colors displayed. To support individuals with visual impairments, the Web Content Accessibility Guidelines (WCAG) recommend a minimum contrast ratio of 4.5:1 between the text (or images of text) and the background.
Image Descriptions	Websites should provide textual descriptions (alt text) for images to ensure that the information is accessible, even if the image does not display properly or is difficult to perceive. Every non-text element, such as images and videos, should have a meaningful text equivalent that conveys the same purpose and context.
Photosensitivity	For individuals with Photosensitive Epilepsy (PSE), visual stimuli like flashing lights can trigger seizures. To prevent harm, websites should avoid using media that includes flashes more than three times per second or sudden changes in brightness, which can pose health risks to viewers with sensitivity to visual stimuli.
Subtitles	Audio-only content can exclude users who are hearing impaired. To make this content accessible, all audio should be accompanied by subtitles or closed captions, as well as audio descriptors for visual elements that contribute to the context. This ensures that all users receive the same information.
Text-to-Speech (TTS)	Text-to-Speech technology converts digital text into spoken words, providing an audio version of on-screen content. This feature is particularly useful for individuals who have difficulty decoding and understanding written text, as TTS can improve comprehension and remove barriers by allowing text to be read aloud.

Healthcare provider organizations serve diverse audiences with varying needs and abilities. It is crucial that representatives clearly communicate which messages are intended for each audience group to avoid confusion and ensure effective outreach. Given that digital health tools often cater to users with different physical and cognitive abilities, usability becomes a key factor in patient engagement. If aspects like visibility, readability, and navigation are not optimized, these tools can create barriers rather than support accessibility. To enhance usability, developers must carefully consider these elements to provide a seamless experience for all users.

2. Issues Guiding Content Management

The second set of Form-related issues providers need to consider centers on the management of the digital tool's content. The subjects under this heading are divided into four sub-categories:

(a) Content Credibility

The internet contains a significant amount of unreliable information. Patients searching for health-related content through digital tools are particularly concerned about the credibility of what they find, as it can directly impact their health decisions. To strengthen trust and confidence, healthcare organizations must clearly communicate the accuracy and reliability of the information provided on their digital platforms to assure patients that the content is trustworthy and evidence based.

(b) eCommerce

Ecommerce, also known as electronic commerce or internet commerce, refers to the buying and selling of goods or services using the internet, and the transfer of money and data to execute these transactions. As many healthcare organizations leverage their websites to facilitate these types of transactions, organizations should clearly articulate their policies related to the provision of electronic services on the website or other web-based tools they provide to patients.

(c) Editorial Policy

The editorial policy details the set of procedures used to select and publish web-based content. In doing so, digital tool owners provide patients with the information they need to make their own judgments about the veracity of the information included in the tool.

(d) Transparency and Honesty

Patients should be offered some basic assurances regarding the intentions of the healthcare organization offering the digital tool.

7 Conclusions and Perspectives

Patient engagement has long been understood to play a pivotal role in ensuring quality care and positive health outcomes. Enabled through a wide array of digital health solutions, it is the position of this author that digitally mature healthcare organizations should be defined in part by the degree to which they offer and support digital patient engagement tools. Yet, with the digital front door encompassing every touchpoint along the patient's journey, it's possible for providers to offer patients an array of disparate digital tools without fully addressing key parts of a patient's empowerment needs.

The digital patient empowerment framework presented in this chapter can be used by providers to guide the organization's selection of digital health tools to advance patient engagement goals. It is a framework that encourages providers to consider both the digital health tool's function and design.

References

Becker KD, Lee BR, Daleiden EL et al (2015) The common elements of engagement in children's mental health services: Which elements for which outcomes? Journal of Child and Adolescent Psychology 44(1):30–43. https://doi.org/10.1080/15374416.2013.814543

Duncan R, Eden R, Woods L et al (2022) Synthesizing dimensions of digital maturity in hospitals: systematic review. J Med Internet Res 24(3). https://doi.org/10.2196/32994

Harvey AG, Gumport NB (2015) Evidence-based psychological treatments for mental disorders: modifiable barriers to access and possible solutions. Behav Res Ther 68:1–12. https://doi.org/10.1016/j.brat.2015.02.004

Hickmann E, Richter P, Schlieter H (2022) All together now – patient engagement patient empowerment and associated terms in personal healthcare. BMC Health Serv Res 22:1116. https://doi.org/10.1186/s12913-022-08501-5

King G, Currie M, Petersen P (2014) Child and parent engagement in the mental health intervention process: a motivational framework. Child Adolesc Mental Health 19(1):2–8. https://doi.org/10.1111/camh.12015

Kreyenbuhl J, Nossel IR, Dixon LB (2009) Disengagement from mental health treatment among individuals with schizophrenia and strategies for facilitating connections to care: a review of the literature. Schizophr Bull 35(4):696–703. https://doi.org/10.1093/schbul/sbp046

Longtin Y, Sax H, Leape LL et al (2010) Patient participation: current knowledge and applicability to patient safety. Mayo Clin Proc 85(1):53–62. https://doi.org/10.4065/mcp.2009.0248

Pettit L (2020) Cultivating a Digital Culture for Effective Patient Engagement - A Strategic Framework and Toolkit for Health-Provider Websites. CRC Press

Staudt M (2007) Treatment engagement with caregivers of at-risk children: gaps in research and conceptualization. J Child Fam Stud 16(2):183–196. https://doi.org/10.1007/s10826-006-9077-2

Yatchmenoff DK (2005) Measuring client engagement from the client's perspective in nonvoluntary child protective services. Res Soc Work Pract 15(2):84–96. https://doi.org/10.1177/1049731504271605

Yu SH, Gumport NB, Mirzadegan IA et al (2019) Addressing the challenges of recruitment and retention in sleep and circadian clinical trials. Behav Sleep Med 18(1):23–34. https://doi.org/10.1080/15402002.2018.1518230

Digital Technologies for Climate Change Resilient Health Systems

Toni J. Laracuente, Monica L. Nakielski, Chethan Sarabu, and Philip Bradley

Abstract

This chapter explores the critical role of digital maturity, technology adoption and data analytics as hospitals and health systems seek to adapt to the physical and operational stressors imposed by the impact of climate change. As environmental conditions change and extreme weather events become more frequent and severe, healthcare services face significant disruptions that can exacerbate known and yet-to-be defined challenges in delivery systems, care of populations and public health services. Factors such as air and water pollution, food production and distribution, shifts in prevalence and geographies of acute, chronic and infectious diseases, social conflict, resource availability and population health burden place additional strain on healthcare infrastructure and demand innovative adaptation, mitigation and resilience strategies.

The integration of digital technologies in healthcare emerges as a critical tool in maintaining continuity, quality of care, and cost efficiency amid these challenges. In this chapter, we will review use of digital solutions, including

T. J. Laracuente (✉)
College of Health Information Management Executives, Ann Arbor, MI, USA
e-mail: toni.laracuente@outlook.com; Tlaracuente@chimecentral.org

M. L. Nakielski
Harmeda, Hartland, WI, USA
e-mail: mnakielski@harmeda.com

C. Sarabu
Director of Clinical Innovation, Cornell Tech, New York, NY, USA

Climate Health Innovation & Learning Lab, New York, NY, USA
e-mail: chethan@cornell.edu

P. Bradley
HIMSS, Chicago, IL, USA
e-mail: pbradley@himss.org

A. Scheuer, J. Studzinski (eds.), *Digital Maturity in Hospitals*,
https://doi.org/10.1007/978-3-031-80704-6_13

telehealth, clinical health information systems, and mobile health applications that facilitate real-time health data collection, dissemination and utilization at the patient and population level, and their role in health systems adaptation, strengthening and resilience. This chapter also highlights the role of artificial intelligence and predictive analytics in modeling climate impacts on health, enabling proactive public health interventions and resource planning.

Moreover, the authors address the implications of climate change for global mortality rates and economic costs, emphasizing the importance of resilient health systems in mitigating these impacts. By examining case studies and emerging practices, this paper provides insights into how health systems worldwide are adapting to the realities of climate change through technology and data utilization to improve system readiness and response in the new normal of global health challenges.

Keywords

Digital health technology and data analytics · Climate change risk, resilience and adaptation · Digital health maturity · Health system strengthening · Population health

Abbreviations

EHVI	Extreme Heat Vulnerability Index
LCA	Life Cycle Assessment
SCM	Supply Chain Management
SDoH	Social Determinants of Health
ERP	Enterprise Resource Planning
EHR	Electronic Health Record
NOAA	National Oceanic and Atmospheric Administration
CDC	Centers for Disease Control and Prevention

1 Introduction

Climate change has profound implications for every dimension of human health, from nutrition to public safety. Health systems worldwide are grappling with escalating demands for care within resource-limited settings. Strengthening, adapting, and enhancing the resilience of these systems are critical to addressing the multifaceted effects of climate change on human health and existing care delivery models.

Climate change introduces formidable challenges to healthcare systems globally. The increasing frequency and severity of extreme weather events disrupt healthcare services and exacerbate public health issues through elevated pollution levels, extreme heat, and other environmental risks. Furthermore, alterations in supply chain, food production, and environmental conditions induced by climate change may contribute to higher costs and increase potential for chronic diseases and greater morbidity and mortality, placing additional strain on healthcare infrastructure.

In response, integration of digital technologies into healthcare can be leveraged to bolster system resilience and adaptability. Such technologies can augment resources and enable meaningful outcomes in system security and stability, continuity of care, equitable access, and the sustained performance of health systems. They can also prepare health systems to cope with both current and future climate-related disruptions and adverse conditions. This article examines the real and potential impact of climate change on healthcare delivery systems and traditional care models. It reviews digital technologies that can be employed to fortify health system resilience against such impacts and discusses their role in preparing these systems to operate effectively in the 'new normal,' where populations globally are increasingly affected by climate-related health challenges.

For the purposes of this discussion, health system adaptation measures are specifically focused on the use of digital technologies and data analytics to design and build new, innovative care delivery models that position health systems for future challenges, strengthening their resilience to external forces and the disruptions they may cause to care delivery. Resilience in this context refers to a state of strengthened capability and capacity to provide comprehensive care for populations served, mitigating the impact of potential risks while maintaining continuity, quality, and equity of care delivery in a manner that achieves positive consumer and service provider experience in a cost efficient way.

2 Resilience in Essential Health Services: Protecting Vulnerable Lives

As climate-related health impacts intensify, the need for health system mitigation, adaptation, and resilience is escalating. Specific geographies face place-based risks, such as regions with high heat and air pollution experiencing more hurricanes, wildfires, or floods. These compounded risks increase health impacts on vulnerable communities. As place-based risk factors grow, people in these areas become more vulnerable, facing high health sensitivity to environmental conditions and increased physical and mental health risks from frequent disasters.

The National Academies of Sciences, Engineering, and Medicine (2024) report provides examples of the most vulnerable populations, including pregnant women, ethnic/racial minorities, females, children, elderly adults, Indigenous people, low-income communities, undocumented immigrants, unhoused populations, prisoners, people with disabilities, LGBTQ populations, unprotected occupational groups, medically high-risk patients, and the politically marginalized. Essentially, everyone at some stage in their lives falls into one or more of these cohorts. People in these at-risk groups may experience multiple intersecting health risk factors, further compounding disease burden, negative impact on quality of life, and increasing rates of disability.

Health systems must ensure equitable access to care for high-risk populations while managing inpatient beds, resources, and workforce demand. Climate-related illnesses and chronic diseases are becoming more prevalent, leading to surges in emergency department (ED) presentations during disaster events. For instance, ED visits rise during extreme weather and poor air-quality events, necessitating on-location emergency care. Data-informed, proactive bed management balances service, workforce and physical resource supply and demand during surges. This process requires diverse datasets and advanced analytics.

During predicted surge events, early and timely discharge of patients can free up beds for anticipated admissions. However, discharging high-risk individuals to homes without air conditioning, air filtration, or adequate access to food and water can lead to their return to the ED, necessitating readmission. This negatively impacts the individual's recovery and quality of life, strains hospital resources, blocks beds for higher-acuity patients, increases staff workload, and exacerbates health equity issues for vulnerable populations.

Pregnant and postpartum women, newborn babies, individuals with chronic diseases such as diabetes, asthma, kidney, lung, and heart disease, and the frail elderly are among the most vulnerable populations requiring special consideration as health systems develop adaptation and continuity of care strategies to mitigate climate change impact. Although the drivers of vulnerability differ among these patient cohorts, the climate factors impacting their health are similar. Extreme heat, high pollution days, food insecurity, wildfires and wildfire smoke, and weather disasters like hurricanes and floods pose significant health risks and increase demand for inpatient and emergency department services.

2.1 Impact of Climate Change on Healthcare Delivery Systems

Healthcare systems consist of complex, interdependent subsystems. Disruptions to any subsystem can have cascading effects on overall system performance and economic efficiency. According to the World Health Organization (WHO 2023), climate change is projected to cause an annual mortality rate of 250,000 additional deaths by 2030. This mortality is a stark indicator of how climate change multifariously impacts healthcare delivery, manifesting through various direct and indirect channels:

- **Extreme Weather Events:** The frequency of catastrophic weather events—hurricanes, floods, heatwaves—has surged from an average of two per year in the 1980s to over twenty recently. These events disrupt healthcare services, damage infrastructure, and disproportionately affect vulnerable populations, increasing the need for medical and mental health support. Health systems must develop reserve resources and alternative care models to maintain service delivery during such disruptions (WHO 2021).

For example, Hurricane Beryl in July 2024 knocked out power for millions in Houston, Texas, followed by a heatwave with temperatures over 108 °F. It was unsafe to discharge patients to homes without air conditioning, leading to access blocks, delays in care, and the expense of opening field hospitals (Taft et al. 2024). Additionally, over 90 people in western US states died from heat-related illnesses during the same heatwave. In Sri Lanka, since the early 1990s, there has been a sharp rise in acute kidney disease among fishermen and farmers aged 30 to 40, linked to extreme heat, dehydration, and water contamination (Mandavilli 2024). These men, often on subsistence incomes, cannot afford to stop working in dangerously hot conditions. Although healthcare is free, the impact on their income means they often seek treatment only in late-stage kidney failure, highlighting a significant health equity issue for this vulnerable population.

- **Pollution and Health:** In 2019, a young girl in the UK developed asthma due to high levels of smog and vehicle emissions near her urban home. Similarly, in 2021, a 70-year-old woman in British Columbia was hospitalized with asthma, dehydration, and heat stress, all exacerbated by climate change. Factors like fossil fuel emissions, prolonged pollen seasons, airborne mold, rising temperatures, wildfire smoke, and elevated ground-level ozone are increasing the prevalence and severity of respiratory ailments, particularly among young children.
  Globally, thousands of children under five die annually from lung diseases or face persistent health challenges. This escalating health burden strains healthcare systems, especially under value-based care models, degrading quality of life, and increasing societal burdens. Affected children suffer impacts on education and future employment, leading to reduced community vitality and tax revenue as Disability Adjusted Life Years (DALYs) rise and Quality Adjusted Life Years (QALYs) fall.

The World Health Organization estimates air pollution causes around seven million deaths annually, many linked to climate change. The World Economic Forum and Wyman (2024) predict six climate-related weather events by 2050 will result in 1589 million DALYs, 417 million years of life lost, 14.5 million deaths, and $12.5 trillion in economic losses. These events include wildfires, sea level rise, heat waves, tropical storms, droughts, and floods.

One of the most addressable sources of healthcare's contribution to climate change and air pollution is the use of volatile inhaled anesthetic gasses and nitrous oxide. Many health systems are eliminating the use of Desflurane, the most volatile anesthetic gas with the longest atmospheric life, 14 years, and Nitrous Oxide, which contributes to Ozone depletion and has an atmospheric life of 114 years (ASA Committee on Environmental Health 2024). Project SPRUCE, a consortium of 18 hospitals around the world, has a mission of reducing the GHG emissions and waste from anesthesia services by 50% within one year. Seattle Children's Hospital was the founding site of Project SPRUCE, initiated and led by Dr. Elizabeth Hansen. Seattle Children's Hospital has achieved resounding success, with an 87% reduction in GHG emissions and an annual cost saving of $180,000 on inhaled anesthetic gasses. Project SPRUCE demonstrates the critical importance of robust clinical data analytics in climate adaptation and the financial return on investment associated with investing in health system sustainability.

- Vector and Waterborne Diseases: Rising global temperatures and shifting precipitation patterns are altering the incidence and distribution of vector-borne diseases such as malaria, dengue fever, Zika virus, Lyme disease, and chikungunya. A 2% rise in global average temperatures could increase dengue fever occurrences by 37% (Romanello et al. 2023). Waterborne diseases like cholera, typhoid, giardiasis, and E. coli are also spreading to new regions, affecting populations with little or no immunity (van Daalen et al. 2024).
  The emergence of these diseases in previously unaffected areas strains health systems, requiring robust prevention, response, and treatment services. This includes educating healthcare workers and launching public health campaigns tailored to the new realities. The US Centers for Disease Control and Prevention (CDC) has developed the National Public Strategy to Prevent and Control Vector-Borne Diseases, outlining measures to mitigate these impacts (U.S. DHHS, CDC 2024).
- **Food Production and Nutrition:** Climate change significantly impacts agricultural and aquatic food sources, especially grain crops and fish. Environmental stressors like drought, floods, sea level rise, wildfires, freshwater shortages, salinization, and increased CO2 levels are reducing crop yields and nutritional content. Rising ocean temperatures and acidification are disrupting marine ecosystems, particularly in equatorial regions reliant on fish for protein.
  These changes escalate malnutrition and famine risks, forcing reliance on less nutritious and more expensive imported and processed foods. This situation increases the risk of chronic diseases like diabetes, heart disease, obesity, and mental health disorders. The Lancet Report on Climate Change projects that a 2% rise in global temperatures could result in an additional 525 million people experiencing food insecurity (Romanello et al. 2023).

The economic repercussions include reduced workforce productivity, increased community burdens, and added strain on healthcare systems, raising healthcare costs and demand for services. Climate change drives health inequity, disproportionately affecting vulnerable populations who contribute the least to the problem. Low-income countries and small island developing states are highly susceptible to severe effects, with death rates from extreme weather events fifteen times higher than in more resilient countries (WHO 2023).

Global Annual Mortality Rates and Economic Costs Due to Climate Change

Mortality and Morbidity Rates: Climate change is expected to cause approximately 250,000 additional deaths annually between 2030 and 2050 from malnutrition, malaria, diarrhea, and heat-related illnesses (WHO 2023). This figure excludes deaths from other climate-exacerbated causes like vector-borne diseases, starvation, mental health disorders, and heart and kidney diseases. The WEF and Wyman (2024) project that six climate-related disasters (floods, drought, heat waves, tropical storms, wildfires, and rising sea levels) will result in 14.5 million additional deaths and expose 500 million more people to diseases like malaria, dengue fever, and Zika virus.

Economic Costs: Currently, the global economic burden of climate change is about $143 billion per year (Bennett 2023), likely an underestimate. Projections suggest this could rise to $12.5 trillion annually, with an additional $1.1 trillion in direct health system costs by 2050 (WEF and Wyman 2024). Low-income countries face the highest economic risks despite contributing minimally to climate change.

Specific National Impacts:

- United States: Health-impacting events in 11 states in 2012 had associated health costs of $10 billion, mortality costs of $8.4 billion, and lost wages of $1.6 billion (Limaye et al. 2019). The Natural Resources Defense Council estimates direct health expenses related to climate change at $820 billion annually (De Alwis and Limaye 2021). Extreme weather events cost the U.S. $150 billion annually (National Oceanic and Atmospheric Administration [NOAA] 2023). Every 1 °F increase in global temperatures accelerates economic harm, with a 2 °F increase causing more than double the damage of a 1 °F increase.
- European Union: The economic burden of climate change over the past decade is estimated at 145 billion euros, with a forecast loss of around 11% of GDP by 2050 (Eurostat 2022). Heatwaves significantly contribute to both mortality and increased healthcare costs.
- Asia: Economic losses from a 2 °C increase in temperature are forecast at 14.9% of GDP, and 26.5% in a 3.2 °C worst-case scenario by 2050, with direct health expenses of $1.8 trillion and 588 million DALYs (Ellerbeck 2022).
- Africa and the Middle East: These regions bear the greatest share of the climate change burden, with forecasts of 1.27 million DALYs and $2.7 trillion in economic losses (WEF 2024).

2.2 Digital Health Technologies: The Building Blocks for Digital Transformation

Digital health technologies have emerged as critical tools in enhancing the resilience of health systems worldwide. These technologies provide innovative solutions to maintain healthcare delivery and health system performance, particularly in times of significant stress, such as during public health crises, pandemics, or environmental disasters. The World Health Organization's Health System Resilience Indicators Toolkit[1] provides a detailed, comprehensive reference for achieving a state of resilience to external forces.

[1] https://www.who.int/publications/i/item/9789240088986

This article focuses only on the technology and data analytics tools that contribute to achieving greater resilience, particularly to withstand the impact of climate change related stressors. It should be noted that there are several critical success factors to successful adoption of technology to achieve high value outcomes, including executive sponsorship, organizational change management, structured governance, and communication strategies.

Table 1 based on the author's experience, highlights some of these essential technologies, their applications in a resilient healthcare delivery model and, where available, real-world case studies demonstrating how they contribute to robust and adaptive healthcare systems. It is important to note that as technology rapidly advances and new use-cases emerge, the list of readily adopted technology and tools will continue to evolve. This list is a starting point, not considered exhaustive.

Case Studies and Examples

Telehealth: became the most rapidly adopted digital technology in healthcare history due to the COVID-19 pandemic in early 2020. Countries like the United States relaxed regulations and improved reimbursement rates, removing barriers to adoption. Telehealth allowed access to healthcare while isolating, and its use has persisted post-pandemic due to patient convenience and improved access to distant services. Many countries have enacted or are enacting permanent regulatory and funding changes, better preparing global health systems for future pandemics.

Since Russia's invasion of Ukraine in February 2022, telehealth has been crucial for maintaining healthcare capacity and continuity. Ukraine's health system[2] exemplifies national telehealth adoption. Telehealth was also vital during Hurricane Maria in Puerto Rico, providing critical consultations and support when healthcare facilities were inoperative, ensuring continuous care, reducing emergency room pressure, and reaching remote areas (Health Matters 2018).

Artificial Intelligence (AI): has shown significant promise in disease prediction and management. In Brazil, AI-driven models using combined EHR, socioeconomic and climate data are utilized to predict the spread of dengue fever. With Dengue fever primarily affecting impoverished, highly vulnerable populations, particularly young children, these predictive capabilities have enabled targeted interventions and the efficient deployment of preventative measures. As a result, timely vector control actions being implemented are leading to reduced transmission rates of the disease (Hemenway 2023).

Centralized Operational Command Center: Intermountain Health, based in Utah, USA, has implemented a data-informed, hybrid command center for managing patient air and ground transport, telehealth, and virtual care. Using predictive analytics for resource and capacity management, tele-ICU, and a federated care partner model, the health system works with community hospitals and rural services to provide care in place, reducing the need for patients to travel for access to

[2] https://www.lhssproject.org/our-work/ukraine

Table 1 Key digital health tools for health system resilience

Digital technology and tools	Application for use in care delivery
Telehealth, Telemedicine, and Virtual Care (Norris 2023)	Telehealth and virtual care platforms provide tools for remote consultation and treatment. Telehealth systems, integrated to the wider digital health ecosystem, provide a fast and convenient solution for specific patient care scenarios that don't require in-person access. During times of disruption to in-person care, such as in a disaster event, conflict situation, or when physical infrastructure is compromised, telehealth is a highly valuable tool for triage and treatment, enabling continuity of access. During the COVID-19 pandemic, telehealth became a lifeline for many patients, providing continuity of care despite lockdowns and severe hospital capacity issues (Children's Hospital of Philadelphia 2023).
Hospital in the Home	An extension of acute care inpatient services, enabling those needing a higher level of care than the usual 'home health' service, to receive care in their home with care supported and augmented by remote monitoring, patient communications, engagement and patient reported outcomes, and predictive analytics or AI to identify potential and actual changes in patient health status (American Hospital Association 2024).
Centralized Operational Capacity Command Center	A centralized command center for monitoring and management of multiple services and resources. Utilizes predictive analytics, AI and ML, and real-time monitoring with decision support for operational management. May include services such as air and ground transport, inpatient bed management, transplant and ICU services, surgical services and operating rooms, emergency department, diagnostic imaging, community, and sub-acute resource management. Used across regions and health networks to predict and monitor resource utilization and load-balance workload across facilities (Small et al. 2024).
Diagnostic imaging and laboratory systems	Portable diagnostic imaging and laboratory services for use in remote locations with 'offline mode' when cellular or Wi-Fi service is unavailable. Vendor neutral archive and universal viewer platforms.
Real-time and predictive capacity management for resource and workforce allocation and demand management.	Hospital capacity management systems with integrated RTLS provide predictive forecasting of demand for services, enabling managers to plan bed resources, staffing, and procurement such as food, linen, medical supplies, budgets, and services.

(continued)

Table 1 (continued)

Digital technology and tools	Application for use in care delivery
Health Information Systems (HIS)	Robust HIS facilitate the efficient management of patient demographic and scheduling data, resource allocation, and coordination of care.
EHR	Electronic Health Records (EHRs) enable healthcare providers to access patient histories and treatment plans quickly, improving response times in crisis situations. More than 95% of US health systems utilize an EHR, yet several barriers to integrated, comprehensive EHR adoption still persist. These barriers include procurement and ownership costs, system usability issues that result in low user adoption among providers and patients, and a lack of interoperability for data sharing. Digital maturity assessment tools such as the CHIME Digital Health Most Wired Survey (https://www.dhinsights.org/most-wired) and HIMSS EMRAM (https://www.himss.org/what-we-do-solutions/maturity-models-emram) are used to measure EHR adoption. Standardized criteria to measure adoption can assist in the prioritization of investments in digital technologies.
Social care referral management systems integrated with EHR.	Personalized wellness and social care platform for secure, closed-loop community referral management. Consumer engagement platforms and portals that can connect consumers in need with resources and services for social care, maternal and newborn care, food banks, suicide prevention and mental health resources, housing and transport, etc. For example: Unite Us (https://uniteus.com/)
Mobile Health (mHealth) Applications	Mobile apps can provide real-time, trusted health information, track disease outbreaks, and offer advice to populations affected by environmental health factors (Park et al. 2022). For example, apps like *mHero* connect healthcare workers with real-time data and communication tools during emergencies to improve response and coordination. Similarly, the *Google Maps Platform* offers consumers and healthcare providers personalized, real-time local data on health risk factors. User-friendly mobile apps, including patient portals and health metrics trackers, enhance consumer engagement in healthcare. mHealth apps help individuals monitor high-risk health days, such as those with high pollution and pollen levels that affect asthma and lung disease. Secure mHealth apps also promote

(continued)

Table 1 (continued)

Digital technology and tools	Application for use in care delivery
	health equity by providing access via smartphones and tablets, with AI-enabled listening and translation services to improve communication for diverse populations (Health Systems Strengthening Practice Spotlight 2022). Mobile app data tracking and digital social behavior change (SBC) analysis can also assist with health system strengthening. Refer to the below case study of SBC analysis for targeted public health prevention and intervention measures.
Robust data analytics for real-time reporting, decision support, and operational management	Robust and trustworthy data analytics services are an essential tool for building high-reliability, proactive healthcare delivery systems. Analytics may be: descriptive, predictive, prescriptive, or diagnostic (Cote 2021). User requirements and the intended purpose of the data analysis can guide which type(s) of analytics to use. For example, predictive analytics, a form of machine learning and a subcategory of artificial intelligence, tools can forecast and model the impacts of climate change on health, enabling proactive measures, education and engagement, and better resource planning (Parker 2021). Predictive models can forecast outbreaks of diseases like Zika virus and Malaria based on climate data, allowing for preemptive public health interventions.
Vaccine and disease registries, integrated with EHR and public health disease surveillance applications	Vaccine and disease registries, integrated with the EHR, provide valuable tools for preventive health, quality improvement, and population health services (HealthIT.gov 2013). Digital registries, integrated with or interfaced to EHRs, enable healthcare providers to identify and engage with at-risk patient cohorts, delivering care tailored to their needs. For instance, a registry of immunocompromised cancer patients helps providers to advise on vaccinations according to best practices and track vaccination status for follow-up actions. Similarly, asthma patients can receive warnings about high pollen and pollution days to avoid symptom exacerbation. Extracting data from these registries to identify and report on specific patient cohorts enhances care quality and population health outcomes.

(continued)

Table 1 (continued)

Digital technology and tools	Application for use in care delivery
	As climate change affects infectious disease patterns and spread, registries for conditions like diabetes, cancer, and asthma are crucial for tracking and managing care for disease cohorts. Knowing who is vaccinated or vulnerable to environmental conditions is vital for health systems implementing adaptation and resilience strategies. Structured, interoperable registries are essential for data sharing. In the USA, state-level Healthcare Information Exchanges (HIEs) and global initiatives like IHE International facilitate standardized healthcare data exchange. Maintaining these registries helps health systems mitigate climate change impacts on public health by enabling preemptive notifications to vulnerable patients.
Wearable Devices and IoT	Wearables monitor health metrics in real-time, providing valuable data for managing chronic conditions exacerbated by environmental factors. Example: Wearable devices that monitor air quality, heart rate, blood pressure, and respiratory function to support patients with asthma and other pulmonary diseases in managing their conditions more effectively and can transmit data to the EHR for healthcare provider alerts and monitoring.
Energy usage submetering, monitoring, and analytics	Energy submetering systems monitor and analyze energy usage to the department level, enabling energy efficiency strategies to be managed at the cost center level. When planning for health system resilience, organizations can gain cost savings on energy during periods of low utilization such as turning off lights and adjusting air conditioning temperatures when operating rooms are not in use. During surge periods or disaster situations, load balancing of energy usage enables service provision to be delivered efficiently, without expense overruns or unplanned stress on energy supply (UtiliVisor n.d.).
Environmental health risk surveillance, monitoring, and communications	Water quality, heat surveillance, infectious diseases, SDOH risk monitoring, air quality with pollen & pollution monitoring, location tracking for guidance on accessing fresh food/food choices.

(continued)

Table 1 (continued)

Digital technology and tools	Application for use in care delivery
Generative AI bots for in-home care and mental health support	Interactive AI applications and robotics for physical and mental health support. These systems can enable virtual monitoring for physical and mental status, provide patient engagement and communications between healthcare providers and patients, and be used to provide care support such as medication or blood glucose testing reminders.
Communications and RTLS systems	System-wide communications systems to alerts and updates during mass casualty incidents.
Carbon Accounting or GHG Accounting Systems	Carbon accounting and Greenhouse Gas (GHG) information systems track, report, and monitor Scope 1, Scope 2, and Scope 3 emissions in healthcare. Carbon accounting systems enhance data transparency to enable actions such as environmental impact assessments, regulatory compliance reporting, and performance and progress monitoring in GHG reduction projects. They can also quantify the return on investment in sustainability initiatives by applying monetary value to energy usage (Scope 2) or reduction, waste generation and management, and GHG emissions generated from care delivery (Scope 1), supply chain (Scope 3).
Unmanned aerial delivery systems (drones) for clinical supply logistics	Drone technology distribution systems offer a cost and energy efficient alternative to traditional transport methods. Particularly useful in rural, remote, disaster, pandemic and conflict affected areas, drone systems are deployed for rapid delivery of clinical supplies, pathology testing samples, blood bank products, medications and vaccines, enabling faster time-to-treatment, improved continuity and experience of care, and lower supply chain expense. Unmanned aerial delivery systems, when deployed at scale, offer substantial cost savings over road truck transport, and have a significantly lower carbon footprint. While the adoption of drone delivery systems is rapidly expanding in healthcare around the world, one of the earliest and most impactful use cases is the deployment of aerial delivery for blood supplies, vaccines and medications in rural Kenya (Maina 2024; She and Ouyang 2023; Raghunatha et al. 2023).

https://youtu.be/g3Vom9pRL6U?si=rLoKfupaHAS4L0ZP
https://www.gavi.org/vaccineswork/zipline-drones-wing-vaccines-malaria-prone-western-kenya
https://www.itskrs.its.dot.gov/2024-b01845
https://www.nature.com/articles/s41598-023-38922-z

specialized services such as oncology and high acuity treatment. Load balancing of critical care services between urban and community facilities has enabled more effective utilization of ICU beds at the main campus. This model not only provides an enhanced patient experience with care close to home, but also helps to improve capacity and health system readiness for mass casualty events, disruption to access due to weather disasters and other external factors that impact traditional care delivery models.

Mobile health (mHealth): *ZephAir* is an air-quality monitoring app, launched by the US Department of State in 2020 to send air-quality data and health advisories to US citizens in foreign countries, particularly in cities with significant pollution that presents a potential health risk. App users can personalize the advisory settings to receive alerts of air-quality changes that may put them at risk for exacerbation of lung diseases and breathing problems. By receiving timely alerts and health advice, users can take proactive steps to protect their health, such as staying indoors during periods of high pollution. This real-time information empowers individuals to make informed decisions about their exposure to pollutants, thereby improving their overall health management (U.S. Embassy Tbilisi 2020).

Digital Social and Behavior Change in Health System Strengthening (Health Systems Strengthening Practice Spotlight 2022)**:** RISE II Project, Social Monitoring and Listening. The Resilience in the Sahel Enhanced (RISE) II project focuses on behavior change to enhance service quality, build resilient health systems, and improve maternal, newborn, and child health and nutrition (MNCHN) outcomes among vulnerable populations. The project utilizes social monitoring and listening to identify priority needs and improve programming. By tracking social media engagement, implementers identified key beliefs, attitudes, and social norms prevalent in online conversations. This approach provided candid feedback that traditional surveys or focus groups might not capture, allowing RISE II project staff to enhance MNCHN service quality.

3 The Role of Data in Health System Resilience

This chapter examines essential data types for enhancing preparedness and response to climate-related health challenges, showcasing effective health policies that leverage climate data.

3.1 The Lifeblood of Resilience: Types of Data

Diverse and accurate data helps predict, understand, and mitigate the effects of environmental changes on public health. Categorizing data into environmental, epidemiological, and resource allocation allows health systems to tailor responses to climate impacts, enhancing effectiveness in routine and emergency scenarios. The following sections detail critical data types for bolstering health system resilience:

- **Environmental Data:**
  Temperature, air quality, and humidity serve as critical indicators for anticipating and responding to climate impacts. Monitoring temperature trends, for example, prepares for heat waves or cold spells that disproportionately affect vulnerable populations. Similarly, air-quality indices that measure pollutants like PM2.5 and ozone are vital for assessing respiratory health risks and implementing appropriate health advisories (Tran et al. 2023; Tiotiu et al. 2020; Fuller et al. 2022; EMJ 2024).
- **Epidemiological Data:**
  Epidemiological data encompass disease prevalence, social determinants of health (SDoH), and geographic variations. Analyzing this data allows health systems to understand where diseases are most prevalent and how social factors influence health outcomes. With this understanding, public health interventions can be tailored to meet the specific needs of different communities, enhancing the effectiveness of disease prevention and control measures (NETEC 2024; Mora et al. 2022; National Research Council et al. 2001).
- **Resource Allocation Data:**
  Data on resource allocation creates effective linkages between health systems, public health infrastructure, and disaster preparedness resources, ensuring a coordinated response during emergencies. It helps allocate medical supplies, deploy healthcare workers, and mobilize community resources.

3.2 The Pillars of Effective Data: Accuracy, Accessibility, and Timeliness

Effective data management in healthcare is underpinned by three qualities: accuracy, accessibility, and timeliness. Below is a detailed exploration of each pillar:

- **Accuracy:** Reliable data ensures that decisions are based on sound information, which is paramount for protecting lives and managing resources effectively. While minor inaccuracies in less critical data, such as temperature or air quality, may be permissible, high precision is needed for effective long-term healthcare planning and emergency response.
- **Accessibility:** Data must be universally accessible to all members of the healthcare ecosystem, from front-line clinicians and health administrators to patients themselves. Ensuring broad accessibility to health data empowers all stakeholders to participate in informed decision-making. It facilitates community engagement by providing public access to health data, while also safeguarding patient privacy through controlled access to more sensitive health information.
- **Timeliness:** Immediate access to real-time data is crucial for responding swiftly to ongoing health emergencies, while the availability of predictive models supports proactive planning for future healthcare needs. Additionally, historical data plays a vital role in assessing past health interventions and outcomes, allowing health systems to adjust and improve their strategies.

Together, these pillars support a robust health data infrastructure that enhances the resilience and responsiveness of health systems to environmental challenges and public health emergencies.

3.3 From Collection to Action: Harnessing Data for Climate Resilience

By systematically collecting and analyzing data, health systems can transform raw information into actionable insights, which in turn supports strategic planning and response to environmental challenges. The following outlines the process from data collection to analysis:

- **Data Collection:** Reliable data is sourced from both public and private entities. Public organizations such as the CDC, NOAA, and EPA provide extensive information on environmental conditions and health trends. Additionally, private companies like Weather Channel/IBM and BreezoMeter/Google contribute valuable data on weather patterns and air quality. These diverse sources ensure a comprehensive dataset that supports robust health system planning and response.
- **Data Analysis:** Utilizing advanced analytical techniques, health systems can convert collected data into meaningful insights. Predictive models and machine learning algorithms enable the anticipation of health trends and the development of targeted interventions. Technologies like digital twins create virtual replicas of health systems, allowing for the simulation of various scenarios to test responses and strategies. Personalized predictive models further enhance this process by tailoring recommendations to individual health profiles, thereby optimizing patient care and resource allocation.

3.4 Transforming Data into Resilience

Effective use of data ensures that health interventions are timely, targeted, and efficient. The following sections outline how different systems can leverage data to build resilience:

- **Community Participation**: Engaging communities in data collection ensures relevance and benefits. Robust governance and collaborative policies enhance resilience through broad participation.
- **Healthcare Systems**: Providers use environmental risk data for preventive care alerts, advising patients during poor air quality to reduce health complications and lower care costs.
- **Insurance Systems**: Insurers use risk data to recommend preventive measures, such as advising pregnant women to avoid pollution, reducing preterm births, and benefiting families.

- **Public Health Systems**: Data guides public health strategies and interventions, enhancing resilience by understanding environmental risks and health outcomes to protect populations.

4 Protecting Populations with Climate Data and Digital Technologies

2023 was the hottest year on record, and 2024 is on track to beat 2023. Average global temperatures from January to June, 2024, were 1.29 degrees Celsius warmer than the same period in any of the ten hottest years previously recorded.[3]

Heat Health Action Plans (HHAP): As heatwaves become more frequent, HHAPs have been adopted by many countries and regions (WHO n.d.-b). These plans use innovative technologies, integrating climate and health data to mitigate risks. HHAPs allow health systems to forecast heatwaves, identify vulnerable populations, allocate resources, and implement early warning systems. Typical measures include public advisories, cooling centers, and mobilizing healthcare workers to assist vulnerable populations during extreme heat.

Heat-related illnesses cause approximately 1600 deaths annually in the United States, making heat health action planning a strategic priority for the Department of Health and Human Services (HHS). The HHS has allocated substantial resources to combat extreme heat and has published numerous publicly available tools to support prevention strategies.[4]

The US Extreme Heat Vulnerability Index (EHVI) exemplifies the effective use of diverse data sources to support heat-related public health mitigation strategies. Figure 1 illustrates the EHVI integrated heat data from NOAA[5] with social vulnerability data from the CDC[6] to identify communities most at risk for extreme heat-related health impacts. This integration enables local government agencies, health systems, and community services to plan and implement proactive risk mitigation measures.

Some US health systems are adopting a holistic approach to managing patients with heat-associated health risks. These innovative, prevention-focused programs include prescribing air conditioning or assisting patients with energy expenses and developing policies to protect the health of outdoor workers (Tonn et al. 2021, 2014; Breysse et al. 2014). Such measures not only deliver substantial health benefits but also reduce emergency department utilization and readmission rates. Consequently,

[3] https://www.ncei.noaa.gov/access/monitoring/monthly-report/global/202406/supplemental/page-1

[4] A comprehensive reference list of tools can be found on the HHS website: www.hhs.gov/sites/default/files/hhs-resources-extreme-heat-health-2024.pdf

[5] https://www.noaa.gov/

[6] https://www.atsdr.cdc.gov/placeandhealth/svi/index.html

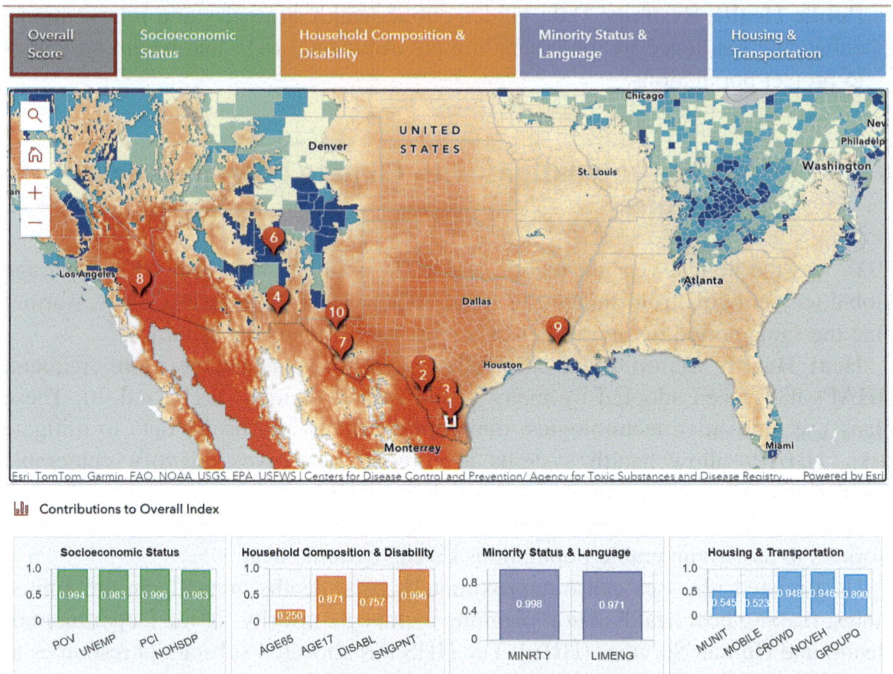

Fig. 1 US Heat Vulnerability Index (Adapted from Climate Mapping for Resilience & Adaptation 2018)

these strategies lower the overall cost of care by improving patient wellness, reducing the risk of heat-related illnesses, and minimizing unnecessary inpatient bed days.

4.1 Selected Regional Strategies to Address Climate Change Impact

As referenced above, 2023 and the January to June period of 2024 have been the hottest average global temperatures on record. The number of billion dollar disaster weather events has also steadily increased in recent years from 3 per year in the early 1980s to more than 20 per year since 2019 (Smith 2024). Healthcare delivery systems have experienced major disruption and increases in costs, while in population health, issues such as access to care services, health equity, morbidity and mortality rates, and the risk of infectious diseases have also compounded.

Various nations employ broader health strategies and tools to address the multifaceted impacts of climate change. The following sections outline the strategic approaches of the United States, the European Union, and Australia in integrating climate data into health policy planning and public health resilience initiatives.

United States: Health and Human Services Strategies and Tools

The impact of climate change on health is a significant health equity issue. The Office of Climate Change and Health Equity (OCCHE),[7] established in 2021 under the Biden Administration as part of the US Department of Health and Human Services (HHS), aims to protect Americans from the effects of climate change (Executive Order 14008, section 222(d)) and to improve health equity for populations disproportionately affected by climate change. OCCHE leads the development and implementation of climate-related strategies, actions, and planning to enhance health outcomes, reduce health risks, and strengthen public health resilience to climate impacts.

European Union: Climate Adaptation Strategy

The European Union's National Energy and Climate Plans (NECPs) for 2021–2030, introduced by Regulation (EU) 2018/1999, are a key part of the Clean Energy for All Europeans policy. These plans aim to meet the Paris Climate Agreement goals by reducing greenhouse gas emissions by 55% by 2030 and achieving climate neutrality by 2050. The five dimensions of the strategy are decarbonization, energy efficiency, energy security, the internal energy market, and research and innovation.

The process involves coordination across government departments at both national and EU levels, and extensive public consultation to facilitate investment and stakeholder engagement. Draft NECPs were submitted in December 2018, analyzed by the Commission, and recommendations published in June 2019. Every 2 years, countries report on their progress and update their NECPs, with the European Commission assessing the plans and monitoring EU progress. The latest updates were published in June 2023, with assessments and recommendations released in October and December 2023.

Australia: National Climate Resilience and Adaptation Strategy

Australia's national strategy emphasizes the integration of climate data into health policy planning to enhance resilience against climate related health risks across four key domains[8]: Natural Environments, Built Environments, Social and Community, and Economic. Although country and regional strategies may differ in focus and execution according to local needs and resources, the universal goal is to build person-centered healthcare ecosystems capable of withstanding the impacts of climate change. Achieving this requires close collaboration among diverse stakeholders, including patients, providers, and payers, new models of care, and the integration of advanced data capabilities and innovative technologies.

[7] https://www.hhs.gov/ash/ocche/about/index.html

[8] https://www.dcceew.gov.au/climate-change/policy/adaptation/strategy/domains

5 Energy Resources for Digital Health Technologies and Resilient Healthcare

Healthcare facilities consume substantial energy due to the need for maintaining sterile environments, operating medical devices, and providing continuous care. Greenhouse gas (GHG) emissions generated in the course of healthcare delivery are classified as Scope 1 (direct) emissions. Effective energy and resource management within healthcare delivery necessitate robust digital capabilities and data analytics. Energy, whether in the form of electricity, gas, heat, or steam, is categorized as Scope 2 (indirect) emissions[9] (Rade 2024).

Conducting energy usage audits and employing predictive analytics can identify areas where energy is wasted and forecast peak demand times for proactive management (U.S. Department of Energy 2011). Implementing energy-efficient lighting (e.g., LED), optimizing HVAC systems,[10] and utilizing energy-efficient medical equipment can lead to significant energy savings, thereby reducing both costs and carbon emissions (Klein and Hostetter 2023). By switching to renewable energy sources, for instance, installing solar panels on hospital rooftops or using geothermal energy for heating and cooling can drastically cut emissions and operational costs. While these solutions may require substantial initial investment, the cost–benefit demonstrates a positive return. Boston Medical Center, a 54 bed urban safety net hospital, began their infrastructure investment in renewable energy and resilience to weather disasters in 2012. The project was focused on their goals of reducing operational costs, reducing carbon footprint, and enabling the health system to continue to serve all patients, regardless of their ability to pay. With their investment supplemented by clean energy grants, issuing of green bonds and offsetting operational costs, BMC is now realizing more than $one million in annual energy savings, forecast to reach $150 million in expense reduction over 10 years. They have achieved net zero energy supply, greater energy security and financial resilience while patient volume has increased by almost 30% (Klein and Hostetter 2023).

As governments promote renewable energy policies, options for financing such as grants,[11] low-interest funding, tax rebates, and financial incentives to promote adoption are expanding.

Building Management Systems (BMS), submetering systems and energy plant management systems monitor and control energy usage in real-time. These systems can significantly enhance energy efficiency, reduce energy consumption, GHG emissions and reduce energy costs. Hackensack University Medical Center, a 775 bed academic hospital, is the largest healthcare provider in the US state of New Jersey. With implementation of submetering, energy monitoring and analytics

---

[9] https://www.sustain.life/blog/scope-emissions

[10] https://www.ashrae.org/professional-development/all-instructor-led-training/catalog-of-instructor-led-training/healthcare-facilities-best-practices-for-hvac-design-and-operation

[11] https://betterbuildingssolutioncenter.energy.gov/financing-navigator/explore

for load balancing, they achieved a break-even point on their investment within 1 year and are now realizing energy savings of $473,000 per year (utiliVisor Team 2014).

Renewable energy improving energy security for at-risk people, a case study:

- **Boston Medical Center Clean Power Prescription:** Energy insecurity has a serious potential health impact for those using therapies such as dialysis and ventilators at home and people who are susceptible to extreme heat. BMC has implemented a program of supporting their patients who are affected by energy insecurity with clean energy credits generated by their solar power array that was installed on the hospital roof utilizing Inflation Reduction Act (IRA) funding for climate sustainability investment.[12]

6 IT and Data Services & Systems in Healthcare

Robust IT and data services contribute significantly to the carbon footprint. Healthcare facilities should optimize data centers to be more energy efficient. This can be achieved in part through server virtualization, more efficient devices, recycling of retired technologies, more efficient data center cooling systems, and utilizing cloud services where possible that prioritize sustainability.

6.1 Health System Preparedness and Resilience for Disaster Events

Healthcare systems must be prepared for and resilient to disasters to maintain operational services, reduce impact on patient care during crises and for readiness to return to normal business as quickly as possible following the event. While it is standard procedure to be equipped with backup power sources such as emergency power generators, renewable energy storage systems such as solar plus storage, provide a cleaner alternative to diesel generators, reducing air pollution. In many older hospital facilities, emergency power generators, electrical circuitry and energy distribution systems are housed in the basement. There are several known instances of backup power systems being rendered inoperable due to basement flooding, therefore they are best positioned on building rooftops or other elevated locations, including all systems components and controls.

Of course, flooding is not the only threat requiring preparedness, mitigation and engineering investment. Healthcare facilities must also plan and mitigate for extreme events with potential to impact on-campus operations, workforce capacity and the

[12] https://www.bmc.org/news/boston-medical-center-health-system-announces-first-nation-program-clean-power-prescription

health of surrounding populations such as acute exacerbation of respiratory diseases from wild fire smoke and high air pollution days, air or water borne chemical or biological contamination, extreme heat, transportation disruption and mass casualty events.

Ethically, economically, and strategically, modern disaster preparedness and resilience planning should incorporate sustainability principles such as renewable, clean energy and life cycle assessment to help mitigate the environmental impact of emergency incident response. This includes training staff on energy-efficient practices during emergencies and using sustainable materials in emergency kits.

Disaster preparedness doesn't have to be a trade-off between risk mitigation and sustainable practices. By prioritizing sustainability principles alongside other critical organizational mandates such as care quality, financial efficiency and governance, we can lessen the impact of healthcare on climate change while we prepare for its impact on healthcare.

Technology Infrastructure for Climate Resilience

Building resilience to climate change impacts requires significant investment in modern technologies and data analytics. These investments enable health systems, providers, and governments to proactively plan and respond to evolving healthcare needs, ensuring high-quality, equitable care during daily operations and disasters like hurricanes, floods, and wildfires. Table 2 provides examples of technology investments that help to increase infrastructure resilience to withstand the potential impact of climate-related disaster events.

Given that all technology consumes energy, water, and natural resources, integrating energy efficiency and renewable energy strategies into the health system's climate adaptation plan is crucial. Effective technology procurement involves more than just buying equipment; it must include considerations of energy efficiency, life cycle assessment, total cost of ownership, disposal, and recycling.

Table 2 Infrastructure technology investment and adoption to build resilience in a climate-ready healthcare delivery system

Technology adoption checklist	Description
Broadband Infrastructure and Reliable Internet Connectivity	Investment in high-speed internet access for effective communications, mHealth connectivity, telehealth services, real-time data exchange, remote monitoring, and disease surveillance.
Cloud Computing	Utilizing cloud platforms enhances scalability, reliability, time-to-recovery and accessibility of health information systems, improves data security, and increases resilience. Cloud computing is up to four times more energy-efficient, signifcantly reduces carbon emissions (up to 99% with renewable energy powered data centers) and high performing Power Usage Effectiveness (PUE) scores (International Organization for Standardization 2016).

(continued)

Table 2 (continued)

Technology adoption checklist	Description
Interoperability Standards	Developing and adopting standards like HL7 FHIR ensures seamless communication and data exchange between health information systems, improving data quality, efficiency, and reliability.
Advanced Analytics and Business Intelligence	Analytics platforms offer data analysis, reporting, and predictive insights. AI and ML promise significant transformation, though they are energy-intensive and still in early adoption stages.
Cybersecurity, Data Privacy, and Security	Implementing robust cybersecurity measures, including encryption, secure access protocols, surveillance, and regular audits, protects sensitive health data from breaches and cyberattacks, ensuring system availability during emergencies.
Data Integration Platforms	Platforms integrating diverse datasets (public health, EHR, wearable tech, environmental and weather systems, climate data) provide comprehensive insights for decision-making and predictive analytics in patient care and operations.
Telehealth Infrastructure	Investing in telehealth platforms, hardware, and software supports virtual care, reduces disease spread and frees emergency resources. Post-pandemic, telehealth reduces physical infrastructure demands and aids chronic disease management and emergency care during disasters.

7 Use Cases and Digital Strategies to Bolster Resilience in Health Care

Advanced technologies and data analytics enhance health service accessibility, equity, and effectiveness through innovative care models. Drawing on experiences in Australian and US health systems, the authors present scenarios and digital strategies to improve care delivery resilience.

Acute Care Physical Resource Management: During surge events, proactive, data-informed bed and physical resource management constitutes sound operational practice in routine healthcare delivery.

Digital Strategy:

- Integrated bed management, staffing, and supply chain systems with predictive analytics support decision-making and include early warning and escalation protocols for routine and surge events.
- Region-wide capacity management systems centralize monitoring and resource prioritization during emergencies.

- Integrating virtual care, telehealth, and emergency services with facility-level systems and EHRs allows proactive management of emergency and acute care arrivals.
- EHR integration across care settings creates a digital health ecosystem, ensuring seamless patient data flow throughout the care journey.

Infectious diseases: In many countries, including the United States, one of the greatest set-backs in the COVID-19 early response strategy[13] was a lack of real time, accurate data on disease spread, infection rates, and other relevant clinical data. As COVID-19 was a novel virus, access to real-time clinical and demographic data as infection rates rose and disease patterns spread across geographies was critical to recognizing and responding to the crisis. In the USA, like the very fragmented healthcare system itself, public health data is still very disparate, rarely interoperable, siloed in various databases and in many cases, still being sent by fax when third parties need the information.

Digital strategy:

- A federal public health surveillance system utilizing advanced data analytics for data capture and analysis.
- Modern state and county public health information systems, registries, and technologies interoperable with the federal system (no more faxes).
- Robust data governance legislation and policies addressing data sovereignty, privacy, security, and access before public health events.

Renal Dialysis: For patients with kidney failure, uninterrupted access to renal dialysis is a matter of life and death. Climate-related disruptions can pose severe risks to these patients.

Digital Strategy:

- Mobile and home-based dialysis units that can be deployed to disaster affected locations with integrated access to the EHR.
- Remote clinical monitoring of dialysis patients with deteriorating patient status early warning and clinical decision support analytics.
- EHR integrated with emergency services and weather forecasting data, and advanced analytics to identify populations and individuals at risk for or impacted by environmental factors such as weather disasters, extreme heat, floods, pollution, wildfires, or wildfire smoke.

Maternal and Newborn Care: Protecting maternal and newborn health during climate events such as extreme heat, high pollution days, and weather disasters (hurricanes, floods, wildfire, etc.) reduces mortality and morbidity.

[13] https://www.nytimes.com/2022/09/20/us/politics/covid-data-outbreaks.html

Digital strategy: Implement mobile health units, telehealth for prenatal and postnatal care, and ensure access to emergency obstetric care. Identify at-risk mothers and newborns, implement early warning systems and evacuation planning, and prescribe air conditioning, and air and water filtration units.

Blood Bank and Pathology Services: Maintaining a stable supply of blood, blood products, and laboratory services during disasters.

Digital Strategy: Developing regional blood banking networks, using digital tracking systems for inventory and storage management, drone delivery and robust logistics management and establishing mobile blood donation centers can ensure supply continuity.

Pharmacy dispensing and distribution: Ensure supply management, distribution, and access to essential medications, controlled drugs, and IV fluid supplies.

Digital Strategy: Decentralized pharmacy distribution, drone deliveries, predictive analytics for stock management, on-site manufacturing, robotic stock management, closed-loop electronic prescribing, integrated EHR with pharmacy management, and dispensing systems.

Chronic Disease Management: Continuity of care for chronic conditions, such as diabetes and hypertension, is vital to prevent complications and preventable ED presentations.

Digital Strategy: Implementing telehealth consultations, remote monitoring using wearable devices, and ensuring access to medications can support chronic disease management.

Mental Health Services: The psychological impact of climate events can be profound, necessitating robust mental health support. Climate Anxiety is highly prevalent among younger population age groups in all countries (Fig. 2), with more than 50% of people 16 to 25 years of age reporting significant distress, negative thoughts and emotions, and 45% reporting climate-related thoughts and anxiety impacting their daily functioning.

Digital Strategy:

- Deploy tele-mental health and virtual care screening services with access to treatment available in scheduled and on-demand options. Provide access options that meet the digital literacy and accessibility levels for the target population.
- Implement public education campaigns on social media platforms to raise awareness of climate anxiety and the mental health impacts of climate change, including in-person and virtual access options for treatment, available digital resources for services and education, and data privacy and security.
- Establish community-based crisis intervention teams with specific training on adolescent mental health and climate impact.
- Equip primary care providers and community health workers with specialist training in climate change communication, digital health screening, and support tools to integrate mental health and climate anxiety screening into care delivery models to support affected populations.
- Include SDOH data collection in the electronic health record and utilize clinical decision support and analytics to identify at-risk patients.

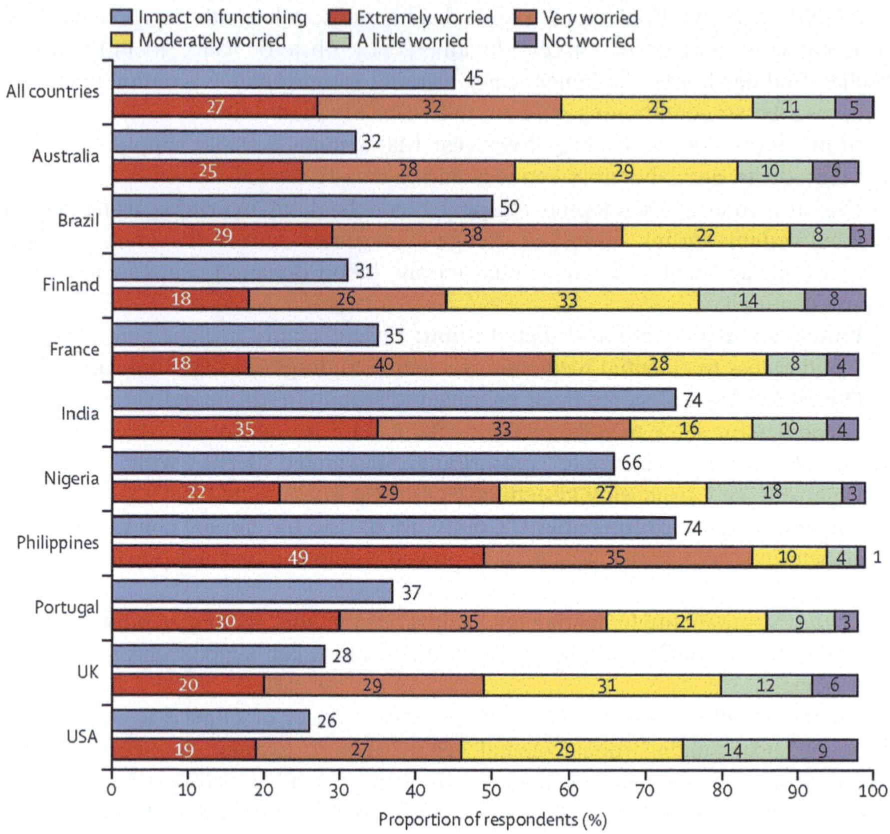

Fig. 2 Prevalence of climate anxiety among populations aged 16 to 25 (Hickman et al. 2021)

Emergency and Disaster Response: International initiatives such as the Global Digital Health Partnership[14] (GDHP) and the International Patient Summary[15] are examples of modern disaster event response strategies and capacity.

Digital Strategy:

- Use of predictive analytics capacity management and surveillance systems at the facility, region, and state level.
- Virtual and mobile care, tele-mental health, tele-ICU, virtual surgical services and consultations.
- Disaster response planning with surge capacity and risk mitigation, cross-training healthcare workers in emergency care, establishing field hospitals and using

[14] https://gdhp.health/

[15] https://international-patient-summary.net/

simulation training events are critical to emergency care delivery, health system preparedness, and disaster response.
- Data interoperability, record sharing, and systems integration.
- Data sharing policies and workforce capacity building for cross-border care in times of widespread disruption of care delivery such as major natural and man-made disasters, deprivation of service cybersecurity incidents, social conflict and war.

Adopting innovative care models, technologies, strategies, and governance can strengthen healthcare systems to deliver high-quality, safe, and equitable care as climate change impacts population health. Proactive planning and investment in resilience can safeguard the health and well-being of all populations, especially those most vulnerable to climate change.

Clinical Supply Chain: Extreme weather events, disease outbreaks, wildfires, and pollution can disrupt healthcare supply chains by damaging infrastructure, affecting logistics, delaying transportation, and reducing manufacturing capacity, leading to critical shortages. Robust data analytics and supply management systems are essential in such scenarios.

The COVID-19 pandemic exposed these vulnerabilities globally, highlighting the need for resilient supply chain technology. Despite efforts by governments and health systems, shortages of PPE, ventilators, and other essentials persisted, affecting workforce retention, patient outcomes, and budgets.

The clinical supply chain is complex, requiring multidimensional investment strategies to strengthen systems against disruptions in manufacturing, transportation and logistics. ERP resilience relies on surveillance, prevention, preparation, response, and recovery, necessitating investments in digital infrastructure, predictive analytics, and advanced ERP systems (Chawla n.d.).

8 Governance, Board Sponsorship and Funding

Effective organizational governance and board sponsorship are crucial for successful digital health implementation for health system strengthening (WHO n.d.-a). Beyond technology acquisition, digital transformation demands investment in stakeholder engagement, cultural change, budget allocation, risk management, workforce development, digital literacy, policy formulation, process design, staff education, benefits realization, and more. Governance frameworks and board sponsorship ensure oversight, authority, and mandates for sustained adoption.

Key Strategies for Effective Governance:
- Strategic frameworks with prioritized deliverables, justification for investment and anticipated outcomes.
- Organizational Change Management: Implement significant behavioral change through consulting, communicating, consensus-building, and training. Recruit Board members with climate change expertise.

- Benchmarking and Metrics: Compare climate policies against peers, regulations, and industry guidelines. Include climate action as a regular board agenda item with clear goals and metrics.
- Governance and Compensation: Establish a Board committee for climate issues. Link executive pay to achieving emissions reduction targets and sustainability metrics.

Gaining Board Sponsorship:
Develop the business case: To gain board sponsorship for resilience strategies, the business case must include real or projected financial data, risk assessments, and benefits. A cost–benefit analysis can demonstrate financial advantages, such as a $500,000 telehealth investment saving $one million annually by reducing ER visits and readmissions (NGPX 2024).

Engage stakeholders: Engage and align key stakeholders including clinical staff, patients, and community leaders, to educate the Board members on climate risks and business opportunities to gain consensus on the business case (Kaiser Permanente 2022).

Monitoring and evaluation: Metrics should measure the impact of resilience strategies and be provided in regular board reports (Mass General Brigham 2024). Frame climate adaptation as risk management to emphasize the board's responsibility for risk mitigation.

Transitioning to a Low-Carbon Economy: Developing green products or services and increasing efficiency presents business opportunities (GSK 2023). Enterprise resource management (ERM) is significantly impacted by climate change, offering opportunities to optimize value from constrained resources. Healthcare payers have an opportunity to impose pay-for-performance measures tied to climate sustainability, presenting revenue opportunities and financial risks.

Funding Climate Adaptation Initiatives:
Funding strategy options include:

- Payment Incentives: Programs like the US CMS TEAM[16] decarbonization incentive payment program and private insurer incentives.
- Internal Budgeting: Flexible capital planning policies and green revolving funds, such as UC's $200 million fund, Cleveland Clinic's $seven million fund, and Yale New Haven Health's $five million fund.
- Government Grants: Funding from agencies like the EPA, DOE, and HHS. For example, Hackensack Meridian Health leveraged $50 million from the Inflation Reduction Act for an energy efficiency project.

[16] https://www.cms.gov/team-decarbonization-and-resilience-initiative

- Philanthropic Funding: Support from foundations like Robert Wood Johnson Foundation, The Kresge Foundation, The Rockefeller Foundation, The Wellcome Trust, PATH, and Digital Square.[17]
- Green Bonds: Issuing green bonds and environmental impact bonds to raise capital for sustainable projects.
- Incentives and Rebates: Federal, state, local, and utility company incentives for energy efficiency and renewable energy installations.

Case study: Boston Medical Center, Greening the Health Care Safety Net.[18]

9 Conclusion: E Pluribus Unum, From Many, One

Digital technologies enhance the resilience and adaptability of healthcare systems to climate change. Utilizing telehealth, HIS, mHealth applications, advanced analytics, and standardized data exchange, healthcare providers can ensure continuity of care, better manage health risks, and improve population health outcomes amid climate-related disruptions. Maximizing the potential of digital health technologies requires investment in energy-efficient infrastructure, training, community engagement, governance, and policy. This holistic approach will lead to greater health equity, better population health, and a resilient healthcare system ready to face future climate challenges.

References

American Hospital Association (2024) Hospital-at-Home. https://www.aha.org/hospitalathome. Accessed 5 July 2024

ASA Committee on Environmental Health (2024) The environmental impact of inhaled anesthetics. https://www.asahq.org/about-asa/governance-and-committees/asa-committees/environmental-sustainability/greening the operating-room/inhaled-anesthetics. Accessed 10 July 2024

Bennett P (2023) Climate change is costing the world $16 million per hour: study. https://www.weforum.org/agenda/2023/10/climate-loss-and-damage-cost-16-million-per-hour/. Accessed 5 July 2024

Breysse J, Dixon S, Gregory J et al (2014) Effect of weatherization combined with community health worker in-home education on asthma control. Am J Public Health 104(1):e57–e64. https://doi.org/10.2105/AJPH.2013.301402

Chawla A (n.d.) Supply chain resiliency at Stanford. Stanford Medicine. Internal document

Children's Hospital of Philadelphia (2023) A lifesaving call: Ellis' journey with single ventricle heart disease. https://www.chop.edu/stories/lifesaving-call-ellis-journey-single-ventricle-heart-disease. Accessed 5 July 2024

[17] https://globalgoodsguidebook.org/

[18] https://www.commonwealthfund.org/publications/case-study/2023/aug/greening-health-care-safety-net

Climate Mapping for Resilience & Adaptation (2018) Future heat events and social vulnerability 2018. https://resilience.climate.gov/apps/nationalclimate::future-heat-events-and-social-vulnerability/explore. Accessed 10 July 2024

Cote C (2021) 4 Types of data analytics to improve decision-making. https://online.hbs.edu/blog/post/types-of-data-analysis. Accessed 10 July 2024

De Alwis D, Limaye SV (2021) The costs of inaction: the economic burden of fossil fuels and climate change on health in the United States. NRDC. https://www.nrdc.org/sites/default/files/costs-inaction-burden-health-report.pdf. Accessed 10 July 2024

Ellerbeck S (2022) Climate change has cost the EU €145 billion in a decade. https://www.weforum.org/agenda/2022/12/climate-europe-gdp-emissions/. Accessed 10 July 2024

EMJ (2024) Climate change and air pollution: how healthcare providers can help mitigate the risks to respiratory. Health 9(2):31–41. https://doi.org/10.33590/emj/JUYJ2358

Eurostat (2022) Losses from climate change: €145 billion in a decade. https://ec.europa.eu/eurostat/web/products-eurostat-news/-/ddn-20221024-1. Accessed 5 July 2024

Fuller R, Landrigan PJ, Balakrishnan K (2022) Pollution and health: a progress update. The Lancet Planetary Health 6:6. https://doi.org/10.1016/S2542-5196(22)00090-0

GSK (2023) GSK announces major step towards sustainability ambitions with advancement of low carbon Ventolin programme to Phase III trials [Press release]. https://www.gsk.com/media/10711/gsk-announces-major-step-towards-sustainability-ambitions.pdf. Accessed 10 July 2024

Health Matters (2018) How New York-presbyterian used telemedicine to expand care in Puerto Rico. https://healthmatters.nyp.org/newyork-presbyterian-used-telemedicine-to-expand-care-in-puerto-rico/. Accessed 10 July 2024

Health Systems Strengthening Practice Spotlight (2022) Digital social and behavior change in HSS: strengthening health systems through social and behavior change interventions that use digital technologies. https://www.usaid.gov/global-health/healthsystems-innovation/health-systems/resources/practice-spotlight-series. Accessed 10 July 2024

HealthIT.gov (2013) Frequently Asked Questions: The Basics. https://www.healthit.gov/faq/what-diseaseimmunization-registry. Accessed 10 July 2024

Hemenway J (2023) Children's health using artificial intelligence to fight dengue. https://www.unicefusa.org/stories/using-artificial-intelligence-fight-dengue. Accessed 10 July 2024

Hickman C, Marks E, Pihkala P et al (2021) Climate anxiety in children and young people and their beliefs about government responses to climate change: a global survey. The Lancet Planetary Health 5(12):e863–e873. https://doi.org/10.1016/s2542-5196(21)00278-3

International Organization for Standardization (2016) ISO/IEC 30134–2:2016: Information technology – Data centres – Key performance indicators – Part 2: Power usage effectiveness (PUE). https://www.iso.org/obp/ui/#iso:std:iso-iec:30134:-2:ed-1:v1:en. Accessed 10 July 2024

Kaiser Permanente (2022) Building a better world 2022 sustainability & responsibility report. https://about.kaiserpermanente.org/content/dam/kp/mykp/documents/reports/kaiser-permanente-esg-report_2022_ADA.pdf. Accessed 10 July 2024

Klein S, Hostetter M (2023) Greening the health care safety net. https://www.commonwealthfund.org/publications/case-study/2023/aug/greening-health-care-safety-net. Accessed 10 July 2024

Limaye VS, Max W, Constible J et al (2019) Estimating the health-related costs of 10 climate-sensitive U.S. events during 2012. GeoHealth 3(9):245–265. https://doi.org/10.1029/2019GH000202

Maina J (2024) Zipline drones wing vaccines to malaria-prone western Kenya. https://www.gavi.org/vaccineswork/zipline-drones-wing-vaccines-malaria-prone-western-kenya. Accessed 10 September 2024

Mandavilli A (2024) The killer stalking Sri Lanka's men: climate change and contaminated water have combined to create an epidemic of kidney disease. https://www.nytimes.com/2024/07/08/health/sri-lanka-kidney-disease.html?unlocked_article_code=1.6U0.62ol._nY-Am4lhDP3&smid=url-share. Accessed 10 July 2024

Mass General Brigham (2024) Sustainability impact report. https://www.massgeneralbrigham.org/content/dam/mgb-global/en/about/newsroom/impact-reports/documents/mgb-sustainability-impact-report-2024.pdf. Accessed 10 July 2024

Mora C, McKenzie T, Gaw IM et al (2022) Over half of known human pathogenic diseases can be aggravated by climate change. Nat Clim Chang 12:869–875. https://doi.org/10.1038/s41558-022-01426-1

National Academies of Sciences, Engineering, and Medicine (2024) Compounding disasters in gulf coast communities 2020-2021: impacts, findings, and lessons learned. The National Academies Press, Washington, DC. https://doi.org/10.17226/27170

National Oceanic and Atmospheric Administration (2023) Climate change impacts are increasing for Americans. https://www.noaa.gov/news-release/climate-change-impacts-are-increasing-for-americans. Accessed 10 July 2024

National Research Council, Division on Earth and Life Studies, Board on Atmospheric Sciences and Climate et al (2001) Under the weather: climate, ecosystems, and infectious disease. https://nap.nationalacademies.org/catalog/10025/under-the-weather-climate-ecosystems-and-infectious-disease. Accessed 10 July 2024

NETEC (2024) Climate change and infectious diseases. https://netec.org/2024/03/25/climate-change-and-infectious-diseases/. Accessed 10 July 2024

NGPX (2024) How UPMC is using remote patient monitoring and telehealth. https://patientexperience.wbresearch.com/blog/upmc-remote-patient-monitoring-and-telehealth-strategy. Accessed 10 July 2024

Norris A (2023) Study: blue shield virtual care pilots reduced carbon emissions, water use. https://news.blueshieldca.com/2023/07/19/study-blue-shield-virtual-care-pilots-reduced-carbon-emissions-water-use. Accessed 10 July 2024

Park Y, Lee C, Jung JY (2022) Digital healthcare for airway diseases from personal environmental exposure. Yonsei Med J 63(Suppl):S1–S13. https://doi.org/10.3349/ymj.2022.63.S1

Parker J (2021) What's the difference between AI & predictive analytics?. https://www.techadv.com/blog/whats-difference-between-ai-predictive-analytics. Accessed 10 July 2024

Rade A (2024) Scope 1, 2, and 3 emissions explained. Sustain Life. https://www.sustain.life/blog/scope-emissions. Accessed 5 July 2024

Raghunatha A, Lindkvist E, Thollander P et al (2023) Critical assessment of emissions, costs, and time for last-mile goods delivery by drones versus trucks. Sci Rep 13:11814. https://doi.org/10.1038/s41598-023-38922-z

Romanello M, Napoli DC, Green C et al (2023) The 2023 report of the Lancet Countdown on health and climate change: the imperative for a health-centred response in a world facing irreversible harms. Lancet 402(10419):2346–2394. https://doi.org/10.1016/s0140-6736(23)01859-7

She R, Ouyang Y (2023) Analysis of drone-based last-mile delivery systems under aerial congestion: a continuum approximation approach (Report No. ICT-23-014). Illinois Center for Transportation. https://doi.org/10.36501/0197-9191/23-014

Small T, Baker L, Egizi AM et al (2024) Assessing the value of a command center model in large health care delivery systems. https://www2.deloitte.com/us/en/insights/industry/public-sector/imagining-virtual-command-center-for-federal-health-system.html. Accessed 10 July 2024

Smith AB (2024) 2023: a historic year of U.S. billion-dollar weather and climate disasters. https://www.climate.gov/news-features/blogs/beyond-data/2023-historic-year-us-billion-dollar-weather-and-climate-disasters. Accessed 10 July 2024

Taft I, Jones J, Blinder A (2024) Hospitals in Houston 'backed up' after hurricane, as millions in U.S. swelter. https://www.nytimes.com/2024/07/10/us/heat-power-outage-houston.html. Accessed 10 July 2024

The U.S. Department of Health and Human Services and the U.S. Centers for Disease Control and Prevention (2024) The National Public health strategy to prevent and control vector-borne diseases in people. https://www.cdc.gov/VBD. Accessed 5 July 2024

Tiotiu AI, Novakova P, Nedeva D et al (2020) Impact of air pollution on asthma outcomes. Int J Environ Res Public Health 17(17):6212. https://doi.org/10.3390/ijerph17176212

Tonn B, Rose E, Hawkins B et al (2014) Health and household-related benefits attributable to the weatherization assistance program. https://weatherization.ornl.gov/wp-content/uploads/pdf/WAPRetroEvalFinalReports/ORNL_TM-2014_345.pdf. Accessed 10 July 2024

Tonn B, Rose E, Hawkins B et al (2021) Health and financial benefits of weatherizing low-income homes in the southeastern United States. Build Environ 197. https://doi.org/10.1016/j.buildenv.2021.107847

Tran HM, Tsai F, Lee Y et al (2023) The impact of air pollution on respiratory diseases in an era of climate change: a review of the current evidence. Sci Total Environ 898:166340. https://doi.org/10.1016/j.scitotenv.2023.166340

U.S. Department of Energy (2011) Hospitals pulling the plug on energy wasting electric equipment and procedures. https://www1.eere.energy.gov/buildings/publications/pdfs/alliances/hea_plugloads.pdf. Accessed 10 July 2024

U.S. Embassy Tbilisi (2020) New U.S. air-quality app helps users avoid pollution. https://ge.usembassy.gov/new-u-s-air-quality-app-helps-users-avoid-pollution/. Accessed 10 July 2024

UtiliVisor (n.d.) Submetering for hospitals. https://www.utilivisor.com/submetering/hospitals-research.html. Accessed 10 July 2024

utiliVisor Team (2014) Hackensack University Medical Center saves $473K with energy monitoring. https://www.utilivisor.com/news/utilivisor-delivers-customized-energy-solution-to-new-jerseys-premier-hospital.html. Accessed 10 July 2024

van Daalen KR, Tonne C, Semenza JC et al (2024) The 2024 Europe report of the Lancet Countdown on health and climate change: unprecedented warming demands unprecedented action. Lancet Public Health 9(7):e495–e522. https://doi.org/10.1016/S2468-2667(24)00055-0

World Economic Forum, Wyman O (2024) Quantifying the Impact of Climate Change on Human Health. https://www.weforum.org/publications/quantifying-the-impact-of-climate-change-on-human-health/. Accessed 10 July 2024

World Health Organization (2021) Fast facts on climate change and health. https://cdn.who.int/media/docs/default-source/climate-change/fast-facts-on-climate-and-health.pdf?sfvrsn=157ecd81_5&download=true. Accessed 10 July 2024

World Health Organization (2023) Climate change. https://www.who.int/news-room/fact-sheets/detail/climate-change-and-health. Accessed 10 July 2024

World Health Organization (n.d.-a) Health systems governance. https://www.who.int/europe/health-topics/health-systems-governance. Accessed 10 July 2024

World Health Organization (n.d.-b) Planning heat–health action. https://www.who.int/europe/activities/planning-heat-health-action. Accessed 10 July 2024

Green IT to Reduce Healthcare's Climate Impact: The French Approach

Rudy Chouvel, Laurie Marrauld, Emeline Flinois, Marie-Claire Viez, Nathalie Baudiniere, and Brigitte Seroussi

Abstract

Climate change poses a significant threat to human health. While the healthcare sector addresses these challenges, it is also a major contributor of greenhouse gas emissions. Although digital health improves care efficiency, it also increases the sector's carbon footprint. To mitigate this impact, promoting sustainable practices on the path toward digital maturity is crucial. France is

R. Chouvel
Département de la Transition Ecologique en Santé, Fédération Hospitalière de France, Paris, France
e-mail: r.chouvel@fhf.fr

L. Marrauld
Institut du Management (IDM), Université de Rennes, EHESP, CNRS, INSERM, Rennes, France
e-mail: laurie.marrauld@ehesp.fr

E. Flinois
Agence Nationale de la Performance Sanitaire et Médico-Sociale (ANAP), Paris, France
e-mail: emeline.flinois@anap.fr

M.-C. Viez
Département de la Stratégie, Fédération de l'Hospitalisation Privée, Paris, France
e-mail: marieclaire.viez@fhp.fr

N. Baudiniere
Agence du Numérique en Santé, Paris, France
e-mail: nathalie.baudiniere@esante.gouv.fr

B. Seroussi (✉)
Digital Health Delegation, French Ministry of Health, Paris, France

Sorbonne Université, AP-HP, Tenon Hospital, Public Health Department, INSERM, Université Sorbonne Paris Nord, Limics, Paris, France
e-mail: brigitte.seroussi@aphp.fr

A. Scheuer, J. Studzinski (eds.), *Digital Maturity in Hospitals*,
https://doi.org/10.1007/978-3-031-80704-6_14

committed to reducing the environmental impact of digital health on a national scale. The country has implemented measurement tools and ecoscores to help software developers assess and reduce the ecological footprint of their health solutions, including those of health applications, hospital information systems, and teleconsultation services. Additionally, Green IT initiatives have been launched by public and private hospitals, including eco-responsible purchasing, reusing heat from data centers, utilizing green search engines, saving energy by shutting down unused digital equipment, adopting conservative email practices, and managing the second life and end of life of electronic materials. Political support has facilitated legal and financial measures to create a favorable environment for green innovation and the adoption of environmentally-friendly technologies. By coordinating actions at all levels, France aims to make digital health a model of sustainability while continuing to enhance the quality and accessibility of care for everyone.

Keywords

Digital health · Carbon footprint · Sustainability · Ecoscores · Green IT · Digital maturity

Abbreviations

COP	Conference of the Parties to the Convention
CSR	Corporate social responsibility
EHDS	European Health Data Space
EPR	Extended producer responsibility
GGHH	Global Green and Healthy Hospitals
GHG	Greenhouse gas
IPCC	Intergovernmental Panel on Climate Change
LCA	Lifecycle analysis
MTE	Ecological Transition Ministry
NHS	National Health Service
NIS	Network and Information Systems
PCR	Product Category Rules
PHR	Personal health record
SONS	Open and Non-Selective System
SPO	Second Party Opinion

1 Anthropocene and Global Warming

1.1 The Impact of Climate Change on Health

At COP26 in Glasgow, approximately 50 countries pledged their commitment to a resilient, low-carbon health system under the auspices of the United Nations. At COP28 in Dubai, old and new signatory countries, including France, have renewed this commitment.[1] This action marks an initial stride toward a unified and global approach to manage the vulnerabilities of healthcare systems and their impacts on the environment. Indeed, the sustainability of healthcare has emerged as an urgent imperative.

The World Health Organization (WHO) has underscored the importance of health resilience as a top priority in the forthcoming years.[2] Sustainability serves as a pathway to fortify this resilience, ensuring the continuity of care well into the future without imposing undue strain on finances, ecosystems, or natural resources (Schroeder et al. 2013; McGain and Naylor 2014). Furthermore, it addresses social imperatives by striving to reduce inequalities and enhance the quality of life, both of which are intricately linked with health and environmental concerns.

Scientists warn that climate change poses the most significant threat to human health in the twenty-first century (Costello et al. 2009). Projections by the Intergovernmental Panel on Climate Change (IPCC) suggest that without a reversal in current greenhouse gas emission trajectories, global temperatures could rise by as much as 4 °C on average by 2100,[3] which represents an existential peril to humanity. Even with concerted efforts, a 2 °C increase in average temperature is anticipated by 2050, driven by existing carbon emissions and the inertia of economic systems. Such a rise would precipitate global food insecurity and exacerbate air and water pollution, vector-borne diseases, and mental health challenges, as documented by reports from the Lancet Countdown in 2019 and 2020 (Watts et al. 2019; Watts et al. 2021). Presently, the most vulnerable populations endure the most of these health risks, yet the entire global populace already feels the repercussions of global warming.

The healthcare sector and its organizational frameworks are not immune to these climate and environmental transformations. Globally, healthcare activities contribute to 4.4% of greenhouse gas emissions, with estimates suggesting this figure could exceed 8% of national emissions in the United States (Health Care Without Harm Europe 2019). Given its singular position in the economy, the healthcare sector's vulnerability is a pertinent concern, both as a provider of care and as a complex entity reliant on diverse resources such as energy and materials.

[1] https://www.cop28.com/en/cop28-uae-declaration-on-climate-and-health

[2] https://intranet.euro.who.int/__data/assets/pdf_file/0005/351284/resilience-report-20171004-h1635.pdf

[3] https://www.ipcc.ch/sr15/

However, people pay scant attention to the real impact of health systems. To date, the carbon footprint of digital technology stands at approximately 3.5% of global emissions (Ferreboeuf et al. 2021). In healthcare, digital technologies already account for over 4% of the sector's carbon footprint (The Shift Project 2023).

In France, according to the last estimate of the Think Tank "The Shift Project ," healthcare activities are responsible for 8% of the national carbon footprint, mirroring trends observed in the United States. In France, although greenhouse gas reporting has been mandatory for most healthcare establishments since 2010, few have published reports on the dedicated national platform (ADEME)[4] (République Française and Bilans n.d.). Achieving ecological transition objectives and combating climate change necessitates transforming the modes of production, consumption, operations, and, consequently, the governance and management of healthcare institutions. Collectively, institutions wield significant influence in delineating and implementing short-, medium-, and long-term strategic objectives geared toward ecological transition.

1.2 The Impact of Digital Health on the Environment

Digital companies and users often heralded digitalization as a "dematerialized" solution, replacing physical processes and tools with digital alternatives. However, this perception overlooks the material and environmental implications inherent in digitalization. French sociologist Bruno Latour (2011) challenges the notion of dematerialization, contending that digital technologies, despite their apparent immateriality, rely on extensive material infrastructures, including data centers, servers, electronic devices, and networks. Latour's critique underscores the intricate entanglement of material and immaterial aspects in modern digital technologies, urging a more nuanced understanding of their environmental impacts.

Digitalization's material footprint manifests across various dimensions:

- **Material manufacturing**: The production of digital devices, from smartphones to desktop computers, and to servers, entails the extraction of physical materials such as metals, plastics, and rare earth elements, resulting in environmental impacts such as habitat destruction, pollution, and resource depletion (Fig. 1).
- **E-Waste**: Digital devices have finite lifespans, contributing to over 50 million tons of electronic waste annually (UN Environment Programme 2019). Improper disposal of e-waste poses significant environmental hazards due to toxic components, and much of it ends up in landfills or is incinerated, exacerbating pollution.
- **Energy consumption**: Digital technologies require energy for manufacturing, usage, and data storage in data centers. The energy consumption associated with

[4] ADEME-Agence de l'Environnement et de la Maîtrise de l'Energie (French Agency for Environment and Energy Management).

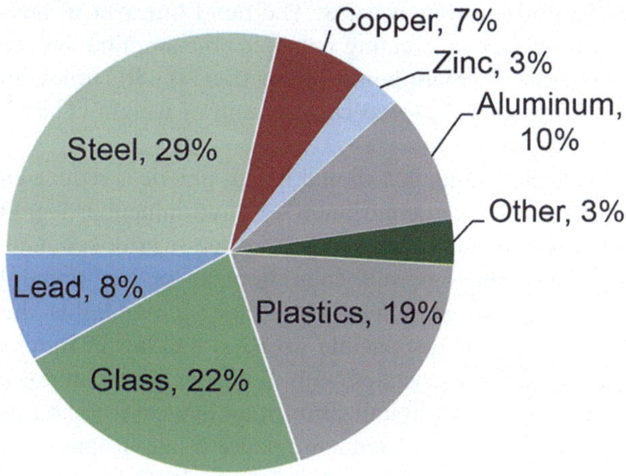

Fig. 1 Composition of a desktop computer (U.S. EPA 2016 cited in Center for Sustainable Systems 2023). Documentation for Greenhouse Gas Emissions and Energy Factors Used in the Waste Reduction Model)

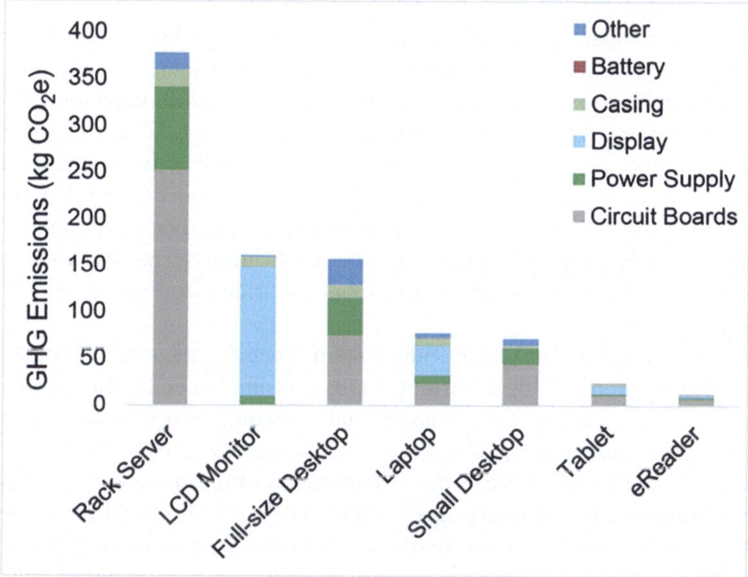

Fig. 2 Embodied greenhouse gas emissions: computing and electronics products (Teehan and Kandlikar 2013 cited in Center for Sustainable Systems 2023)

data centers and Internet infrastructures contributes to greenhouse gas emissions and strains energy resources. Unrestricted use of AI can greatly increase energy consumption.

- **Supply chain impacts**: The global supply chain for digital technologies involves complex processes, spanning multiple countries. Each step, from manufacturing components to assembling devices, involves transportation, energy consumption, and waste generation.

- **Resource intensiveness**: The rapid turnover of devices drives demand for new products, exacerbating resource consumption and waste generation (Fig. 2). For example, smartphone production is 80 times more energy-intensive than combustion-powered cars per unit of weight (The Shift Project 2019).

Despite claims that digital health practices reduce emissions, the entire lifecycle of digital devices contributes to environmental degradation, with concerns about e-waste and energy consumption often overlooked. Green IT provides a framework to reduce the environmental impacts of traditional IT by emphasizing energy efficiency and sufficiency of usage. It highlights the importance of awareness, regulation, and responsible decision-making in minimizing the negative environmental effects associated with digital health technologies.

Hence, while digitalization may decrease certain physical materials in specific areas of life, such as reducing paper usage, it fails to eradicate the overall environmental impact. It is imperative to consider the entire lifecycle of digital technologies, spanning from raw material extraction to disposal, to comprehend accurately their environmental footprint.

E-health, encompassing digital health technologies, yields numerous environmental repercussions despite its potential benefits in healthcare. The healthcare sector's contribution to greenhouse gas emissions due to digitalization is on the rise (Chevance et al. 2020). While some advocate that digital health practices like telemedicine reduce emissions, some researchers contend that the complete lifecycle of digital devices fosters environmental degradation, often disregarded by users. Concerns regarding e-waste and the substantial energy consumption of digital technologies are highlighted, projecting potential future scenarios wherein these technologies significantly contribute to global electricity consumption and greenhouse gas emissions.

Green IT, standing for Green Information Technology, and denoted as the practice of creating and using environmentally-sustainable computing resources, is recognized as a strategy for creating and utilizing eco-friendly information technologies. According to the Center for Sustainable Systems at the University of Michigan, Green IT can alleviate the environmental impacts associated with conventional information technologies (Corbett 2010). Traditionally, "examples of Green IT include energy-efficient hardware and data centers, server virtualization, and monitoring systems. Green IT focuses on mitigating the material and energy burdens associated with conventional IT while meeting our information and communication demands" (Center for Sustainable Systems and University of Michigan 2023). More recently, the concept of Green IT has embraced the notion of usage sufficiency. Studies indicate that solely emphasizing energy efficiency is insufficient to counterbalance the growing environmental impact of IT, prompting reflection on the genuine needs of organizations and the total cost of deploying these devices (Ferreboeuf et al. 2021; The Shift Project 2019).

To address the significant impact of digitalization on their carbon footprint, hospitals must establish comprehensive methodologies for measuring this impact and devise targeted decarbonization strategies. These strategies should focus on minimizing carbon emissions arising from both the manufacturing processes and the operational use of digital technologies.

In this chapter, we advocate for principles fostering a sustainable digital health ecosystem: moderation in digital consumption, enhancement of the lifecycle management of digital products, and adoption of interdisciplinary approaches to comprehend and mitigate the environmental impacts of digitalization in healthcare. This underscores the necessity for awareness, regulation, and responsible decision-making at individual, community, and regulatory levels to mitigate the adverse environmental effects of digital health technologies. This leads to considering digitization's carbon footprint in digital maturity concepts for hospitals with two aspects:

- ***Green4IT*** aims to implement eco conception and sobriety to reduce direct carbon emissions of energy-intensive digital solutions, such as hospital information systems and artificial intelligence.
- ***IT4Green*** aims to leverage the potential of digital solutions to reduce emissions across the entire health ecosystem, e.g., through telemedicine, remote patient monitoring, and virtual wards which can help hospitals cut down on transportation-related emissions or to use digital solutions to manage anesthetic gases more efficiently.

This chapter describes the implementation of Green4IT solutions for hospitals in France.

2 Hospital Information Systems

The computerization of hospitals offers tangible benefits, yet it also has some drawbacks that necessitate a thorough evaluation of the benefit/risk ratio.

2.1 Benefits of Hospital Information Systems

In the healthcare sector, understanding the environmental impact of a system holds value only when coupled with knowledge of the benefits it provides, following the benefit/risk principle. Digital technology enables the overcoming of spatial and temporal constraints, particularly crucial in the healthcare sector where time is of the essence and specialized skills are paramount for efficient care. The following list outlines several advantages of digital systems in healthcare facilities:

- **Enhanced care coordination** through the instantaneous and secure sharing of patient records, imaging results, and blood tests accelerates and improves healthcare professional coordination.

- **Implementation of telemedicine** widens access to healthcare, particularly in remote areas, thereby improving medical service availability and patient remote care.
- **Telemedicine or telemonitoring** can contribute to hospital stays reduction.
- **Algorithms and artificial intelligence** facilitate diagnostic decision support and early diagnosis.
- **In-depth analysis of healthcare data** with AI methods aids researchers and healthcare professionals in identifying trends, developing patient-specific treatment strategies, and enhancing chronic disease management.
- **Alert systems** for drug interactions increase patient safety and reduce medical errors by simplifying allergy management and ensuring medical information accuracy.
- **Healthcare professional education through** virtual simulations and e-learning enables continuous skills upgrading.

Overall, computerized systems enable more efficient (i) access to patient information, (ii) appointment management, (iii) secure prescriptions, and (iv) examination results availability, thereby improving care coordination and optimizing hospital resource utilization. Moreover, they play a pivotal role in reducing medical errors, notably by mitigating prescription risks and granting rapid access to patients' medical histories.

2.2 Risks of Hospital Information Systems

To realize these benefits, an accelerated deployment of digital technology is crucial. It's the expansion of digital technology and the interoperability of tools that allow accessibility of patient health information across all care settings and even in patients' homes, facilitating seamless information flow to enhance patient care. This alleviates constraints such as time for healthcare professionals to access patient-specific relevant information and location. However, hospital digitization also brings inherent risks.

- **Data security**: Computerized hospitals must implement robust security measures to protect patients' medical data from threats such as cyberattacks.
- **Training and adaptation**: Implementing computerized systems necessitates thorough training of personnel, and some healthcare professionals may find it challenging to adapt to new technologies.
- **Initial costs and maintenance**: The installation of computerized systems requires substantial investments in both hardware and software, with ongoing maintenance also representing a significant expense.
- **Loss of human connection**: Increasing automation can reduce human contact between healthcare professionals and between healthcare professionals and patients, potentially affecting the quality of care.

- **Environmental impact**: The production of electronic devices, servers, and other hardware components for health information systems has environmental implications, and digital health systems involve significant energy consumption, particularly in data centers.

In summary: The computerization of hospitals boosts efficiency and care quality but also presents challenges such as data security risks, personnel training needs, and environmental impacts. A well-planned implementation that focuses on security, training, and sustainability can help maximize the benefits of hospital computerization while minimizing its risks.

2.3 The French Approach to Managing the Benefit/Risk Ratio of Hospital Information Systems

France's healthcare sector has been undergoing an accelerated digital transformation, driven by "Ma Santé 2022"[5] program. In 2018, a report had drawn a gloomy picture of digital health, stating that the patient had been overlooked in the digital health transition, that healthcare professionals were facing a fragmented digital health landscape making usage complex in daily practice, and that the national digital health strategy was unclear to software developers and difficult to adhere to.

A French national digital health roadmap was elaborated by the Digital Health Delegation (Délégation au Numérique en Santé or DNS) of the French Ministry of Health with proposals to fix all these different issues, and address the expectations and concerns of all stakeholders of the ecosystem, e.g., patients, healthcare professionals, healthcare facilities federations, and software developers.

The implementation of the French national digital health roadmap was supported by an unprecedented investment of 2 billion euros, France's "Ségur Digital Health" program which served as a powerful accelerator for leveraging digital technology in healthcare. The aim was to facilitate the seamless and secure sharing of healthcare data between professionals and with patients to enhance prevention, care, and follow-up. This program represents a new approach to public policy within a State-platform framework, characterized as follows:

- The deployment of shared infrastructures on a national scale, particularly through "*Mon espace santé*"[6] (L'Assurance Maladie 2024). *Mon espace santé* is a digital health platform launched in France in 2022 to provide citizens with a secure, centralized, and user-friendly way to manage their health information. *Mon espace santé* is designed as a comprehensive personal health record (PHR) that consolidates various aspects of an individual's health information. The aim is to

[5] https://sante.gouv.fr/systeme-de-sante/masante2022/

[6] Mon espace santé- My Health Space (https://www.monespacesante.fr/).

improve the quality of care, enhance coordination among healthcare providers, and empower patients with better access to their health data;
- The continuous co-development of national policies together with patient associations, representatives of professionals and facilities in the health, social, and medico-social sectors, and digital health industry stakeholders;
- Unprecedented government involvement to support professionals' software updating and the development of associated digital uses through two funding programs:
  - The "Open and Non-Selective System" (SONS) program, which provides funding on a time-limited regulatory framework, for the update of Ségur Digital Health – referenced health software. Such software is capable of feeding *Mon espace santé* and enabling exchanges via secure messaging (MSSanté) between healthcare professionals and with patients, in compliance with stringent specifications. This funding is paid directly to healthcare software developers on behalf of healthcare institutions and professionals, with no remaining costs (excluding infrastructures).
  - The "Ségur Digital Health in Healthcare facilities" (SUN-ES)[7] program, which offers financial support to healthcare facilities that promote the use of *Mon espace santé* and secure health information exchange between professionals and citizens through messaging.

2.4 Carbon Footprint of Health Information Systems

Healthcare facilities host a wide array of digital equipment, ranging from conventional IT gear to specialized connected devices used for therapeutic or diagnostic purposes (such as imaging and radiology systems, network-connected cardiology equipment, automated biology systems, and monitoring systems in operating rooms or intensive care units), along with shared IT gear and networks situated outside the facility, often developed using cloud technologies and architectures.

To pinpoint the primary areas of impact, equipment can be categorized as follows:

- User workstations
- Printing devices
- Network equipment and services
- Server rooms (on-site or off-site)
- Cloud services
- Specific diagnostic or treatment equipment unique to the healthcare sector, referred to as connected medical devices: technical and operating room equipment, imaging, biology, and radiotherapy equipment.

[7] SUN-ES-Ségur Usage Numérique en Établissements de Santé

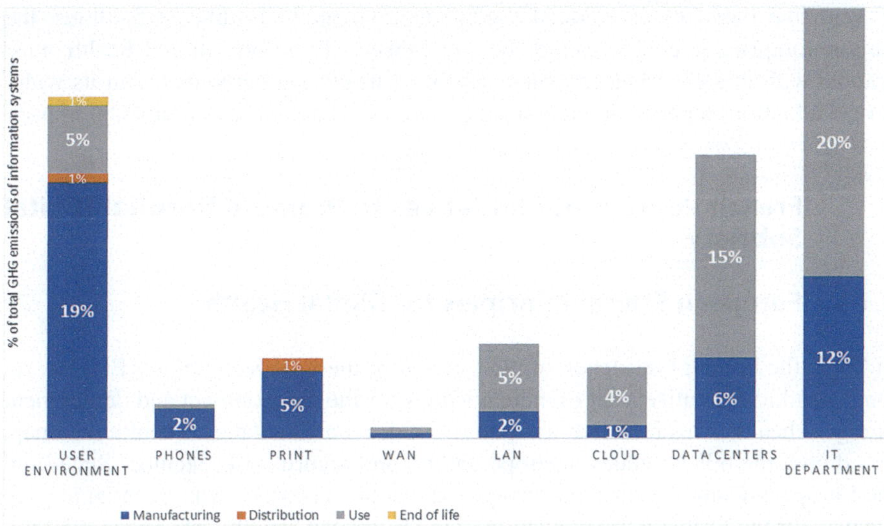

Fig. 3 2023 greenhouse gas emissions breakdown by IT domain and lifecycle stage (Anne and Resilio 2023)

For instance, the number of computer workstations solely used in France's public healthcare facilities is estimated to exceed 500,000.[8] If hospitals were standard businesses (without specialized equipment), user workstations, server rooms, cloud, and IT department operations would account for more than 80% of hospital digital CO_2 emissions (Fig. 3). IT department operations encompass physical premises, staff travel, and intangible services required for the IT department's functioning.

Greenhouse gas (GHG) emissions from internal IT (excluding connected medical devices and external IT solutions like those hosted in the cloud) for all public healthcare facilities (875 in 2018)[9] are estimated to exceed 190,000 tons of CO2eq per year,[10] equivalent to over one million round trips by plane between Paris and Marseille.[11] Imaging equipment alone potentially accounts for 4% of a hospital's energy consumption.[12]

---

[8]2020 Atlas of Hospital Information Systems https://sante.gouv.fr/IMG/pdf/atlas_2020_-_vf.pdf

[9]2018 Atlas of Hospital Information Systems - https://solidarites-sante.gouv.fr/IMG/pdf/dgos_atlas_sih_2018.pdf

[10]Estimates based on the IT inventory of healthcare institutions, published by the DGOS (Atlas de SI Hospitaliers 2018- https://solidarites-sante.gouv.fr/IMG/pdf/dgos_atlas_sih_2018.pdf) and the DNS calculator developed by F. Bordage.

[11]https://impactco2.fr/comparateur

[12]French Society of Radiology- Livret blanc - version web 10 dec 2021.pdf (radiologie.fr).

With the dawn of accelerated digital deployment in healthcare facilities, the carbon footprint it entails cannot be overlooked. Therefore, digital health must grapple with this equation: the value of digital health is intertwined with its widespread adoption, representing a potential source of increased environmental impact.

3 French Policies and Initiatives to Promote Hospital Digital Sobriety

3.1 European Ethical Principles for Digital Health

In 2022, the French Presidency of the Council of the European Union (EU) has set the ground to formalize a set of ethical values for the development and deployment of digital health. This initiative was part of France's national digital health roadmap, based on three pillars: ethics, interoperability, and security. EU Member States and the European Commission soon recognized that this approach was essential to pave the way for the forthcoming regulation of the European Health Data Space (EHDS). Building on a proposal from the French EU Presidency, European ethical principles for digital health were developed and discussed during the very first weeks of the French EU Presidency by the eHealth Network representatives and were unanimously adopted on January 26, 2022. The aim was to articulate simple, clear, and self-explanatory principles in a direct and straightforward manner, ensuring they could be understood by all EU citizens.

These European ethical principles for digital health are organized into four dimensions with four principles each. The fourth dimension, named "Implement eco-responsible digital health," urges all stakeholders to commit to moving toward eco-friendly digital health and to undertaking concrete actions in this area. The four principles of this dimension are:

- "Environmental impacts of Digital Health are identified and measured."
- "Digital Health services are developed in compliance with eco-design best practices."
- "Re-use and recycling of Digital Health equipment is ensured."
- "Digital Health stakeholders are committed to reducing their ecological footprint."

3.2 Engaging in the Ecological Transition

Healthcare players are caught between the imperative to act in favor of ecological transition and the need to cope with often insufficient financial resources. However, recent years have seen the emergence of numerous sustainable development initiatives covering a wide range of topics: circular economy, increased waste recovery, combating food waste, developing short distribution channels, promoting soft mobility, sustainable purchasing, eco-designing care products, and improving indoor air quality. Before taking action, professionals have become aware of the need to preserve resources that are no longer considered infinite. This was illustrated by the record participation to the webinars organized by ANAP[13] on energy savings (5000 registrants) and water savings (3000 registrants).

Pioneering hospitals have made their CSR strategy—the comprehensive plan companies and founders use to design, execute, and analyze their corporate social responsibility initiatives—a lever of attractiveness. In a context where there are lasting tensions in terms of human resources, CSR has become a differentiating factor for professionals in choosing their place of work. In a French survey conducted by the EHESP,[14] in 2022, 76% of healthcare managers surveyed said they were extremely sensitive to sustainable development. They consider a sustainable development policy at both the facility and regional levels an important factor in attractiveness. Faced with this growing awareness in the field, national players are committed to supporting professionals and providing them with the means to take action. One example is the deployment of a network of Energy and Ecological Transition Advisors in the healthcare sector, where some 150 professionals are working to implement energy performance plans and the ambitions of the tertiary eco-energy decree, helping facilities decipher various regulatory obligations and financial aid available to set up their projects.

The growing mobilization of facilities is also reflected in the latest national "My Sustainable Development Observatory"[15] campaign results. Conducted by ANAP every year since 2017 as part of the sustainable development agreement signed by the French government and federations in the healthcare and medico-social sectors, this survey provides a current snapshot of the maturity of facilities in different areas: governance, societal, social, environmental, and economic.

After 2 years of low-key surveying due to the COVID pandemic in 2022, more than 2600 healthcare facilities and medico social structures responded to the survey, more than twice as many as in 2019, with strong insights into the work being carried out by healthcare players in favor of the ecological transition, which still needs to be strengthened:

- Facilities are structuring their actions, and the movement is underway: 44% of respondents have appointed one or more professionals as sustainable development and CSR referents; 50% have set up a sustainable development steering committee; and 65% have established working groups.

[13] ANAP: Agence Nationale de la Performance Sanitaire et Médico-Sociale—The French national agency to support the performance of healthcare and medical-social facilities (https://www.anap.fr/)

[14] EHESP: Ecole des Hautes Etudes de la Santé Publique, School of High Level Public Health Studies

[15] https://anap.fr/s/article/rse-publication-1644

- The picture is more mixed when it comes to the environmental footprint, with 47% of facilities having undertaken work to improve their energy efficiency, and 45% having installed water-saving equipment: tensions over energy prices and availability, as well as 2023 summer's drought, undoubtedly played an accelerating role.
- On the other hand, the entry into force of regulations has not always had a leverage effect: more than half have never assessed their greenhouse gas production, despite the regulatory obligation for structures with more than 250 employees to draw up a balance sheet (Marrauld et al. 2021).

These results, along with the feedback received, indicate that facilities are aware of their environmental impact and their responsibility in the ecological transition of the healthcare system and are taking action to address it.

3.3 Green IT Initiatives Already Implemented at the French National Level

The ethics unit of the DNS, under the French Ministry of Health, spearheads efforts in digital health ethics, overseeing all digital health transformation projects. This unit endeavors to cultivate ethical digital health practices that uphold patient rights, nurture humanized care relationships, instill user trust, engage stakeholders, and encourage adoption of digital health for the benefit of all. The initial working step involved crafting the Digital Health Ethics Framework,[16] anchored in five principles including eco-responsibility, alongside beneficence, non-maleficence, autonomy, and justice.

Aligned with the French government's "Digital and Environment" roadmap,[17] the DNS ethics unit embarked on raising awareness among healthcare professionals, industry stakeholders, and public authorities regarding the environmental impact of digital health. This endeavor culminated in the publication of the inaugural sectorial report on the environmental impact of digital health in June 2021.[18] The aim was to balance the rapid advancement of digital technology with environmental responsibility, acknowledging the potential of digital technologies to improve patient care despite the challenges of their widespread implementation. Moreover, the report conveyed a compelling message to digital stakeholders in France: effective disease prevention and health promotion necessitate curbing our carbon footprint. The solution lies not in curtailing digital health usage but in eliminating unnecessary

[16] CENS -Cadre de l'Ethique du Numérique en Santé (https://esante.gouv.fr/produits-services/referentiel-ethique).

[17] https://www.economie.gouv.fr/environnement-numerique-feuille-de-route-gouvernement

[18] https://esante.gouv.fr/sites/default/files/media_entity/documents/RAPPORT_GT6_VF.pdf

digital healthcare practices and embracing responsible digital health approaches. This necessitates a deliberate assessment of our needs. The healthcare sector must exemplify environmental stewardship by reducing the carbon footprint of digital healthcare solutions. This dedication also enhances the quality of patient care, aligning environmental goals with healthcare excellence.

Subsequently, in collaboration with the National Agency for Digital Health,[19] the DNS ethics unit developed various ecoscores,[20] offering them to software developers as free and simple tools for gauging the environmental impact of their digital health solutions. The ecoscore for health apps (Baudinière et al. 2024) evaluates the environmental impact of app usage and suggests areas for minimizing this impact. Referencing an application in the "*Mon espace santé*" store requires an ecoscore. Presently, approximately 70 ecoscores for digital health apps have been published, with ratings ranging from 40/100 (requiring improvement) to 86/100 (good), and an average score of 62/100. Additionally, an ecoscore for teleconsultation has been developed, mandated for teleconsultation companies to receive reimbursement from the French health insurance system.[21]

EvalCarbone SIH,[22] a national carbon footprint calculator for hospital information systems, was launched in May 2024. Available online, EvalCarbone SIH utilizes the calculation engine advocated by the MTE,[23] adhering to the "Product Category Rules" (PCR[24]) for digital services[25] of the ADEME agency. In this methodology, the hospital information system is delineated by the hardware assets and virtual machines managed by the hospital. EvalCarbone SIH calculates the carbon footprint based on the impact factors[26] of each digital asset at each stage of its lifecycle.[27] Additionally, it offers functionalities that allow for simulations to assess decarbonization strategies.

---

[19] ANS—Agence du Numérique en Santé (https://ue.esante.gouv.fr/).

[20] Ecoscores—https://ecoscore.esante.gouv.fr/

[21] Ecoscores, initially promoted within France, demonstrate potential for international adoption, as evidenced by their inclusion in the fourth chapter of the European Ethical Principles of Digital Health and active discussions within the eHealth Network (eHN).

[22] https://eval-carbone-sih.esante.gouv.fr/

[23] MTE: Ministère de la Transition Ecologique (French Ministry of Ecological Transition).

[24] The objective of the Product Category Rules (PCR) is to provide a precise method for evaluating environmental impacts in order to standardize processes.

[25] https://librairie.ademe.fr/produire-autrement/6022-referentiel-methodologique-d-evaluation-environnementale-des-services-numeriques.html

[26] Impact factors used for the calculations are those made available by the MTE, under a Creative Commons license, from the Resilio database, with scope for expansion (few impact factors are currently available). The categories of materials and proposed models are subject to change as this reference is enriched.

[27] The general formula for the calculation is as follows: Unit Impact (with, for example, the impact unit expressed in kgCO2 eq) = Activity (activity unit = km, kWh…) x Impact Factor (impact unit/activity unit).

Ecoscores have also contributed to quantifying the environmental impact of pathology digitalization. Pathology, experiencing significant changes supported by authorities, is undergoing a transformation of its practices,[28] leveraging slide digitization and PACS-like tools for viewing. As part of these advancements, an experiment was conducted to estimate the environmental impact of viewing virtual slides. Utilizing the ecoscore platform provided an approximation of the CO_2 cost for processing a slide at around 2 g of CO2 per slide. While these results are indicative, they prompted dialogue with software developers on mitigating their impact.

3.4 Green IT Initiatives Already Implemented in French Public and Private Hospitals

A number of Green IT initiatives have been launched in public and private hospitals to reduce the environmental and social impact of IT and digital equipment and hardware (servers, computers, storage devices, network equipment, medical equipment and devices, etc.) used to provide care to the public.

Implementing Green IT in healthcare facilities involves a range of practices such as server virtualization, data center pooling, efficient energy management, extending the lifespan of workstations and servers, integrating sustainable purchasing criteria for IT equipment, recycling IT equipment and managing electronic waste, and implementing good storage practices. This includes limiting mailbox sizes, encouraging users to delete unnecessary files from network drives and workstations, reducing paper consumption through printing policies and electronic information systems, and training teams in digital responsibility.

3.4.1 Eco-Responsive Purchasing

Hospitals are increasingly attentive to the procurement of materials and equipment, which constitute a significant portion of greenhouse gas emissions, with digital technology being no exception. French regulations mandate certain purchasing obligations for the State and local authorities, including a reparability index from January 1, 2023, and a durability index from January 1, 2026. Although these purchasing obligations do not apply to private and public hospitals, some hospital purchasers use the indices, along with labels such as the European Ecolabel,[29] Blue Angel,[30] Nordic Swan,[31] and Energy Star.[32]

[28] https://sante.gouv.fr/actualites/presse/communiques-de-presse/article/francois-braun-lance-une-mission-sur-la-pathologie-digitale

[29] https://www.ecolabel.be/fr

[30] https://www.blauer-engel.de/en

[31] https://www.nordic-swan-ecolabel.org/

[32] https://www.energystar.gov/

Beyond purchasing, several facilities have taken steps to reduce the environmental impact of their IT equipment. For example, the *Vivalto Santé* Group,[33] with its 50 private hospitals, initiated a structured approach to digital sobriety in 2022, driven by the Information Systems Department and based on the initial carbon footprint assessment conducted in early 2023. Since then, several actions have been deployed or are in progress within the Group: changing IT hardware purchasing criteria to include more reconditioned equipment (over 50% in Brittany for digital workplaces, with a similar process starting for data center hardware), including the carbon footprint as a selection criterion for recurring equipment (such as screens), introducing an inventory process for key equipment, setting policies for the replacement and lifespan extension of user equipment and data centers, rolling out waste sorting for electrical and electronic equipment, and raising awareness about green IT and responsible IT consumption among healthcare professional users during the European Sustainable Development Week. This approach aligns with the Group's environmental commitment, having been a mission-driven company since 2021. Extra-Financial Performance is measured, tracked, and audited annually by a Second Party Opinion (SPO).

3.4.2 Best Practices for Using and Building Digital Health Tools and Services

In 2022, Rennes University Hospital, in close partnership with Philips, conducted a lifecycle analysis (LCA) of a recently purchased biplane image-guided therapy system (Azurion) dedicated to neuroradiology. This study showed that the system contributes up to 0.007% of the annual carbon footprint, or 6 tons of CO_2 equivalent, of Rennes University Hospital, equivalent to the emissions of 1.3 French citizens. The main environmental impact came from raw material extraction and component production, particularly related to printed circuit boards, which represent 43% of the total impact. The energy consumption associated with using the system had a relatively low impact (8%) due to the large share of nuclear power in the French energy mix, resulting in approximately 4.5 times less CO_2 emissions compared to the European average. On-site energy measurements revealed the opportunity to decrease total energy consumption by up to 22% by raising professionals' awareness about switching off the system between procedures. This translates into a potential avoidance of 176 kgCO2eq per year. Based on these observations, a concrete action plan was jointly established, both at the operational and strategic levels, including co-creation sessions on energy efficiency and resource optimization, a 360-degree analysis of the Cath lab[34] carbon emissions, and circular strategies such as system upgrades and refurbishment.

[33] https://www.vivalto-sante.com/

[34] A catheterization laboratory, commonly referred to as a cath lab, is an examination room in a hospital or clinic with diagnostic imaging equipment used to visualize the arteries and the chambers of the heart, and treat any stenosis or abnormality found.

Energy Aspects

The energy aspect of hospital data centers is a significant lever that institutions are using to reduce their carbon footprint and energy costs. Large hospital data centers are modernizing to better insulate these spaces, thus cooling accounts for 40% of data center expenses, and incorporating airflow capture logic. For example, the University Hospital of Nancy and the Grand Est region are constructing a mutualized data center designed to collect heat.

The University Hospital of Bordeaux (Nouvelle Aquitaine region) is modeling the energy consumption of its entire fleet (10,500 machines). The University Hospital of Poitiers (Nouvelle Aquitaine region) has programmed the shutdown of machines at 11 p.m. if no activity has been detected 30 min prior. This initiative currently covers approximately 500 computers and is being extended, requiring the localization of all computers. In addition, at the University Hospital of Poitiers, free cooling has been implemented in two data centers: 80% of the cold is obtained for free without running the cold groups, with an additional investment cost of around 10% (15,000 to 20,000 € on a 200,000-€ investment for 80% energy savings). The hospital is still looking for how to use the heat generated by the data centers.

As part of its facilities, Mipih,[35] a public structure of hospital cooperation specialized in the development and hosting of hospital and healthcare information systems, implements air conditioning regulation systems to optimize cold production. The excess heat produced by the data center in Toulouse (France) is recycled to heat other buildings. The data center in Amiens (France) is connected to the city's district heating and cooling network. The heat produced by the data center is used for the city's heating network, and the cooling is produced by geothermal energy. Amiens Energies has committed itself to injecting energy into the Intercampus network, a student federation in Champagne-Ardenne, as part of the development of its networks, through information notes and awareness campaigns about eco-friendly practices.

A lot of facilities are implementing various cost-saving measures with promising results: a 3 °C increase in data center temperature, the widespread use of LED lighting, and the systematic distribution of equipment protection to increase longevity. In the Auvergne region, the reprography service of the University Hospital of Clermont-Ferrand has opted for a cold printing machine with water-based ink, which is less polluting, reduces paper jams, and consumes less energy in a ratio of 1 to 12.

Search Engines

Search engines also have an impact, and several facilities have opted for alternatives to Google or Bing. For example, the University Hospital of Clermont-Ferrand (France) chose the French search engine Lilo,[36] which compensates for its

[35] https://www.mipih.fr/

[36] https://fr.wikipedia.org/wiki/Lilo_(entreprise).

greenhouse gas emissions and allows self-financing through a points system ("water drops") for each search. These water drops can be converted into money to finance university hospitals' projects. The University Hospital of Clermont-Ferrand can then be supported by Lilo users from outside the hospital who wish to donate their points to the hospital rather than another association. Subsequently, the Necker Hospital (*Assistance Publique – Hôpitaux de Paris*), the University Hospital of Lille in the Hauts de France region, and the University Hospital of Montpellier in the Occitanie region have joined Lilo.

The University Hospital of Bordeaux (France) in the Nouvelle Aquitaine region chose to use Ecosia also named "the green search engine"[37] from Germany where searches fund tree planting).

Email Management

Recognizing the significant energy consumption in digital storage, various institutions have initiated original actions. For example, the Moulins-Yzeure Hospital Center (Allier) launched its first email deletion initiative ("Vendretri," which could be translated as "SortingFriday") in 2020, and repeated it in 2021. Over a period of 2 weeks, nearly 2.4 million emails (including backups) were deleted. Hospital staff were made aware of the initiative, and the IT department deleted unused mailboxes and removed emails based on a list of recurring emails such as press reviews, software updates, and newsletters.

French Hospitals Federation, in collaboration with DNS and the National Digital Health Agency (ANS), launched a similar national challenge that was taken up by about twenty public hospitals, resulting in 2 terabytes of storage being deleted in these institutions in October 2022. Some hospitals have continued these efforts beyond the period of the challenge. For instance, the University Hospital of Nantes (Pays de la Loire region) deleted four million emails and 960 mailboxes, representing 640 GB of storage and accounting for 3% of the total volume and 5% of emails in 2022. The University Hospital of Bordeaux deleted 1.4 million emails, freeing up 1 terabyte of data in October 2023. Although this initiative may seem minor, it raises awareness among hospital staff, reallocates digital storage resources, and engages digital departments in the ecological transition of their institutions. Many hospitals, like the University Hospital of Clermont-Ferrand, have streamlined their digital sharing practices to reduce storage volume.

3.4.3 Actor Awareness and Transparency

The Rive Gauche hospital in Toulouse (Occitanie region) has been participating in the Digital CleanUp Day since 2020. This day aims at raising awareness of hospital staff about the environmental footprint of digital technology, encouraging people to take the first concrete steps by cleaning up data and giving digital equipment a second life.

[37] https://www.ecosia.org/

3.5 Second Life and End-of-Life Management

Considering that IT equipment manufacturing has the highest environmental impact in the digital sector, the end of life of IT equipment should receive special attention. France has established "Extended Producer Responsibility" (EPR) programs that require manufacturers to manage the end-of-life process for their equipment. Regarding electronic waste, "eco-organizations," funded by a fee included in purchase invoices, collect and recycle this waste as efficiently as possible.

Therefore, many healthcare facilities have integrated reconditioning channels into their renewal processes so that old equipment, such as PCs, infrastructure equipment, network cores, or servers, is donated, refurbished, and redistributed. For example, the University Hospital of Bordeaux has facilitated the reuse of more than 13,000 computers recycled in associations and schools since 2006, with its internal procedure anticipating around 5 years of use at the University Hospital followed by a donation to an association. This hospital will also organize a campaign to collect used smartphones with the Emmaus Connect association.[38]

The Angoulême Hospital Center also donates its non-functional equipment to a local association located less than 10 km from the healthcare facility, which repairs and sells them at a modest price to people in need. In 2022, they donated 9 servers, 97 central units, 144 screens, 37 keyboards, 4 printers, and 8 mice.

At the initiative of its employees, the University Hospital of Lyon recently created a "Second Life" program for the 3000 computers replaced annually in its digital park of 24,000 units. At least half of the used computers are now donated to Lyon associations for refurbishment and resale at low prices to people in need. The University Hospital of Clermont-Ferrand also donates a few dozen PCs to schools each year, giving them a second life. Additionally, this hospital maximizes computer refurbishment based on technical possibilities and capacities by integrating SSDs and RAM into central units to preserve the maximum functionality of the machines. Finally, the use of eco-organizations for non-repairable equipment is systematic in these healthcare facilities.

The ELSAN group,[39] which operates 140 private hospitals in France, has integrated several key initiatives to enhance its corporate social responsibility (CSR) and promote digital sustainability. Ten percent of their computer purchases are reconditioned devices. The lifespan of digital equipment has been extended to about 6 years for workstations and to around 8 years for servers. Additionally, 60% of their data center operations are shared, with a target to increase this to 80%. ELSAN also donates end-of-life computers, ensuring they are reused rather than discarded. These efforts are part of a broader CSR approach that aligns with ecological and economic goals.

In the context of hospital digital sobriety, actions such as the deletion of emails may seem minor, but they are significant in raising awareness and reallocating digital

[38] https://emmaus-connect.org/

[39] https://www.elsan.care/fr

resources. These initiatives play a crucial role in the ecological transition of healthcare institutions. By integrating resource-related indicators such as energy, storage, CPU usage, and terminals, hospital digital services management can effectively undertake actions contributing to cost savings and a reduced carbon footprint. Overall, these measures reflect a growing trend in healthcare toward more sustainable practices, balancing technological advancement with environmental responsibility.

4 Discussion

4.1 Green IT in Healthcare System, a Worldwide Recognized Imperative

The health sector is responsible for about 5% of European greenhouse gas emissions and thus contributes to an increasingly unhealthy future unless we take action quickly. In line with the commitments of the signatory countries of the Paris Agreement in 2015 COP to limit the temperature increase well below 2 degrees, new carbon neutrality policies have emerged worldwide with a focus on the healthcare sector or a focus on IT, but rarely on health IT.

For example, the Council of the European Union (EU) adopted the 2030 policy program "Path to the Digital Decade,"[40] which ensures that the EU meets its objectives for a digital transformation in line with EU values. A study on digital health implementation in the EU was carried out in April 2022. Among the 29 countries within the scope of the study, the majority have included the necessity of reducing the carbon footprint of the digital sector in their development policies, but digital health is not specifically addressed. Since then, initiatives to reduce the environmental impact of health IT have been launched in a few European countries, including Hungary and Spain.

In the UK, the NHS has set out to become the world's first net zero national health service.[41] On July 1, 2022, the NHS became the first health system to embed net zero into legislation, through the Health and Care Act 2022. Two primary targets have been established: first, to optimize emissions directly controlled by the NHS, the NHS Carbon Footprint, reaching net zero by 2040, and second, to optimize emissions influenced by the NHS, the NHS Carbon Footprint Plus, achieving net zero by 2045. With over a decade of experience in sustainable healthcare and strong support from staff, the NHS is well-positioned to undertake this challenge. However, NHS efforts to reduce carbon impact have not yet extended to reduce the environmental impact of health IT.

[40] https://www.consilium.europa.eu/fr/press/press-releases/2022/12/08/path-to-the-digital-decade-council-adopts-key-policy-programme-for-eu-s-digital-transformation/

[41] Greener NHS » Delivering a net zero NHS (england.nhs.uk)

In the Pacific area, the Global Green and Healthy Hospitals (GGHH[42]) members have set their own public carbon emissions reduction targets in the healthcare sector.[43] For example, the Australian Government has committed itself to reduce national emissions by 43% below 2005 levels by 2030, and to achieve net zero national emissions by 2050. These targets apply to all parts of the Australian economy, including the health system.

Amidst the increasing momentum of governments worldwide setting targets for achieving net zero emissions, none is as consequential as China's pledge. In September 2020, it was announced that China will strive to peak CO_2 emissions before 2030 and achieve carbon neutrality by 2060.[44] China is striving to establish green and low-carbon data centers with higher computing efficiency and less power consumption (Li et al. 2023). As the world's largest energy consumer and carbon emitter, accounting for one-third of global CO_2 emissions, China's progress in reducing emissions will significantly influence global efforts to limit global warming to 1.5 °C.

While there are international efforts to reduce the environmental impact of digital technology and healthcare, only a few countries are actively engaged in eco-friendly digital health, and in this regard, France aims to play a pioneering role globally.

4.2 IT for Green

Beyond the potential to reduce its own carbon footprint, as discussed in this chapter, digital technology is often heralded as a crucial solution to the climate crisis, capable of lowering greenhouse gas emissions across various sectors. This concept, known as "IT for Green" (IT4Green), leverages digital tools to optimize physical processes and minimize their environmental impact. Energy management systems in buildings and eco-driving technologies in vehicles exemplify this approach.

In healthcare, digital technology is similarly viewed as a means to support decarbonization and enhance resilience (The Shift Project 2023). Teleconsultation systems, which allow patients to stay at home, can save between 0.70 and 372 kg CO2e per consultation (Purohit et al. 2021). More broadly, digital solutions for care coordination, remote monitoring using IoT, and early diagnosis powered by AI aim to improve health outcomes, which can inherently reduce the reliance on emergency care and curative procedures, thus lowering their carbon footprint. Additionally, digital solutions can help avoid unnecessary and redundant healthcare procedures as

[42] GGHH is a collaborative network of health institutions that are committed to reducing their ecological footprint and promoting public and environmental health.

[43] https://assets.nationbuilder.com/caha/pages/3280/attachments/original/1708045627/GGHH-Annual-Report-2022-2023.pdf?1708045627

[44] https://iea.blob.core.windows.net/assets/9448bd6e-670e-4cfd-953c-32e822a80f77/AnenergysectorroadmaptocarbonneutralityinChina.pdf

those caused by a lack of shared patient information, further reducing their environmental impact.

To fully realize the environmental benefits of these digital health solutions, it is essential to methodically assess their positive impact. This involves defining a baseline scenario and considering potential rebound and indirect effects (Vasselin and Benedini 2024). Only rigorous assessment can ensure that digital innovations contribute effectively to a more sustainable healthcare system.

The environmental benefits of digital health technologies, such as telemedicine, are evident in the reduction of carbon emissions associated with travel. For example, a single teleconsultation can save substantial CO_2 emissions by eliminating the need for patients to physically travel to healthcare facilities. However, the production, usage, and disposal of digital devices used in these technologies contribute to environmental issues. The manufacturing processes, energy consumption during usage, and e-waste generated at the end of their lifecycle all add to their environmental footprint, leading some researchers to argue that the complete lifecycle of digital devices can lead to significant environmental degradation. This aspect is often overlooked by users who focus on the immediate benefits of digital health solutions.

4.3 Cybersecurity for Green IT

In the healthcare sector, where cybersecurity and data protection are major strategic issues, reducing the environmental footprint of information systems may seem incidental or even contradictory to some cybersecurity measures. For example, the high availability and redundancy recommended in certain cybersecurity contexts can lead to hardware complexity and duplication, which may increase the environmental footprint. In healthcare, French facilities classified as OSE (Opérateur de Service Essentiel) and OIV (Opérateur d'Importance Vitale), following the adoption of the European "NIS2" directive[45] into the French law, have strict legal obligations in terms of cybersecurity. Cybersecurity issues will take primacy over environmental issues.

However, cybersecurity also plays a crucial role in reducing the environmental impact of information systems in several ways. Firstly, the principle of least privilege limits user rights to the bare minimum on client devices, resulting in fewer installed software applications and preventing unnecessary installations. This not only reduces the attack surface but also potentially decreases energy consumption and premature hardware obsolescence caused by memory limitations or disk space saturation. Another aspect of digital sovereignty involves controlling supply chains and even equipment production, intersecting with cybersecurity, environmental, and social concerns, including human rights considerations.

Furthermore, eco-design, cybersecurity, and data protection share specific frameworks and regulations tailored to each domain. Decommissioning, the

[45] https://eur-lex.europa.eu/legal-content/EN/TXT/?uri=CELEX:32022L2555

cessation of digital services accompanied by data, software layer, and associated resource management, is another practice that intersects with eco-design, cybersecurity, and data protection. It helps reduce attack surfaces, limit security vulnerabilities in obsolete systems, decrease the volume of data to protect, free up storage space and resources, and reduce energy consumption. Overall, decommissioning contributes to reducing the environmental footprint and financial costs associated with digital services, including storage and maintenance expenses.

4.4 Green IT French Policy in Hospitals, What Is Next?

The rise in global temperature has accelerated and has significant impacts on all ecosystems and human health, spanning across all sectors. It is imperative to accelerate the ecological transformation of all activities, particularly healthcare. To this end, for the first time in the history of France, the President of the Republic has entrusted ecological planning directly to the Prime Minister.

The healthcare and medico-social sector, which accounts for 8% of national greenhouse gas emissions (nearly 50 million tons of CO_2 equivalent), must contribute to the necessary reduction of greenhouse gas emissions by 5% annually until 2050, to adhere to France's national commitment to stay below the +1.5 °C target. The sector must transform to pave the way for a more sustainable, circular, and less fossil fuel-dependent system.

Many local initiatives for ecological transformation have emerged across the territory in recent years. Building on these initiatives, launched by the Prime Minister under the "France Nation verte" brand,[46] and coordinated by the General Secretariat for Ecological Planning, a novel steering committee, structured around eight themes, was established on May 22, 2023, for the healthcare and medico-social sector to control its environmental impacts and ensure the sector's long-term ecological transformation. The workforce has identified eight action areas to ensure the ecological transformation of the sector.[47]

1. Building and energy management
2. Industries and health products
3. Sustainable procurement
4. Eco-responsible healthcare
5. Sector waste management
6. Education and research in ecological transformation
7. Sustainable mobility
8. Environmental impact of digital technologies

[46] https://www.info.gouv.fr/france-nation-verte

[47] https://sante.gouv.fr/IMG/pdf/dp_pess.pdf

The aim of action 8 is to develop and implement a roadmap for greening digital healthcare, some of whose objectives have been defined:

- Promoting the purchase of digital equipment and materials from the circular economy through the implementation of an ecolabel.
- Encouraging the purchase of eco-designed devices. Modularity of devices is particularly advantageous as it allows for part replacement instead of total renewal (in line with sustainable procurement).
- Combatting "bloatware" and promoting the uninstallation of unused services and cleaning of servers and computers to prevent the accumulation of digital waste that unnecessarily consumes energy.
- Promoting eco-design approaches and encouraging the development of less energy-intensive software by altering coding practices through professional software training and introducing the concept of "energy bugs" for any code consuming energy without functional necessity.
- Calculating and monitoring the environmental impacts of digital healthcare services (providing ecoscore calculation services for health applications, software for managing outpatients, telehealth services including teleconsultation, healthcare facilities, and social and medical-social sectors).
- Combating obsolescence by proposing not to renew functional hardware, favoring reuse within solidarity initiatives, and systematizing the recycling of devices at end of life.
- Promoting actions to control the energy consumption of data centers and certain medical devices (only one-third of the energy is used to operate servers, with the remainder used to power a cooling system), such as regulating air conditioning to optimize cold production and reusing excess heat produced to heat other buildings.
- Measuring and controlling the volumes of data exchanged. Not all personal data may be relevant in clinical or population studies.
- Evaluating the actual health benefit of an innovative digital device before implementation, compared to the environmental risk it represents and any potential rebound effects.

These national targets will eventually be used to assess the environmental performance of digital health devices. Associating Green IT with the performance of healthcare facilities offers numerous benefits. It helps reduce operational costs by improving energy efficiency and decreasing waste. Advanced and reliable technologies enhance the quality of care and minimize interruptions, while adhering to environmental standards avoids penalties and secures valuable certifications. A commitment to sustainability strengthens the facility's brand image, attracting eco-conscious patients and staff, and a healthier work environment boosts staff satisfaction and productivity. Ecological practices foster innovation and provide a competitive edge. Ultimately, these practices contribute to public health by reducing pollutant emissions and position the facility as a leader in environmental health.

5 Conclusion

Globally, healthcare activities contribute to 4–5% of worldwide greenhouse gas emissions. On the other hand, the digital sector is responsible for approximately 4% of global greenhouse gas emissions. Some initiatives have been launched in different countries to reduce the environmental impact of IT, or reduce the environmental impact of the healthcare system, but very few to reduce the environmental impact of health IT. In this regard, France represents an exception as it has engaged in the reduction of the impact of health IT since 2020. However, even in France, the digital component of ecological transition efforts in healthcare facilities is still underdeveloped, in contrast to waste reduction and, more recently, the advancement of eco-friendly healthcare practices.

Despite the fact that local initiatives for ecological transformation have emerged, several obstacles hinder the implementation of these initiatives: financial barriers in an underfunded context with inextricable budget constraints, resistance to change, and a lack of awareness and training among stakeholders.

While healthcare professionals recognize the urgency of addressing climate change and ecological issues, they need support in this transition, which is more akin to a revolution. The ecological transition requires investment in training, changes in practices and organization, and funding to scale up. Implementing Green IT measures often requires substantial initial investment.

Beyond financial issues, other levers need to be collectively addressed to continue developing Green IT initiatives: comprehensive training for managers of healthcare establishments and healthcare professionals on the challenges of Green IT and the associated best practices, and support for changing habits and practices.

Lastly, healthcare facilities face a dual constraint: they must participate in the strategy to accelerate digital health promoted by public authorities as part of various national programs for the development of digital health, while also meeting the ambitions of the ecological planning of the healthcare system, which requires healthcare professionals to contribute to greener IT. For healthcare facilities, this means reconciling the development of digital technology with the challenge of sustainability in the healthcare sector.

All of these practices, which involve rethinking the use and management of IT and digital tools, contribute to better IT resource management, the sustainability of digital healthcare, reduced operating costs through more efficient energy use, and the limitation of greenhouse gas emissions as part of a broader social responsibility approach. By considering these aspects in the planning and implementation of their digital maturity strategies, hospitals and other healthcare providers can not only contribute to reducing climate change, but also achieve solid financial and performance benefits.

Acknowledgments The authors would like to thank Camille Farrando, Management Engineer at the University Hospital of Rennes (France), Paul Mayaux, advocacy officer at Agence Nationale de la Performance Sanitaire et Médico-Sociale (ANAP), and Frédéric Duvignaud, Project Manager at Agence du Numérique en Santé (ANS). Their contributions have been instrumental in the writing of this paper.

References

Anne R, Resilio (2023) Benchmark Green IT report. https://www.greenit.fr/wp-content/uploads/2023/12/2023-10-Benchmark_Green_IT-2023-report_v1_EN.pdf. Accessed 25 June 2024

Baudinière N, Philippot O, Leboucq T et al (2024) Reducing the environmental impact of digital health: development of an ecoscore for health apps. Stud Health Technol Inform 310:1598–1602. https://doi.org/10.3233/SHTI231333

Center for Sustainable Systems, University of Michigan (2023) Green IT factsheet. Pub No CSS09–07. https://css.umich.edu/publications/factsheets/built-environment/green-it-factsheet. Accessed 25 June 2024

Chevance G, Hekler EB, Efoui-Hess M et al (2020) Digital health at the age of the Anthropocene. Lancet Digit Health 2(6):e290–e291. https://doi.org/10.1016/S2589-7500(20)30130-8

Corbett J (2010) Unearthing the value of Green IT. In: Proceedings of the international conference on information systems, Saint Louis, MI. https://aisel.aisnet.org/icis2010_submissions/198

Costello A, Abbas M, Allen A et al (2009) Managing the health effects of climate change. Lancet 373:1693–1733. https://doi.org/10.1016/S0140-6736(09)60935-1

Ferreboeuf H, Efoui-Hess M, Verne X (2021) Environmental impacts of digital technology: 5-year trends and 5G governance. In: The shift project. https://theshiftproject.org/wp-content/uploads/2023/04/Environmental-impacts-of-digital-technology-5-year-trends-and-5G-governance_March2021.pdf. Accessed 25 June 2024

Health Care Without Harm Europe (2019) Annual report 2019. https://noharm-europe.org/sites/default/files/documents-files/6406/HCWH_Europe_Annual_Report_2019.pdf. Accessed 25 June 2024

L'Assurance Maladie (2024) Mon espace santé:ce qu'il faut savoir. https://www.ameli.fr/paris/assure/sante/mon-espace-sante/mon-espace-sante-carnet-sante-numerique. Accessed 25 June 2024

Latour B (2011) Plus elles se répandent, plus les bibliothèques deviennent centrales. Bulletin des Bibliothèques de France 56(1):34–36. https://hal.archives-ouvertes.fr/hal-01024235

Li G, Sun Z, Wang Q et al (2023) China's green data center development: policies and carbon reduction technology path. Environ Res 231(Pt 3):116248. https://doi.org/10.1016/j.envres.2023.116248

Marrauld L, Lefèbure A, Baurès E (2021) Comprendre l'impact environnemental du secteur de la santé: Pour un leadership partagé en faveur d'un système de santé durable et résilient. La Presse Médicale Formation 2(6):628–633. https://doi.org/10.1016/j.lpmfor.2021.10.024

McGain F, Naylor C (2014) Environmental sustainability in hospitals – a systematic review and research agenda. J Health Serv Res Policy 19(4):245–252. http://www.jstor.org/stable/26751577

Purohit A, Smith J, Hibble A (2021) Does telemedicine reduce the carbon footprint of healthcare? A systematic review. Future Healthc Journal 8(1):e85–e91. https://doi.org/10.7861/fhj.2020-0080

République Française ADEME, Bilans GES (n.d.) Bienvenue sur Bilans GES. https://bilans-ges.ademe.fr/. Accessed 25 June 2024

Schroeder K et al (eds) (2013) Sustainable healthcare. Wiley

Teehan P, Kandlikar M (2013) Comparing embodied greenhouse gas emissions of modern computing and electronics products. Environ Sci Technol 47(9):3997–4003. https://doi.org/10.1021/es303012r

The Shift Project (2019) Lean ICT: towards digital sobriety. https://theshiftproject.org/wp-content/uploads/2019/03/Lean-ICT-Report_The-Shift-Project_2019.pdf. Accessed 25 June 2024

The Shift Project (2023) Décarboner la santé. https://theshiftproject.org/wp-content/uploads/2023/04/180423-TSP-PTEF-Rapport-final-Sante_v2.pdf. Accessed 25 June 2024

UN Environment Programme (2019) UN report: time to seize opportunity, tackle challenge of e-waste. https://www.unep.org/news-and-stories/press-release/un-report-time-seize-opportunity-tackle-challenge-e-waste. Accessed 25 June 2024

Vasselin Z, Benedini C (2024) Charte d'engagement NZI for IT. https://www.carbone4.com/charte-engagements-nzi-it. Accessed 25 June 2024

Watts N, Amann M, Arnell N et al (2019) The 2019 report of The Lancet Countdown on health and climate change. Lancet 394(10211):1836–1878. https://doi.org/10.1016/s0140-6736(19)32596-6

Watts N, Amann M, Arnell N et al (2021) The 2020 report of The Lancet Countdown on health and climate change. Lancet 397(10269):129–170. https://doi.org/10.1016/s0140-6736(20)32290-x

Readiness and Digital Maturity: Promoting Innovation, Integration, and Scale in Scotland

Janette Hughes, Michelle Brogan, and Ann Wales

Abstract

Digital maturity and readiness are well-known concepts; however in reviewing the research early breakthrough digital health innovations (DHI) can be challenging to capture and assess as part of the common maturity and assessment models utilized. Even though there are many digital maturity models available today (Schallmo, D.R.A., Tidd, J. (eds) Digitalization. Management for Professionals. Springer, Cham. https://doi.org/10.1007/978-3-030-69380-0_5), he challenge is mainly due to innovation projects' immaturity and usual lack of fidelity, integration, and inconclusive service pathway design at that point, along with most organizations marked at being at "early or intermediate stages of digital maturity" (Ladu L, Koch C, Ashari PA, Technol Soc 77: 102564, 2024), and a recent publication stated that "frontier and emerging technology" was not fully aligned with many of the maturity models used (Zhao, Schalet, Alsalamah, Stud Health Technol Inform 305:257–260, 2023).

The *Oxford Dictionary* defines innovation as "the process through which new products, concepts, services, methods, or techniques are developed" (Oxford definitions—www.oxfordreference.com/display/10.1093/oi/authority.20110803100004376) or as Joseph Schumpeter argues ensuring the system is made ready for "supporting the process of creative destruction" (Hospers 2005 Vol18 page 35—https://link.springer.com/content/pdf/10.1007/s12130-005-1003-1.pdf). Through this innovation lens, it is therefore important to examine how, and if, we can assess new emerging innovations through standard digital maturity models (DMMs), and whether DMA digital maturity assessment (DMA) tools can fully accommodate new and upcoming DHIs that are not yet fully conceived, connected, or stable within mature hospital and healthcare systems.

J. Hughes (✉) · M. Brogan · A. Wales
Digital Health & Care Innovation Centre, Glasgow, Scotland, UK
e-mail: janette.hughes@strath.ac.uk; michelle.brogan@dhi-scotland.com; ann.wales3@nhs.scot

A. Scheuer, J. Studzinski (eds.), *Digital Maturity in Hospitals*,
https://doi.org/10.1007/978-3-031-80704-6_15

This chapter discusses how organizations are assessing their digital maturity, taking into consideration a drive for more integrated care ambitions (UK Gov—https://www.gov.uk/government/publications/a-plan-for-digital-health-and-social-care/a-plan-for-digital-health-and-social-care) and the use of disruptive technologies and innovations to help solve the major challenges. Case studies are used within this chapter to illustrate how a heuristic innovation readiness framework is used to demonstrate that the innovation project is increasing in readiness level terms, and how innovations accumulate evidence through each stage with the ambition to be ultimately handed over for service and technical integration (linking to more mature and anchored systems within hospitals and other associated systems), what we would commonly regard as moving out the research and development (R&D) stage into business as usual (BAU).

Keywords

Digital health innovations (DHI) · Digital maturity models (DMM) · Readiness assessment · Integrated care · Innovation adoption and scale · Digital maturity

Abbreviations

BAU	Business as usual
BRL	Business readiness levels
DHI	Digital health innovation
DMA	Digital maturity assessment
DMM	Digital maturity model
EMRAM	Electronic Medical Record Adoption Model
H@H	Hospital-at-home
HIS	Health Information Services
OECD	Organisation for Economic Co-operation and Development
R&D	Research and development
RDS	Right Decision Service
RWE	Real-world evidence
SRL	Service readiness levels
TRL	Technical readiness levels

1 Introduction

The pursuit of digital maturity and readiness for the digital health sector has never been more acute, mainly due to unsustainable pressures on health systems globally. However, digital maturity is regarded highly complex because of the multifaceted nature of health and care, with many stakeholders, and that it is not limited to technical aspects alone, but involves innovating service and business models to create whole system benefit. In addition, it is well recognized that in parallel the facilitation of the right skills for adoption must also be accounted for if any new system is to be successfully adopted and used. Barriers to adoption include the ability and appetite to change, redesign workflows, and create new business models and future skill sets.

If we acknowledge that innovation is a creative process, we can observe that it is common for many industries (e.g., construction, manufacturing, and finance) to take part in a process of transformation to create and deliver innovation and products ready for handover into normal business (Dimitrios Kafetzopoulos Evangelos Psomas 2015). Innovation in health in principle follows a similar process as in other industries, but due to the complexity of the system, it cannot be limited to technical aspects alone. It is increasingly recognized that the service and business model elements produce changes that are equally imperative (Whitehouse and Lange 2022). As international scholars and experts have suggested, a significant barrier to the adoption and implementation of innovation in health is the ability and appetite to "change" and the impact to redesign workflows (Schlieter et al. 2022), along with the necessary skills 'successful digital transformation of health and care require an appropriately skilled workforce' (Rimpiläinen et al. 2024), to be in place to create a state of readiness and maturity from an organization (macro) perspective down to the individual (micro) level.

Successful projects in this field follow an innovation process that iterates and co-designs with multiple partners' input and when/if it reaches a state of readiness (technical, service (Hughes et al. 2021; Whitehouse and Lange 2022), and business readiness) an appropriate value case is developed, and skills are strengthened to onboard innovations into practice. When or if a value case is approved, the innovation is then prepared for industrial production, implemented into practice (to include interoperating with more stable mature legacy systems as necessary), and then stabilized where improvements will then be factored in as required, under a business as usual setup.

On a global scale, the pursuit of digital maturity and readiness for the DHI sector is regarded as being more complex due to the multi-stakeholder involvement in health and care. Indeed, it is understood that the primary obstacle to successful implementation is the willingness and capacity of the multi-stakeholders to work together, agree, and adapt to change, and those that have the requisite competencies to achieve a state of readiness and advancement from two perspectives. The first is the organizational (macro) perspective and second the individual (micro) perspective. The multi-stakeholders should (when appropriate) signal that they are eager and ready to adopt a new way of working using digital tools, with the organization committed to leading, championing, and building a case for change, which is generally a prerequisite to proceed to BAU status and leading to a state of maturity.

Most governments and prominent health and care institutions, both public and private, acknowledge that innovation provides the opportunity to restructure, update, and transform services by reconfiguring processes and embedding DHIs for the advantage of the workforce and the person receiving or interacting with these services. The advantages must be proven and robustly evidenced, ensuring that the system is confident and ready for such a change.

We understand from various studies and frameworks that digital maturity and readiness should be inclusive and be dependent on active multi-stakeholder involvement of all interested and affected parties. In relation to DHI projects, this in most cases includes the workforce, patients/carers, and industry developmental teams. This collaborative nature of designing and testing digital innovations is said to foster a sense of ownership and facilitate active learning, which in turn should create better systems that are fit for purpose and by following this approach should by its nature dynamically create increased digital maturity and capacity within the system.

It is well known from scholars, and a matter of common sense, that rather than simply adding digital innovation on top of an already ineffective health service will not add value or drive efficiencies, and that systems must work with innovators to reimage the future state, which includes reimagining the "structural, technical, and policy preconditions" to imagine new service models (Abernethy et al. 2022). Extensive evidence supports the notion that integrating digital into new methods of working and work practices is crucial, but as important is taking everyone on that journey of discovery which in turn creates a level of 'readiness by doing' and develops the champions within the system to advocate for change to take place.

The importance of innovation activities that encompass the entire care continuum with an emphasis on a person-centered approach is now key. In addition, more emphasis at a strategic level (by Governments) is being placed on seamlessly extending from hospitals into community settings, clearly stated in most strategies and agendas across the world. In accordance with significant EU and Global strategies, these mandates set out the objective to develop health and care systems fit for the future, which are integrated, are sustainable, and provide better care to communities, with more delivered outside the hospital.

In a recent UK publication, the ambition and shape of integration within the community were summarized through seven success statements on "what good looks like (WGLL)."[1] In this chapter, the "person" is defined as users, including professionals, patients, and the circle of care (e.g., carers) that are active care participants. They aim to be more effective and efficient and in meeting this objective it requires testing the readiness of all involved; inclusive readiness may be a new concept, in respect of developing these new innovations and ways of working, but may prove to be a missing part of readiness and maturity roadmaps.

Digital maturity and readiness for this entire stakeholder group is not always well consulted or considered for new and emerging innovations and technology, and this element is not properly catered for in many of the DMMs (Ladu et al. 2024) used as

[1] UK Gov Transform—https://transform.england.nhs.uk/digitise-connect-transform/what-good-looks-like/what-good-looks-like-publication/

benchmarking/assessment tools, as pointed out in a recent publication that involved the World Health Organization (Zhao et al. 2023).

2 Digital Maturity Structures

DMMs create the scaffolding to self-assess in terms of their level of digital maturity, which have been reported in many papers and articles as growing in popularity as a tool and a good way to self-assess the level of maturity of organizations (Tarhan et al. 2020). It is said that "DMMs provide roadmaps for implementing digital technologies, and the progress on these roadmaps can be measured using Digital Maturity Assessments (DMA) tools, which have been reported in many papers and articles as growing in popularity and effective to self-assess the level of maturity of organizations." The digital maturity test is of course not limited to hospital environments; however from research it is found that in many instances, the DMA collects data mostly from hospital leads and health board/trust stakeholders. This assessment (i.e., DMA) is accomplished by utilizing self-assessment inputs and often focused on services and systems already in place and well established. This well-established system therefore possesses a higher level of maturity (high-fidelity product), with the challenge being that new emerging innovations (breakthrough innovations with low fidelity) that require more time to evolve and be made ready for the sector and system are unlikely to be included within the assessment due to the immature nature of such innovation projects.

When considering digital maturity, the focus is generally more on technical systems. The extent to which the workforce and patients are prepared to utilize the emerging digital service has not been thoroughly investigated, and new innovations seldom surface. While certain digital maturity models, such as the HIMSS C-COMM[2] tool and the CCMM[3] models, are applicable to non-hospital settings, they are less commonly used compared to the HIMSS suite of maturity assessment models, which primarily focus on hospital environments. In the UK, the Electronic Medical Record Adoption Model (EMRAM) is a model used by NHS trusts for digital maturity assessments. As of the summer of 2023, it was reported that 76% of the trusts who had used that model ($n = 132$) have not yet achieved level 5 of 7, showing "that the majority of trusts are at a moderate level of maturity" (Health Tech World 2023).

Previous research has demonstrated that the EMRAM tool, along with other established maturity models such as infrastructure and imaging, prioritizes traditional hospital settings and systems rather than the emerging innovative integrated care community models and related innovations. The emerging importance of

[2]HIMSS—HIMSS C-COMM tool—https://www.himss.org/what-we-do-solutions/digital-health-transformation/maturity-models/community-care-outcomes-maturity-model-c-comm

[3]HIMSS Community of care maturity model CCMM—https://www.himss.org/what-we-do-solutions/digital-health-transformation/maturity-models/continuity-care-maturity-model-ccmm

community and integrated innovation models aims to provide healthcare services closer to patients' homes, specifically within community settings.

Whether it is the transition to telemedicine, telehealth, and/or telecare in a community setting, or in more recent times, the drive to use the home as the extended hospital environment, these innovations often sit outside the mature hospital walls. It is understood that the new advanced hospital-at home (H@H) constructs and associated emergent innovations have the ability to admit patients directly into a hospital-at-home (H@H) therefore extending the hospital estate, in some regions commonly known as the drive for digital virtual wards (UK) which is still gathering evidence (Norman et al. 2023). The H@H model (in most cases) relies on establishing a link between primary and secondary care to effectively monitor and promptly respond to any alarms or alerts generated by the associated H@H sensors all of which requires new business models to be developed.

3 Digital Maturity Assessment and the Existing Readiness Assessments

As touched on previously DMA is not a new concept for healthcare, and readiness is represented in some form in almost all the popular digital maturity models and assessment toolkits as highlighted by a recent international publication (Zhao et al. 2023) with the findings revealing existing maturity models have a heavier focus on "health information systems with gaps highlighted on contextualization and the take up of more emerging technology." Furthermore, other reports touched upon other maturity models developed by several EU projects, Momentum[4] and Sirocco[5] to name but a few.[6] These models all highlight the apparent relationship between maturity and readiness.

NHS England Digital Maturity Assessment includes a readiness section, defined as "Readiness: an assessment of the organization's ability to plan, deliver and optimize the digital systems it needs to operate paper-free at the point of care." However, this readiness definition is not usually marked against the lens of the early digital innovation projects which sit more at a micro level and are often invisible due to their low fidelity and iterative nature. The readiness in these models is more about the environment and people being ready to use a stable technical product, or plan for the introduction of new stable products into the system. New disruptive innovations have different associated factors and therefore don't naturally fit into common DMM models and DMA frameworks easily.

---

[4]Momentum project—http://telemedicine-momentum.eu/

[5]Sirocco—www.digihealthcare.scot/our-work/international-engagement/scirocco-exchange-project/

[6]Momentum—Readiness—https://healthmanagement.org/c/healthmanagement/IssueArticle/a-services-readiness-levels-stage-model-a-roadmap

The next section reviews and describes how Scotland has created an assessment tool that assists in rating the digital maturity of health and care stakeholders, on the quest to look wider than just the hospital environment, with a greater emphasis on care outside the hospital (e.g., integrated care) and the multi-stakeholder involvement.

3.1 Digital Maturity in Scotland

Over the last decade or so, Scotland has placed the integrated care agenda at the heart of policy and strategy, and this has been regarded as a key driver to deliver better health and care services. Digital is well recognized as an enabler for this whole system vision, to realize a more sustainable and seamless health and care service delivery model fit for the twenty-first century. The advantage Scotland had in moving closer to this reality is due to key policy changes in 2016, where the Scottish Government passed legislation for an integrated health and care structure. This new framework legally unified health and social care into a single, integrated system, facilitating closer alignment of systems and financial resources.

This was an important step for Scotland with the Scottish Government stating that *"Integration is the most significant change to health and social care services in Scotland since the creation of the NHS in 1948. Integration aims to improve care and support for people who use services, carers, and families"* (The Scottish Government n.d.).

This policy shift created a mandate and further opportunities to review the entire system including digital maturity, taking into consideration the opportunity for a more joined up and integrated perspective to be included. It is considered that Scotland may be relatively unique in their integrated assessment and approach to digital maturity assessments at a national level.

This consistent annual assessment approach open to of the whole "integrated health and social care system" being undertaken is not just focused on hospitals and health boards/trusts but wider stakeholders are included, like the local government (Councils/Municipalities). This considers the well-recognized relationship and symbiotic nature between primary and secondary care created to safeguard the journey and take into consideration the handovers across the care continuum for the end user—the patient and their families on that journey.

Scottish Health and Social Care is a devolved national government (Scottish Government) matter, with funding transferred to the Scottish Government from the UK government treasury via a block grant.[7] This devolved situation is key as it offers Scotland a level of control and freedom to operate, which allows decisions and policies to be implemented when required. The annual H&C financial envelope is approximately £19.5bn (2024) with a further £2bn for social care (The Scottish

[7]UK Government: Public Spending—https://www.deliveringforscotland.gov.uk/scotland-in-the-uk/public-spending/

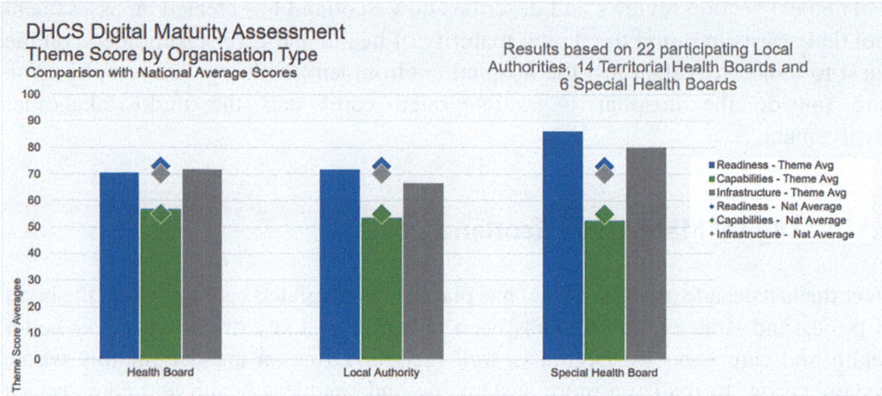

Fig. 1 Scotland Digital Maturity Assessment summary (Digital Health & Care Scotland 2022)

Government 2023). These funds are allocated to meet the needs of the Scottish population,[8] which was 5.4 million in 2022, and with the integration policy, the funds can be utilized across the entire health and care system.

In 2019, Scottish Government undertook the first national DMA survey to establish a Scotland-wide national integrated health and care baseline. This was offered to all "14 territorial Health Boards; 6 Special Health Boards and 22 Local Authorities" (Local Authorities/Council are referred to as municipalities in EU countries) who were asked to complete the assessment (it should be noted that not all organisations took part). These findings were summarized in the annual report (Fig. 1). The results showed that the readiness rating achieved the highest score, closely followed by infrastructure with the capabilities the least mature at that point in time.

Although more recent (2022) results are showing progress, the capabilities within organizations are creating challenges that are slowing down the uptake of digital innovation within the whole system and across the integrated care model. This factor is regarded as key to creating capacity for hospitals to fully optimize performance and shift the balance of care to more sustainable services enabling users to get ready for change.

The various strategies for Scotland, including the digital strategy,[9] digital health and care strategy,[10] and health and social care data strategy,[11] highlight the

[8] Scotland Census 2022—https://www.scotlandscensus.gov.uk/news-and-events/scotland-s-census-first-results/

[9] A changing nation: how Scotland will thrive in a digital world—https://www.gov.scot/publications/a-changing-nation-how-scotland-will-thrive-in-a-digital-world/pages/foreword/

[10] Digital health and care strategy—https://www.gov.scot/publications/scotlands-digital-health-care-strategy/

[11] Health and social care: data strategy https://www.gov.scot/publications/data-strategy-health-social-care-2/

importance of integrated care, person-centered care, and the use of digital technology and data in community settings. These strategies aim to create a more personalized, effective, and efficient integrated service, with the ambition of reducing hospital admissions and providing sustainable and convenient services for the future, facilitated by digital advancements and increased maturity.

Further to the above, the following section attempts to ascertain the difference between the digital maturity of established hospital systems and the readiness for innovation within the wider health and care system. **Disruptive innovations must include the workforce, patients, and those who are designing the new technical systems and infrastructures.**

3.2 Digital Maturity and the Relationship with Innovation Readiness

Digital technology is widely acknowledged as a crucial facilitator for integration, and it is also acknowledged that the process of reaching maturity in this area is not always straightforward. Comprehending a digital baseline position is widely recognized as the initial stage at a large-scale level. However, there is a need for more detailed analysis at an individual level in order to determine the readiness for implementation and expansion.

Digital maturity most often focuses on mature systems that are self-evidently already well established. Health 4.0 shifts the focus onto streamlining hospital settings and practices to remote settings; however taking Industry 4.0 principles, such as Care 4.0,[12] it is much more about placing the person and community at the center. The Care 4.0 aspiration emphasizes the importance of community-led digital health and care innovation that is based on Industry and Health 4.0 capabilities (Li and Carayon 2021) with integration imperative.

The implementation and integration of new emerging systems and skills introduced by work practice redesign must embrace Care 4.0; this requires stakeholders to have time to get ready. While these integrated systems and skills are not yet fully developed, there is a recognition and anticipation that these external innovations and new service models, especially those that are primarily delivered outside of hospital premises, will ultimately need to be integrated into the hospital systems and infrastructures to achieve comprehensive system-wide advantages. This change presents the opportunity to establish a more cohesive healthcare system, but it is often at such an immature stage and needs robust groundwork and preparation to surface the full benefits, and therefore overlooked as part of the baselining assessment activity within maturity frameworks.

[12] Chute C, French T (2019) Introducing Care 4.0: An Integrated Care Paradigm Built on Industry 4.0 Capabilities. International journal of environmental research and public health 16(12), 2247. https://doi.org/10.3390/ijerph16122247

To conclude, the relationship between digital maturity and innovation readiness remains loose and, in some cases, invisible to the normal DMM/DMA process. The gap between early innovations emerging and being considered under normal DMM and DMA is not well accommodated and, in many cases, and would seem not to be on the radar of those completing the assessment process.

3.3 Digital Maturity Is More Than Technology and Mature Systems

The conventional path towards achieving digital maturity typically adopts a narrow technological viewpoint (Cresswell et al. 2019a). Various tools for assessing digital maturity generate an organizational maturity score, which is determined through self-reflection and often answered and aggregated by senior managers. This approach tends to prioritize macro-level considerations, while paying less attention to service innovation, new work methodologies at the micro individual level, and the preparedness for change, as well as the necessary steps to be taken to make it ready to adopt breakthrough innovations.

The most cited literature related to digital maturity is focused on hospital systems like the electronic health record (EHR). While there are other significant hospital systems to consider, EHR usually takes center stage (Duncan et al. 2022). Of course, it is well understood that the EHR is only one main central system; the point is that most emerging innovations and systems are not considered when EHR is assessing an organization's digital maturity level (macro). "When contemplating the micro level, it is crucial for all involved parties to be prepared to utilize the digital functionality and features if they prove useful and a strong case can be made. However, there is often limited consideration of human factors in this process."

However, it is necessary for organizations to understand and embrace how these elements are integrated into new work methods, service pathways, and stable hospital systems. Established systems will often be the backbone of a service; integrating any novel community led innovation into already well-embedded systems tends to make it easier to gain clinical buy-in and leadership and prepare people more generally to make best use of the tools.

3.4 Digital Maturity and Readiness for New Emerging Innovations

Despite the common belief that digital maturity is increasing, previous publications and chapters have shown that there are various groups of stakeholders and individuals who still need to be better prepared and convinced fully to embrace new digital-enabled ways of working. Furthermore, it has been found that when the system is ready in respect of a particular innovation, only then can it transfer into a state of maturity and expand, integrating with other well-established systems, e.g. digitally mature hospital environments.

The next section highlights Scotland's role in advancing cutting-edge DHI (prior to it reaching business as usual status) and driving the expansion of innovation within the sector. Nikkhou et al. (2016) defined maturity as the state of being in a "flawless condition, typically at a macro level." On the other hand, they described readiness as an uncountable unique concept that can be argued to be more at a micro level.

According to the Oxford Dictionary, the term "readiness" refers to the "state of being ready and willing to do something," particularly in terms of being "open to doing things differently or working in a different way." However, it does not necessarily imply that one is fully mature to implement at this stage, and that people must prepare for, and get ready for the change. Put simply, this is primarily focused on making people confident and creating the conditions and a level of readiness to allow all involved to gain a greater understanding. To support the influencing or championing of a new system that then can be implemented as business as usual, with maturity and improvement to then follow.

3.5 Readiness Innovation Journey Case Studies

The UK Future Commission defines "digital health futures readiness as when all people and their communities, the health ecosystems they interact with, and the countries they live in are prepared, equipped, and empowered to use digital technology and data to meet personal health and wellbeing needs" (Governing Health Futures 2030 Commission 2021). According to other prominent literature, the process of innovation in digital health involves various aspects and actors that come together as new services are integrated between the interfaces of hospitals and into the broader communities.

Due to the limitations noted in the last section to early breakthrough innovation being assessed or not properly considered in relation to existing DMM/DMA. Scotland's Digital Health & Care Innovation Centre has utilized an innovation readiness level framework for related DHI research and development projects. This is used at a project level to describe the readiness movements and covers three main innovation strands that allow DHI projects to map advancements. The three innovation strands are technical, service, and business readiness with nine different levels[13] (Fig. 2), this approach mirrors the well-established technology readiness levels (TRL) introduced originally by NASA in the 1970s.[14]

The heuristic readiness innovation level framework employs a logical approach, utilizing triggers, active learning through co-design, simulation, and the activation of real-world evidence (RWE) at a certain level to encourage, in a controlled manner

[13] DHI readiness framework—https://www.dhi-scotland.com/learning/resources/scaling-digital-health-innovation-developing-a-new-service-readiness-level-framework-of-evidence/

[14] Technology Readiness Levels. NASA—https://www.nasa.gov/directorates/somd/space-communications-navigation-program/technology-readiness-levels/

Technical readiness levels	Service readiness levels	Business readiness levels
TR9 – Live implementation proven	SR9 – Service change implemented	BR9 – Commercial sale
TR8 – system complete/qualified	SR8 – Develop Case for scale	BR8 – Reference site, real world test - business model accepted
TR7 - working model demonstrated	SR7 – Evaluation and Evidence concluded	BR7 – procurement route/framework – clarified
TR6 - fully functional prototype	SR6 – Change pilot test - RWE/LL	BR6 – Regulation and Standard check (CE/FDA/MDR/IG/SSP)- interoperable
TR5 - rigorous testing undertaken	SR5 – Future state accepted in principle - Simulated to de-risk	BR5 – Business model review
TR4 – technical validation	SR4 - Future state options co-designed	BR4 – Product fit, tested and adapted and made interoperable etc....
TR3 - proof-of-concept constructed	SR3 – Current state understood/accepted	BR3 – Business plan for industry - developed
TR2 - basic principles studied	SR2 – Market/Gap analysis; best practice (hypothesis dev)	BR2 – Market size and strategy reviewed
TR1 - scientific research (defined)	SR1 – Demand – needs analysis	BR1 -Business idea (defined)

Fig. 2 Technical, service, and business readiness level framework

the innovation enabled by digital to progress to adoption and scale if evidence proves to be positive. The goal is to enhance the overall readiness and maturity of all partners within the care continuum, primarily outside the hospital settings, where possible, and to include all stakeholders including industry partners that are developing the new emergent innovations to reduce adoption and scale friction.

The overarching aim of the framework is focusing on how you move collaboratively and in a considered fashion to the endpoint: adoption, scale, and ultimately the innovation being transferred into BAU. This section demonstrated through key case studies using the technical, service, and business readiness level innovation framework (Fig. 2) to illustrate and describe the readiness shifts that took place.

This framework outlines the progress made of three innovation projects (led by Scotland in partnership with a variety of multi-disciplinary teams) over a given timeframe. It emphasizes how readiness innovation levels shifted, preparing partners to develop innovations and work collaboratively at a project level. The desired result from such projects is that emerging innovations can be evidenced and be ready for the integration into the mature and stable hospital environments and ecosystems when and if necessary.

3.5.1 Case Study #1: The Scottish Capsule Programme (SCOTCAP)

The incidence of bowel cancer is increasing worldwide, putting a strain on traditional endoscopy (endoscopists) services. New innovations are emerging to offer an alternative diagnostic test utilizing Colon Capsule Endoscopy (Pill Cam). This coupled with a shortage of qualified cancer care specialist professionals is exacerbating the demand for hospital services, particularly in departments responsible for cancer investigations. Cancer care and early diagnostics are considered a top priority in most Western healthcare systems. An effective solution centers around improved access to early diagnostic testing to enable quicker interventions to be activated. Research has demonstrated early identification leads to higher survival rates and improved outcomes. The incidence of bowel cancer in Scotland is

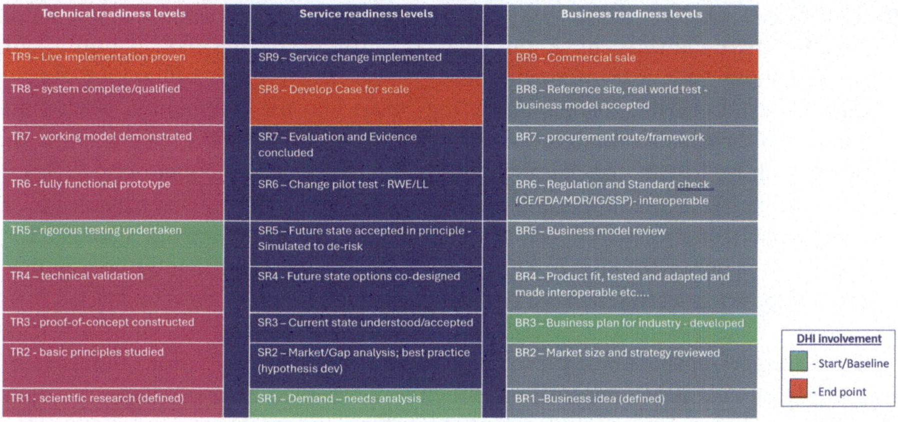

Fig. 3 SCOTCAP innovation project: innovation readiness shifts

increasing, leading to excessively long waiting periods for patients seeking diagnosis and treatment. This innovative project was aimed at enabling earlier diagnosis of bowel cancer symptoms by shifting part of the diagnostic screening service from a conventional, well-established, mature hospital setting to a new patient-centered remote model located within the community.

During the development of the new community care model in Scotland, numerous important stakeholders embarked on a readiness journey to test and establish new service pathways for delivering and testing a comprehensive service offering.

If we examine how the readiness levels shifted (Fig. 3) for this innovation project, we can summarize that the technical readiness focused on aligning and integrating new and legacy systems and testing these to ensure it was acceptable, technically feasible, and worked in practice with more mature established hospital systems (TR5 toTR9). With respect to service readiness, current and future pathways mapping was undertaken, followed by the redesign and integration of new SCOTCAP pathways[15] into the well-established endoscopy hospital systems (SR1 to SR8). This was crucial to the new service to generate interest, acceptance, and trust and to make it more scalable for wider adoption. The business readiness element focused on a review and support for a new business model for industry and initiation of an innovation procurement partnership which allowed faster routes to procurement – https://www.gov.scot/publications/scottish-government-procurement-annual-report-2020/pages/3/ - BR6 (BR4 to BR9).

The project partners highlighted the key opportunities[16] in an article published in 2022:

---

[15] SCOTCAP service model codesigned—https://futurehealthandwellbeing.org/scotcap-service-model-evaluation-report

[16] The Royal College of Surgeons Edinburgh—https://www.rcsed.ac.uk/news-public-affairs/the-rcsed-blog/2022/may/the-future-of-colon-capsule-endoscopy

- CLINICAL LEAD—"The Colon Capsule Endoscopy, which is a pill cam, contains two cameras in a pill that is swallowed. It's not painful, and complications are exceptionally rare. Importantly, this technology takes diagnostics out of the hospital and brings them into the community. The colon capsule is highly acceptable. When asked in the SCOTCAP study, 90% of patients chose to take the colon capsule rather than having colonoscopy (Bond et al. 2023). So, if I look into the future. I could see those low-cost capsules (…) smart, maybe automated reporting of the images could be a game changer for colorectal cancer. Because if we can spot the precursors, the little polyps that grow into cancer early, then we can stop people developing bowel cancer."

The Scottish Capsule innovation project started with a shared vision, enthusiasm, and strong clinical leadership. Effort and early work focused on creating the right conditions to test and iterate with stakeholders the new novel service model.

- INNOVATION PROGRAMME MANAGER—"Groundbreaking in terms of their approach we've taken around innovation partnership. And for me, it's about the collaboration, and it's really allowed us to think differently, bring together industry, bring together innovation centers, bring together NHS partners. (…) I think it allowed us to think out of the box a bit in terms of how we deliver something that's transformational in Scotland. And we've developed strong collaboration and strong working relationships." It is well recognized that the business readiness needs to be fully considered in any digital innovation program. However well documented, you can often end up with an innovation with no home.
- INTERNATIONAL INDUSTRY PARTNER—"If you have a product or service-based innovation that doesn't fit neatly into an existing box, you can't really go through a standard first past the post, buy a widget, procurement process. You need an innovative procurement process, or a competitive dialogue so both sides can identify and realize the value that you're trying to deliver."

The project summary from the organization leading the transformation stated that

- SENIOR SPONSOR—"It's all about innovation. It's about doing things differently, but not just for the sake of doing things differently because we need to make things better."

Projects of this nature necessitate the active participation of multiple stakeholders in the innovation program and readiness process. This encompasses individuals with expertise in clinical leadership, service design, and system integration, including industry collaborators specializing in digital interventions. By combining their efforts with an agile approach, they developed a novel service and business model that provided remote diagnostics, enabling more comprehensive and accessible healthcare for underserved communities. The readiness framework illustrates the shifts and steps made to create the conditions of acceptance and upskilling along

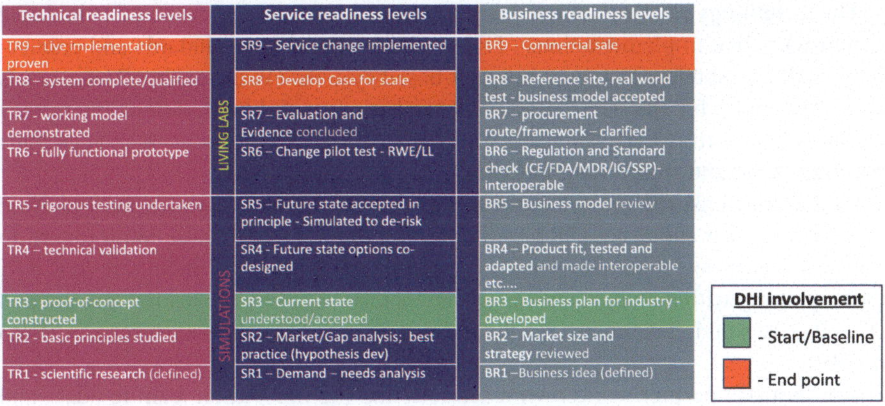

Fig. 4 RDS innovation project: innovation readiness shifts

with joint working and ownership to link the system back into the mature hospital services and systems.

3.5.2 Case Study #2: Right Decision Service (RDS)

The second example is a platform called the Right Decision Service (RDS) that uses digital technology to assist professionals and citizens in making decisions about health and care. The latest innovation strategy in Scotland emphasized the necessity of restructuring public services to address key difficulties and enhance direct interaction, while minimizing inconsistency, through the provision of digital tools to aid more consistent decisions. In addition, the OECD stated that up to 20% of healthcare is of no clinical value, as decisions about interventions are not based on current evidence (Limb 2017).

From the outset, the Right Decision Service[17] was designed with the ambition to scale, positioning it as a "Once for Scotland" service, clearly defining its role as a nationwide innovation project. It is an online source of healthcare digital tools, allowing people to make safe decisions about health and care quickly, and importantly is based on validated evidence. The innovation project objectives were to give health and social care organizations tools to build decision-ready guidance and pathways, risk-scoring tools, shared decision aids, and other decision support resources and by doing so create capacity and better outcomes within the whole system.

The innovation journey to getting the system, service, and business model ready for adoption took over 7 years to reach a level of overall readiness for general adoption and scale. The framework demonstrates the readiness movements over the innovation projects lifetime (Fig. 4).

---

[17] Right Decision Service—https://www.dhi-scotland.com/media/0j5gd50k/brochure-rds-final.pdf

The technology readiness moved from an early proof-of-concept stage to multiple system-wide live implementations (TR3 to TR9) using open standards and connectivity back into stable and mature hospital-based systems (Wales 2020). These levels were influenced by the service activity work that took place within the service readiness levels, from early-stage horizon scanning to confirm the need and demands to current state and future state mapping.

Evaluations ensured the system was safe and effective, which helped build the case for adoption and scale (SR3 to SR8) (Cresswell et al. 2019b). With regard to business readiness, regulations and standards were achieved including UK CA marking along with advanced business models influenced by the testing and iterated using the user and stakeholder feedback from the service innovation activity (BR3 to BR9).

The industry partner worked with services and the innovation team to adapt and revise the business model and achieved a state of readiness with a reference site that covers all of Scotland. This has enabled industry to get feedback and testimonials for collaborative design, leading to the development of over 180 decision support toolkits, supporting 37 health and social care organizations. The process of scaling up these toolkits has been transferred to the national NHS Board Healthcare Improvement Scotland, which leads on evidence and improvement. Furthermore, the RDS has successfully obtained UK CA marking for 12 decision support tools. Several organizations are also integrating the Right Decision Service with digital clinical systems. This includes issuing patient-specific prompts and alerts within electronic health record systems to alert the clinician to address high-risk situations for that patient. It also includes use of a mature clinical document management system (DOCMAN), which captures PDF outputs from decision support tools and makes these available from individual patient records.

The system has now scaled to an annual user base of approximately 1.7 million, as of 2023 statistics, with encouraging feedback that "79% of RDS Leads indicate that RDS tools are already improving patient safety in their organizations and 74% advise that RDS tools have improved quality of patient care, with 65% indicate that they are driving up consistent evidence-based practice."[18]

In 2023, this innovation project was deemed ready to graduate into the next stage of maturity, moving to increased scale within Scotland.[19]

3.5.3 Case Study #3: NWE.CHANCE—Hospital-at-Home (H@H)

This case study demonstrates that all innovations shift their readiness, but do not always get adopted at the end of the readiness process. This European Northwest Europe (NWE) Interreg project involved multiple stakeholders and countries with a focus on innovating for heart failure patients and testing a new digital service model. The project involved multiple stakeholders from the Netherlands and Belgium, in

[18] Wales et al—https://rightdecisions.scot.nhs.uk/

[19] National roll out of RDS—https://www.dhi-scotland.com/news/national-roll-out-for-the-right-decision-service/

collaboration with Scotland and England. The main objective was to assess the system's readiness to implement and develop new care models that could be integrated into hospital operations to deliver a more effective and efficient service and at the same time improve the technical readiness by combining a range of innovations into one single platform.

Even before COVID-19 arrived there were waiting time difficulties for heart failure patients in most Western societies (Ambrosy et al. 2014), and it became evident that the conventional service offering was exacerbated under COVID restrictions (period 2020–2022). Established hospital systems were unable to provide regular services within the hospital settings, as the system aimed to minimize hospital visits and appointments. Once again, like numerous developing nations, the concept of hospital-at home, also known as "Virtual wards" in the UK NHS England, generated expectations of revolutionizing hospital services beyond the confines of traditional hospital walls, with heart failure (HF) patients an interesting cohort (Haywood et al. 2023). This European H@H project consisted of eight partners, including university hospitals, commercial companies, and communication advice firms. In intricate multi-dimensional projects, it is crucial to have a clear understanding of the necessary activities to advance the project to a more advanced and developed state. This project used the Scottish readiness level framework.

The H@H project developed a platform that involved multiple stakeholders. The project started with a low technical readiness level (TR3) and aimed to generate a proof of concept. Over the course of 3 years, the partners successfully created and implemented a functional prototype. This model was showcased and evaluated at three separate locations, including more than 100 participants. The advancement in technology readiness was closely linked to the degree of service readiness. This shift occurred from the stage of being a future concept and option (SR4) to being tested and evaluated, resulting in the creation of early proof. As a result, the service readiness level progressed to level 6 (SR6). The market access plan and business models were examined, enabling commercial partners to deliberate on whether to commercialize the individual components or adopt the platform as the preferred business model for commissioners.

This activity facilitated the preparation of all stakeholders for the implementation of the innovation, by assessing the readiness of commissioners to adopt and utilize this new care model. At this juncture, the project successfully demonstrated the effectiveness of the technology. However, the service model encountered difficulties in terms of resource utilization and the reimbursement models. The perspective of both the commissioner and the insurer was that the iteration of this project proved the technology, reporting an integrated platform that offered training, home access, and integration with the hospital system dashboard. Despite confirming the patient feedback being positive, and the market potential being set to grow, the initiative encountered financial challenges.

Ultimately, despite the promising nature of technology, the business and service innovation model proved to be more expensive with the inclusion of highly paid critical care staff as part of the service model. Overall, the entire system, including staff resources and insurers, was ill-prepared and not ready for an innovation in that

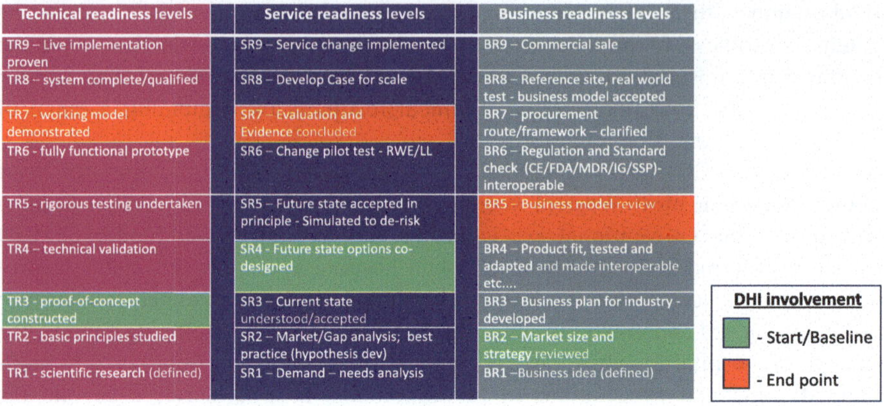

Fig. 5 NWE-Chance innovation project: innovation readiness shifts

format. Despite the hospitals being digitally mature, the readiness for this innovation was still underdeveloped and incurred significant expenses that could not be reconciled as a business case.

The graphic (Fig. 5) illustrates the changes that occurred during the project duration of the Interreg (NWECHANCE) project. This project facilitated the advancement of activities pertaining to a novel technical product and integration plan, H@H service pathway, and refinement of the business model. Additionally, it provided suggestions on the subsequent actions to be taken and the market access for potential commercial partners. The related shifts can be summarized as follows—technical readiness levels (TRL) (from TR3 to TR7), service readiness levels (SRL) from SR4 to SR7, and business readiness levels (BRL) from BR2 to BR5. This tool assisted our European partners in communicating the activity and progress and facilitated the realization that all partners were creating resources and evidence that makes them more ready to scale innovation for digital health and care innovation. This project did not mature to a state of readiness for a value case to be developed as the evidence was not clear that the innovation was feasible or could be sustained to bring service/cost benefits.

These three case studies illustrate the readiness journey outside mature digital systems and demonstrate that innovation readiness for new and emerging projects needs to be better considered in linking with the existing DMM and DMA toolsets.

4 Conclusion

This chapter discussed how digital maturity and readiness in relation to emerging digital health innovation are different and can't be easily assessed at both the macro and micro level using the common existing DMM and DMA tools. This chapter has used case studies to discuss and illustrate how innovation projects progressed their readiness levels, to demonstrate the multifaceted nature of the need to converge technical, service, and business innovation to create value cases for adoption and scale, leading to then mature systems with a level of fidelity.

The chapter also points out that readiness within existing DMM/DMA needs to encompass a greater inclusive aspect for innovation more generally. More emphasis is required on emerging technologies, innovation, and considering the integrated health and care agenda ambitions, along with citizens' preferences.

To promote maturity and innovation readiness in community and integrated settings outside of the hospital it is necessary for individuals and key community Health and Care partners, such as patients, carers, insurers, or charities to undergo a process of upskilling, preparing to get ready to adopt new innovations into service delivery. Additionally, new business and reimbursement models cannot be ignored if scale is to be achieved. As previously indicated, innovative systems that prioritize individual needs over established healthcare models and operate outside of the usual service models must be prepared to adapt and be agile to reap whole system benefits.

This chapter highlighted some of the more common DMM/DMAs that are being utilized worldwide. In addition, it flagged the importance of testing and the development of more integrated health and care DMM/DMA models to create baselines that appreciate the maturity, gaps, innovation opportunity, and skill challenges when dealing with integrated health and care innovations, and how these models remain inconsistent.

Creating the conditions for innovation to land well when ready is paramount, and the overall digital maturity of an organization will have a bearing on the successful transition into business as usual; however, the assessment of whether the innovation is ready to be fully appraised under current DMMs remains challenging.

There is therefore an opportunity to further research and develop DMM and DMA tools to better adapt to the new emerging innovation projects and disruptive technology needs. This adaptation would allow for the creation of a bridge to maturity backed by a community of champions that are ready with the skills and willingness to use in everyday business. Furthermore, all stakeholders need to be prepared for change, and this needs to have a heavier focus with regard to enhanced maturity models (in relation to innovation projects). In many countries, the 'balance of care shift' with the overall goal to better prevent, predict, and create more personalized care outside the hospital walls using emerging technologies could offer a future that is more sustainable, effective, and efficient.

References

Abernethy A, Adams L, Barrett M et al (2022) The promise of digital health: then, now, and the future. NAM Perspect 2022. https://doi.org/10.31478/202206e

Ambrosy A, Fonarow G, Butler J et al (2014) The global health and economic burden of hospitalizations for heart failure: lessons learned from hospitalized heart failure registries. JACC 63(12):1123–1133. https://doi.org/10.1016/j.jacc.2013.11.053

Bond S, Kyfonidis C, Jamieson M et al (2023) Evaluation of an innovative colon capsule endoscopy service in Scotland from the perspective of patients: mixed methods study. J Med Internet Res 25:e45181. https://doi.org/10.2196/45181

Cresswell K, Sheikh A, Krasuska M, Heeney C, Franklin BD, Lane W, Mozaffar H, Mason K, Eason S, Hinder S, Potts HWW, Williams R (2019a) Reconceptualising the digital maturity of health systems. The Lancet Digital Health 1(5):e200–e201. https://doi.org/10.1136/bmjhci-2019-100022

Cresswell K, Callaghan M, Mozaffar H, Sheikh A (2019b) NHS Scotland's decision support platform: a formative qualitative evaluation. BMJ Health Care Inf 26(1):e100022

Digital Health & Care Scotland (2022) Digital maturity assessment results summary. https://www.digihealthcare.scot/app/uploads/2022/03/Digital-Maturity-Assessment-Final-Summary.pdf. Accessed 1 July 2024

Dimitrios Kafetzopoulos Evangelos Psomas (2015) The impact of innovation capability on the performance of manufacturing companies: the Greek case. J Manuf Technol Manag 26(1): 104–130

Duncan R, Eden R, Woods L et al (2022) Synthesizing dimensions of digital maturity in hospitals: systematic review. J Med Internet Res 24(3). https://doi.org/10.2196/32994

Governing Health Futures 2030 Commission (2021) Policy brief: Digital health futures readiness. Geneva. https://www.governinghealthfutures2030.org/pdf/policy-briefs/DigitalHealthFuturesReadiness.pdf. Accessed 1 July 2024

Haywood HB, Fonarow GC, Khan MS et al (2023) Hospital at home as a treatment strategy for worsening heart failure. Circ Heart Fail 16(10):e010456. https://doi.org/10.1161/CIRCHEARTFAILURE.122.010456

Health Tech World (2023) Few NHS trusts have reached top levels of international HIMSS digital maturity benchmark. https://www.htworld.co.uk/news/few-nhs-trusts-have-reached-top-levels-of-international-himss-digital-maturity-benchmark-sb23/

Hughes J, Lennon M, Rogerson R, Crooks, G (2021) Scaling digital health innovation: developing a new service readiness level framework of evidence. Int J Environ Res Public Health 18(23): 12575. https://doi.org/10.3390/ijerph182312575

Ladu L, Koch C, Ashari PA, Blind K, Castka P et al (2024) Technology adoption and digital maturity in the conformity assessment industry: empirical evidence from an international study. Technol Soc 77:102564. https://doi.org/10.1016/j.techsoc.2024.102564

Li J, Carayon P (2021) Health care 4.0: a vision for smart and connected health care. IISE Trans Healthc Syst Eng 11(3):171–180. https://doi.org/10.1080/24725579.2021.1884627

Limb M (2017) A fifth of healthcare spending is wasted, says OECD report. BMJ 356:j215. https://doi.org/10.1136/bmj.j215

Nikkhou S, Taghizadeh K, Hajiyakhchali S (2016) Designing a portfolio management maturity model (Elena). Procedia Soc Behav Sci 226:318–325. https://doi.org/10.1016/j.sbspro.2016.06.194

Norman G, Bennett P, Vardy ERLC (2023) Virtual wards: a rapid evidence synthesis and implications for the care of older people. Age Ageing 52(1):afac319. https://doi.org/10.1093/ageing/afac319

Rimpiläinen S, Bosnic I, Savage J (2024) *How to navigate the digital shift in healthcare? An international review and analysis of frameworks used to support digital working by frontline healthcare staff.* Digital Health & Care Institute, Glasgow.

Schlieter H, Marsch L, Whitehouse D et al (2022) Scale-up of digital innovations in health care: expert commentary on enablers and barriers. J Med Internet Res 24(3):e24582. https://doi.org/10.2196/24582

Tarhan AK, Garousi V, Turetken O et al (2020) Maturity assessment and maturity models in health care: a multivocal literature review. Digital Health 6:1–20. https://doi.org/10.1177/2055207620914772

The Scottish Government (2023) Scottish budget 2024-25. https://www.gov.scot/publications/scottish-budget-2024-25/documents. Accessed 1 July 2024

The Scottish Government (n.d.) Health and social care integration. https://www.gov.scot/policies/social-care/health-and-social-care-integration/

Wales A (2020) Decision support for Scotland's health and social care: learning from an outcomes-focused approach. BMJ Health Care Inform 27:e100124. https://doi.org/10.1136/bmjhci-2019-100124

Whitehouse D, Lange M (2022) A services readiness levels stage model: a roadmap. HealthManagement.org The Journal 22(2):63–66. https://healthmanagement.org/uploads/article_attachment/hm2-servicesreadinesslevelsstagemodelf.pdf

Zhao Y, Schalet D, Alsalamah S et al (2023) From assessment to action: exploring the dynamics between maturity assessments and strategy implementation in digital health. Stud Health Technol Inform 305:257–260. https://doi.org/10.3233/SHTI230477

The Transformative Impact of Artificial Intelligence on the Digital Maturity of Hospitals

John Gachago

Abstract

Artificial intelligence (AI) is revolutionizing the digital maturity of hospitals in the USA across multiple domains, including clinical operations, quality and safety, and consumer engagement. This chapter explores how AI integration enhances digital maturity, particularly through tools such as machine learning, deep learning, and natural language processing, impacting both clinical and administrative processes. The author emphasizes the use of AI in clinical decision support, medical imaging, and personalized medicine, showcasing its role in improving diagnostic accuracy and operational efficiency. Additionally, the chapter discusses the importance of standardized digital maturity frameworks and introduces new AI-specific maturity models, such as the proposed Five-point Scale for hospital AI Implementation; the ADAM Framework which parallels the Society of Automotive Engineers' framework for autonomous vehicles; the proposed five-point scale for hospital AI implementation; Clinical AI Sociotechnical Framework (CASoF); and the Organizational Perspective Checklist for AI (OPTICA) Solution Adoption. These scales provide alternative structured approaches for evaluating and implementing AI tools, ensuring they are integrated safely and effectively into healthcare settings.

J. Gachago (✉)
TTUHSC Institute of Telehealth and Digital Innovation, Lubbock, TX, USA

IFC/World Bank, Washington, DC, USA

National Institutes of Health, Bethesda, MD, USA

American College of Education, Indianapolis, IN, USA

California Baptist University - Strategic Management of Healthcare Organizations and Healthcare Finance & Reimbursement, Riverside, CA, USA

E-Health Solutions LLC/AfriDOKTA, Gainesville, FL, USA
e-mail: john.gachago@ttuhsc.edu

A. Scheuer, J. Studzinski (eds.), *Digital Maturity in Hospitals*,
https://doi.org/10.1007/978-3-031-80704-6_16

The chapter also highlights challenges related to data quality, ethical considerations, and the need for robust governance frameworks. By examining case studies and proposing strategies for overcoming these barriers, the study provides actionable insights for hospitals aiming to achieve full digital maturity through AI adoption, ultimately transforming them into intelligent health systems.

Keywords

Artificial intelligence (AI) · Digital maturity · Clinical decision support · Telehealth · AI ethics in healthcare

Abbreviations

ADAM	AI development and maturity framework
AIA	Artificial Intelligence Act (European Union)
CASoF	Clinical AI sociotechnical framework
CDSS	Clinical decision support systems
CNN	Convolutional neural network
DL	Deep learning
DR	Diabetic retinopathy
EHDS	European Health Data Space
EHR	Electronic health record
HAI	Human-Centered Artificial Intelligence (Stanford University)
HCP	Healthcare professional
HIS	Health information system
HSO	Health service organization
IoT	Internet of Things
ML	Machine learning
MIT	Massachusetts Institute of Technology
NLP	Natural language processing
OPTICA	Organizational Perspective Checklist for AI Solutions Adoption
OR	Operating room
ROP	Retinopathy of prematurity
SAE	Society of Automotive Engineers
SAMD	Software as a medical device
TDM	Therapeutic drug monitoring
UTAUT	Unified Theory of Acceptance and Use of Technology

1 Introduction

The integration of artificial intelligence (AI) in hospitals in the USA is having a significant impact on the digital maturity of hospitals across nine main domains: continuity of care, network and market insights, clinical operations, clinical analytics, quality and safety, value-based care, reimbursement, corporate functions, and consumer engagement. Examples of AI applications in clinical operations include optimizing operating room (OR) schedules and supply chain management. In clinical analytics, AI aids in decision support and treatment recommendations, particularly within radiology. Quality and safety improvements are driven by AI use cases such as early detection of sepsis and prediction of adverse events (Sahni et al. 2023).

Administrative domains such as reimbursement and corporate functions show advanced AI adoption due to their repetitive, manual nature, which AI can streamline effectively (Fig. 1). However, domains such as quality and safety are less mature in AI adoption due to the complexity of creating robust business cases that consider non-financial benefits such as patient experience and access to care (Sahni et al. 2023).

- **Case Study—Corporate Functions:** During the COVID-19 pandemic, a hospital's call center experienced increased call volumes, leading to longer wait times and dropped calls. By implementing an AI-powered virtual agent, the hospital reduced call volume by 30%, improved patient experience, and redeployed staff to more complex tasks.
- **Case Study—Clinical Operations:** A regional hospital used AI to optimize its operating room (OR) block scheduling. Historical data and trends were analyzed to better allocate time slots, increasing OR utilization from 60% to higher efficiency. This change allowed for better management of critical patient cases and improved overall surgical volumes (Sahni et al. 2023).

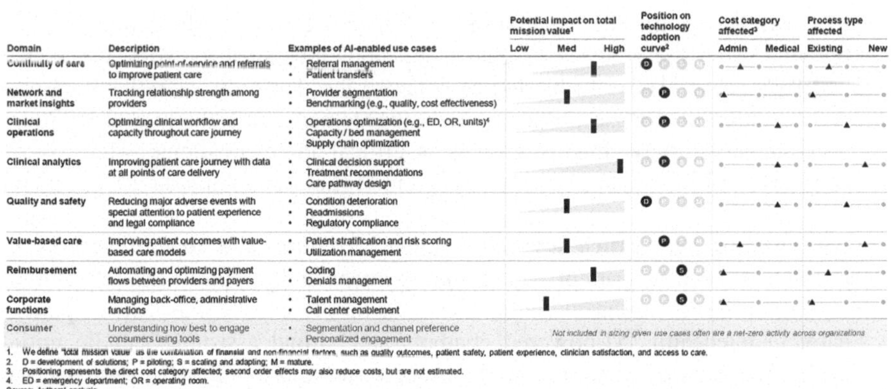

Fig. 1 Hospital AI domains and example use cases

When one looks at AI from an expanded view that encompasses various techniques such as machine learning (ML), deep learning (DL), and natural language processing (NLP) and generative AI, we begin to experience its transformative influence in hospitals and other health service organizations (HSOs) with substantive impact on clinical and administrative processes. AI has become integral to modern healthcare, particularly in telehealth, where it enhances efficiency, accuracy, and accessibility of medical services yet it also has its challenges. Specific areas that we see this digital maturity with beneficial impact include (Alowais et al. 2023):

1. **AI in Clinical Decision Support:** AI algorithms are crucial in diagnosing and predicting diseases. They analyze vast medical data to identify patterns, leading to early detection of diseases like cancer and heart disease. AI's continual learning improves its diagnostic precision, vital for managing chronic diseases. AI techniques like machine learning, deep learning, and natural language processing each contribute uniquely to these advancements. Some key case studies include Cancer Detection, where AI models, such as those developed by Google Health, have shown superior accuracy in detecting breast cancer from mammograms compared to human radiologists; Diabetes Management (machine learning algorithms predict diabetes complications, improving patient outcomes); Heart Disease Prediction (AI models accurately predict heart disease severity, aiding in timely and effective treatment); and Neurological Disorders (AI enhances the diagnosis and prognosis of conditions like Alzheimer's and Parkinson's through advanced imaging and data analysis techniques) (Varnosfaderani and Foruzanafar 2024).
2. **Prognosis and Diagnosis:** ML and DL models are increasingly being used to predict disease outcomes and assist in early diagnosis. For example, these models have shown high accuracy in identifying and classifying cancers, predicting disease progression, and determining patient survival rates. The review highlights the use of ML in analyzing electronic health records (EHRs) and medical images, where AI systems can efficiently extract valuable clinical features and improve diagnostic accuracy (Lalli and Amutha 2020).
3. **Medical Imaging:** The integration of ML and DL in medical imaging has revolutionized the analysis of data from modalities such as MRI, CT scans, and PET scans. These technologies enhance the detection, classification, and segmentation of diseases, assisting radiologists and clinicians in making more accurate and timely diagnosis. This is critical in telemedicine where rapid diagnosis and treatment planning are essential (Lalli and Amutha 2020; Pallavi et al. 2022).
4. **Automation and Efficiency:** AI automates routine and repetitive tasks in healthcare, such as patient record management and appointment scheduling, freeing up healthcare professionals to focus on more critical aspects of patient care. In telehealth, AI-powered chatbots and virtual assistants handle initial patient interactions, streamlining the process and improving patient satisfaction (Pallavi et al. 2022).

5. **Real-Time Monitoring and Treatment:** AI-driven systems are also applied in real-time health monitoring, utilizing wearable devices and IoT sensors. These systems collect and analyze patient data continuously, providing insights that help in managing chronic conditions and improving treatment outcomes. Wearables track vital signs and detect health issues early, while virtual assistants offer 24/7 patient support and improve treatment adherence. AI also enhances telemedicine, expanding healthcare access and enabling remote patient engagement (Lalli and Amutha 2020; Varnosfaderani and Foruzanafar 2024).
6. **AI in Personalized Medicine:** AI tailors medical treatments to individual genetic profiles, improving drug development and customizing treatment plans. AI analyzes genomic data to identify specific markers for targeted therapies, making treatments more effective and reducing adverse reactions. AI also optimizes drug development by predicting patient responses, significantly accelerating the process (Varnosfaderani and Foruzanafar 2024).
7. **Clinical Laboratory Testing:** AI improves the accuracy, speed, and efficiency of laboratory processes, particularly in clinical microbiology. ML systems detect and identify microorganisms, classify diseases, and predict clinical outcomes using genomic data, gene sequencing, and microscopic imaging. AI's role in automating and enhancing laboratory testing contributes significantly to efficient healthcare delivery (Alowais et al. 2023).
8. **AI in Genomic Medicine:** AI's integration with genotype analysis aids in disease surveillance, prediction, and personalized medicine. ML algorithms analyze genomic data to identify genetic markers associated with disease susceptibility, enabling early detection and personalized treatment plans. AI facilitates the identification of novel therapeutic targets and the repurposing of existing drugs, accelerating drug discovery and development (Alowais et al. 2023).
9. **Precision Medicine and Clinical Decision Support:** AI supports personalized treatment by analyzing complex datasets to predict treatment responses and optimize medication dosages. AI-based models improve chemotherapy dosing and therapeutic drug monitoring (TDM), reducing adverse drug events and enhancing patient safety. The personalized approach ensures effective and efficient patient care (Alowais et al. 2023).
10. **Population Health Management:** Predictive analytics and risk assessment in population health management benefit from AI. ML algorithms analyze data to develop predictive models, identifying patients at risk of chronic diseases and hospital readmissions. AI optimizes healthcare delivery by improving the accuracy and efficiency of interventions, reducing costs, and enhancing patient outcomes (Alowais et al. 2023).
11. **Virtual Health Assistants and Mental Health Support:** AI-powered virtual health assistants provide personalized patient care, reduce healthcare provider workload, and improve patient engagement. AI applications in mental health support early detection, diagnosis, and tailored treatment of mental health conditions, offering round-the-clock support and monitoring patient progress (Alowais et al. 2023).

Nonetheless, this digital maturity of hospitals, driven by AI, brings with it inherent challenges and ethical considerations. These include the need for robust data management, the risk of biases in AI algorithms, and the ethical implications of automation. Additional challenges to the widespread adoption of AI in healthcare include the need for large, annotated datasets, and the complexity of integrating AI into clinical workflows. Therefore, to realize the full potential of AI in HSOs it must be implemented in a way that enhances, rather than compromises, the quality of care. AI is rapidly becoming a cornerstone of healthcare and HSO operations, with significant applications in telehealth.

By improving diagnostic accuracy, supporting clinical decisions, and automating administrative tasks, AI has the potential to revolutionize the way healthcare is delivered, particularly in remote settings (Lalli and Amutha 2020; Pallavi et al. 2022).

2 How Hospitals in the USA Are Implementing AI to Facilitate Digital Maturity

Professional associations, such as HIMSS or CHIME, have developed several maturity models that have historically helped US hospitals determine their digital maturity in different areas, such as the use of EMR systems, the adoption of analytics, or infrastructure maturity. However, there is no such standardized framework for classifying AI systems in healthcare based on their risks and benefits. This has led to two main issues. First, many healthcare organizations are reluctant to adopt even low-risk, high-impact AI solutions due to an inability to comprehend and manage the associated risks. Second, some AI systems are implemented beyond their intended scope, leading to potentially unsafe and unethical consequences, such as automated claims denials without adequate human oversight. This led to the introduction of various structured frameworks for implementing AI in healthcare, prompted by President Joe Biden's executive order invoking the Defense Production Act to promote AI adoption while ensuring safety, security, and equitable access. This directive underscores the urgent need to balance the substantial opportunities AI presents with the significant risks it poses (Yasin et al. 2024).

2.1 The Five-Point Scale for Hospital AI Implementation

The proposed five-point scale to classify healthcare AI products was inspired by the Society of Automotive Engineers' (SAE) framework for autonomous vehicles (Fig. 2). This scale is designed to set appropriate validation requirements based on the risk level of each AI product and foster responsible AI adoption in hospitals. It serves four key purposes by:

1. Providing a common vocabulary for strategic discussions around AI interventions in healthcare.

Level	Description	Patient risk	Description	Example use cases	Criteria to implement	Potential risks
1	Back office - administrative	Minimal	• Automation of administrative tasks which do not directly impact medical care. Little or no clinician involvement	• Revenue Cycle Management • Prior authorization submission/Utilization Management review	• Financial • Process integrity	• Financial misadventure • Administrative confusion
2a	Clinical process automation	Low	• Automation of low-level clinical processes and information organization, freeing clinician time to improve satisfaction and efficiency	• Virtual scribes (ambient documentation) • EHR summarization • Facilitated reference access	• Documentation accuracy • Process Integrity • Improved clinician experience/efficiency	• Inaccurate documentation
2b	Patient enablement	Low	• Educating patients on their conditions, recommended, medications, test results etc. • Collecting information from patients (e.g. meds, basics of clinical history)	• EHR inbox management • Patient education/promotional content generation • Patient chatbots • Smart questionnaires for intake • Member services/call centers	• NOT diagnostic or treatment advice • Improved patient experience	• Nontransparency with patients • Misinformation on non-medical issues
3	Clinical Decision Support - non-device CDS	Moderate	• Supporting clinicians in prognosis, diagnosis, or treatment, where clinicians make explicit choices to finalize plan, based on explanations provided by the system	• Computer assisted diagnosis and recommended treatments • Prediction of sepsis • Prior authorization review/approval	• Recommendations explainable and traceable to source • Clinician + CDS perform better than clinician alone	• Incorrect guidance given to physicians which they don't adequately review
4	Clinical Decision Support - SAMD	High	• Supporting clinicians in prognosis, diagnosis, or treatment, where the AI's reasoning is not always transparent, and the AI often provides a specific recommendation • Direct advice to patients through chat, etc w/o HITL	• Automatic pathology and radiology reading • Smart, advice-giving chatbots for patients • Suggestions about ventilator management	• Tightly defined use parameters • Accuracy at least equivalent to expert clinician • Equitable performance tested in a diverse patient population • Speed substantially better than human expert clinician as Gold Standard	• Automated incorrect medical decision making
5	Digital Doctor - SAMD+	Highest	• Autonomous decision making (prescribing, treating, test ordering, diagnosing) by algorithms without human intervention	• Algorithms which would prescribe or make remote diagnoses with no human present	• Clinical outcomes at least equivalent to human expert clinician as Gold Standard • Equitable performance tested in a diverse patient population • Requires new legal/ regulatory framework	• Automated bad medical practice

Fig. 2 Five-point scale of hospital AI implementation

2. Assessing the extent of AI transformation achievable by current or planned interventions.
3. Evaluating the impact and risk associated with specific AI systems.
4. Establishing the necessary controls, validation testing, and safeguards for safe and effective implementation at each level.

1. **Level 1 (Back Office Automation):** These technologies do not involve clinicians or decision-making processes, posing minimal risk of patient harm. They require validation to ensure a positive return on investment and prevent automation of financial fraud.
2. **Level 2 (Clinical Process Improvement):** Divided into:
   - **Level 2a (Staff and Organization-Focused):** These tools improve clinician efficiency and experience, such as documentation aids that must maintain consistent accuracy.
   - **Level 2b (Patient-Focused):** These solutions enhance patient education and navigation of the healthcare system, requiring human review for accuracy but not for real-time outputs.
3. **Level 3 (Clinical Decision Support):** These systems provide diagnostic and treatment guidance to clinicians, requiring explicit user choices and adherence to FDA criteria for non-device clinical decision support systems.
4. **Level 4 (Prescriptive AI):** More prescriptive and less explainable than Level 3, these tools require FDA approval as software as a medical device (SAMD) and must demonstrate performance equivalence or superiority to the standard of care.
5. **Level 5 (Fully Autonomous AI):** These systems would operate without human verification, requiring rigorous pre- and post-market testing to ensure performance exceeding human capabilities and addressing potential data drift.

Most healthcare organizations currently focus on Level 1 and 2 solutions to streamline administration and unburden clinical staff from routine tasks. Some of these solutions include revenue cycle management, prior authorization of drugs, ambient documentation, and electronic health record (EHR) summaries. Level 3 solutions, which hold the greatest potential for improving care quality and outcomes, are underdeveloped. The proposed framework aims to guide developers, purchasers, and regulators in ensuring AI solutions are implemented safely and effectively (Yasin et al. 2024).

2.2 The Organizational Perspective Checklist for AI (OPTICA) Solutions Adoption

Regulatory bodies are currently grappling with the challenge of effectively overseeing AI-driven healthcare solutions, which frequently demonstrate inconsistent performance and produce unexpected outcomes when deployed in new clinical settings. The traditional evaluation frameworks for these technologies often take the form of textual analyses that are not easily translatable into practical, actionable assessments. Additionally, these frameworks frequently overlook the specific needs and data characteristics of the target population and the particularities of the deployment environment (Dagan et al. 2024).

Healthcare organizations, inundated with a growing number of AI solutions, urgently require a pragmatic, actionable framework that can facilitate context-specific evaluations of the suitability of these AI tools. This necessity was particularly evident at Clalit Health Services, a large public healthcare organization with over a decade of experience integrating AI solutions into clinical practice. To address this gap, a comprehensive and practical checklist tool, named OPTICA (Organizational Perspective Checklist for AI solutions adoption), was developed to evaluate AI solutions within healthcare organizations (Dagan et al. 2024).

OPTICA is structured into 13 chapters, each containing between 3 and 12 checklist items, and is designed to guide healthcare organizations through a systematic, end-to-end evaluation process (Fig. 3). The tool identifies five key stakeholders typically involved in the AI adoption process and assigns specific checklist items to each, ensuring a logical order of completion based on dependencies among the tasks. Having been tested in various cases, OPTICA offers a structured, organization-centric approach to assessing the appropriateness and feasibility of AI solutions in new clinical environments, ensuring that evaluations are tailored to the unique characteristics and requirements of the implementing organization (Dagan et al. 2024).

2.3 Clinical AI Sociotechnical Framework (CASoF)

The CASof is introduced as another comprehensive model for incorporating artificial intelligence (AI) systems into healthcare, emphasizing both technical and sociotechnical factors essential for effective deployment. It provides a structured

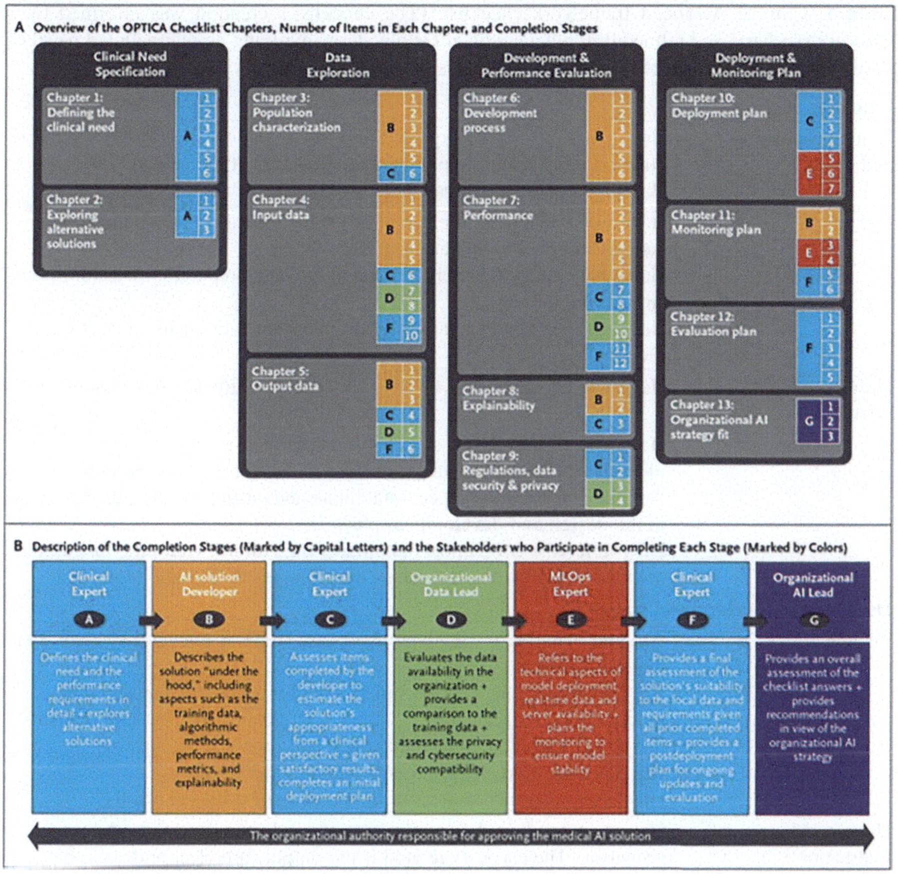

Fig. 3 OPTICA framework checklist components (Dagan et al. 2024)

checklist to facilitate planning, design, development, and implementation of AI tools tailored specifically for clinical settings. It uniquely addresses the gaps in existing frameworks that often overlook sociotechnical dynamics (Owoyemi et al. 2024) (Table 1).

The finalized checklist includes 35 questions grouped into key stages described as the planning stage, design and development stage, and proposed implementation stage and focuses on clarity and practical applicability for users.

The checklist underscores several critical components that include:

1. Value Proposition: Assessing anticipated impacts on both patient outcomes and healthcare efficiency.
2. Data Integrity: Ensuring ethical practices in data collection and maintenance of quality.

Table 1 Clinical AI social framework checklist. (The checklist's creation was informed by a *literature synthesis* and refined through a *modified Delphi study* involving an international panel of healthcare professionals, achieving high consensus on its relevance and clarity)

Planning stage	
Value proposition and utility	Have you outlined the expected impacts on patient outcomes? Have you outlined its expected impact on care provider efficiency and outcomes? Has any economic analysis been conducted for the AI system?
Data	Have you engaged in the use of any ethical data checklist during your data collection and preparation? Have you engaged domain experts in the data preparation, cleaning, and engineering process? Have you delineated an approach to maintain data quality, integrity, and security?
Human–AI interaction	Have you accounted for existing clinical workflows and protocols?
Organization and culture	Have you identified key stakeholders and their needs? Have you identified potential resistance or barriers within the organization? Are there strategies in place to facilitate and ensure end-user engagement in the design and development phase? Do you have a good understanding of the culture within the institution, relevant departments, and teams and changes that might be needed?
Design and development stage	
Technical	Are you planning for hardware/software (EHR if relevant) systems and requirements? Are you creating support documentation for users and management, e.g., model details, explainability details, data details, metrics, and manuals? Have you validated clinical accuracy and reliability? Have you secured any required regulatory approval? Have you taken active steps to mitigate against biased results?
Human–AI interaction	Have you conducted a simulation with end users in real work system scenarios? Have you evaluated if the outputs are clear and understandable for the users? Have you implemented patient and user safety measures? Have you assessed the impact on the delivery of clinical tasks? Have you involved and tested with users? Has resistance to the use of the AI system been identified and addressed? Are you developing strategies to ensure that the alerts from the AI system are relevant, timely, and not overwhelming, to avoid alert fatigue?
Data	Have you tested your method on various types of data to make sure it works well in different situations? Have you planned for data drift and shift (changes in the data over time)?
Proposed implementation stage	
Organization and culture	Have you ensured that this intervention aligns with the existing governance and regulatory frameworks of the organization? Have you prepared necessary training/resources for end users? Have you considered steps to help address end users' questions and alleviate their concerns?
Technical	Have you conducted a real-world evaluation of the AI system? Are you planning for pilot/silent tests? Are you providing user tools for continuous validation and evaluation of the system?
Monitoring and support	Have you created a plan to evaluate the success of the implementation? Have you planned for continuous user feedback on the system? Have you planned for regular audits, reviews, and updates? Have you planned for continuous education and support for users?

3. Human–AI Interaction: Integrating existing workflows to enhance usability.
4. Organizational Culture: Identifying stakeholders and potential barriers to system adoption.

The relevance of the checklist is found it that it is structured into four key stages of AI system deployment, ensuring it is practically applicable and adaptable to the complexities of healthcare settings. Emphasis is placed on the necessity for ongoing evaluation and feedback mechanisms to ensure that AI systems remain effective, user-friendly, and integrated seamlessly into clinical workflows.

The CASoF advocates for a balanced approach that marries technological advancement with a deep understanding of the sociotechnical context in clinical environments. This framework aims to enhance the success rates of AI implementation, ultimately improving patient care and healthcare outcomes. These insights provide essential guidance for healthcare leaders focusing on the transformative role of AI in hospital settings (Owoyemi et al. 2024).

2.4 ADAM Framework: Maturity of AI in Hospitals

The ADAM framework, analogous to the Society of Automotive Engineers' (SAE) levels of driving automation, provides a structured approach to classify and implement AI in medicine (Fig. 4). It allows regulatory bodies and healthcare organizations to set standards and compare the capabilities of various AI technologies. This framework aims to ensure that AI tools are integrated safely and effectively into medical practice while maintaining transparency and patient trust. The ADAM framework categorizes AI in medicine into five levels based on knowledge, safety, emotion, and independence (Kazzazi 2021).

1. **Level 1 (Basic AID Tool):** AI tools with narrow functions that assist human doctors without emotional requirements, such as risk stratification scores.
2. **Level 2 (Adjunctive AID Tool):** AI tools that function independently but still require human oversight, like diagnostic tools used in triage.
3. **Level 3 (Controlled Output AID Tool):** AI tools that provide personalized care, considering patient preferences and characteristics.
4. **Level 4 (Controlled Autonomy AID Tool):** AI tools that can conduct complete doctor–patient interactions but still require verification by human clinicians.
5. **Level 5 (Completely Autonomous AID Tool):** Fully independent AI practitioners capable of managing patient care without human intervention.

Human interaction, empathy, and emotional care are integral to medical practice. AI can augment these interactions by providing information and recommendations, but complete automation might challenge the human aspects of care. The "doctor effect," where friendly and reassuring clinicians improve patient outcomes, emphasizes the importance of human touch in healthcare (Kazzazi 2021).

Level 1: Basic AID tool
Knowledge: narrow-defined tool with single function
Safety: commercial standards (CE mark)
Emotion: no requirement
Independence: reference tool to human doctor care

Level 2: Adjunctive AID tool
Knowledge: narrow-defined tool with single function
Safety: commercial standards
Emotion: no requirement
Independence: adjunct tool to human doctor care

Level 3: Controlled output AID tool
Knowledge: narrow-defined tool with single or multi-function
Safety: specialty-defined standards
Emotion: demonstrate some evidence of holistic care
Independence: outputs selected by clinician

Level 4: Controlled autonomy AID tool
Knowledge: multi-functional tool
Safety: multi-specialty-defined standards for each component of care provided
Emotion: demonstrates emotional and intellectual care that is acceptable to patients
Independence: outcome reviewed by clinician

Level 5: Completely autonomous AID tool
Knowledge: complete specialty knowledge
Safety: Safety: multi-specialty-defined standards for each component of care provided
Emotion: demonstrates emotional and intellectual care that is acceptable to patients
Independence: Independence: requires no human clinician verification

Fig. 4 The five-level AI maturity (Kazzazi 2021)

3 Barriers and Strategies for AI Implementation in Hospitals: An In-Depth Analysis

The implementation and adoption of AI within the hospital context brings its own unique barriers and enablers. To be forewarned is to be forearmed and therefore organizations contemplating integration of AI should be aware of the following barriers and the related strategies to overcome the same.

1. **Barriers to AI Implementation:**
   - **Data Quality and Interoperability:** One of the primary barriers to AI implementation in hospitals is the poor quality and fragmented nature of healthcare data. Insufficient performance assessment, inconsistent data formats, outdated information, and issues with data interoperability across different systems hinder the effective training and deployment of AI models. Additionally, concerns about data privacy and security further complicate data sharing and integration efforts (Rahimi et al. 2024; Nair et al. 2024).
   - **Regulatory and Legal Challenges:** The lack of clear regulatory frameworks and legal guidelines for AI in healthcare poses significant challenges. Issues such as HCPs building trust in AI systems, liability in AI-assisted decision-making, and the regulatory approval process for AI systems create uncertainty and slow down the adoption of AI technologies (Rahimi et al. 2024; Nair et al. 2024). Additionally, the absence of specific regulatory frameworks for AI in healthcare creates uncertainties, making it difficult for both developers and users to navigate legal obligations and compliance requirements.
   - **Workforce and Training Issues:** The successful implementation of AI in hospitals requires a workforce that is well-versed in both clinical practice and AI technologies. However, a significant barrier is the disparate levels of AI literacy among healthcare professionals. Without proper training and education, disruptive integration, and lack of customization to specific hospital needs, there is a risk of resistance to adopting AI systems, as well as improper use of these technologies (Rahimi et al. 2024; Nair et al. 2024).
   - **Ethical Concerns:** Ethical issues such as AI-induced bias, the "black box" nature of AI algorithms, and the potential for AI to depersonalize patient care are critical concerns that need to be addressed. These concerns can lead to skepticism and reluctance among healthcare professionals to fully embrace AI technologies. Significant *ethical concerns, such as privacy, trust, consent, and conflicts of interest, obstruct the implementation of AI in healthcare* (Ahmed et al. 2023; Nair et al. 2024).
   - **Technological Barriers:** Technological limitations are identified as critical obstacles, including challenges related to data quality, access, interoperability, and integration into existing clinical workflows. The review emphasizes that AI systems are only as effective as the data they are trained on, and current issues with data standardization, validation, and infrastructure significantly hinder their practical application in clinical environments (Ahmed et al. 2023).
   - **Workforce Barriers**: The UNESCO review identifies a lack of education and training among healthcare professionals regarding AI technologies as a substantial barrier. There is a pressing need for targeted training programs to enhance AI literacy and competence. Furthermore, concerns about job displacement and the potential impact of AI on clinical roles contribute to resistance among healthcare staff (Ahmed et al. 2023).
   - **Social Barriers:** Social acceptance of AI in healthcare is limited by concerns about digital health inequality and the potential for AI to exacerbate existing biases in healthcare delivery. The review suggests that ensuring diverse

representation in AI training datasets and developing bias-neutral algorithms are essential steps toward mitigating these social barriers (Ahmed et al. 2023).
- **Patient Safety Barriers:** Patient safety remains a paramount concern, with the review highlighting risks such as automation bias and the limitations of AI systems in handling complex, multisystem conditions. The need for robust validation and continuous monitoring of AI systems to ensure their safe integration into clinical practice is underscored (Ahmed et al. 2023).

2. **Strategies for Overcoming Barriers to Digital Maturity with AI:**
   - **Leadership and Change Management:** Effective leadership is crucial for fostering a supportive environment for AI implementation. Leaders need to establish a clear vision, communicate the benefits of AI, and actively engage with stakeholders throughout the implementation process. Change management strategies that include continuous training, clear communication, and the involvement of key stakeholders are essential for overcoming resistance and ensuring the successful adoption of AI (Nair et al. 2024).
   - **Engagement and Collaboration:** Engaging stakeholders early in the planning process and fostering interdisciplinary collaboration are vital for the successful integration of AI systems. Involving clinicians, IT professionals, and AI developers in the design and implementation phases helps ensure that AI solutions are tailored to the specific needs and workflows of the hospital (Rahimi et al. 2024; Nair et al. 2024).
   - **Workflow Integration and Usability:** To facilitate the adoption of AI, it is important to design AI systems that integrate seamlessly into existing hospital workflows. AI tools should be user-friendly, with interfaces that align with the needs of healthcare professionals. Additionally, the implementation process should involve iterative testing and feedback to ensure that the AI system enhances, rather than disrupts, clinical workflows (Rahimi et al. 2024; Nair et al. 2024).
   - **Sustainability and Continuous Improvement:** Sustaining the use of AI systems in hospitals requires ongoing support, including the establishment of maintenance protocols, regular performance evaluations, and continuous training for staff. Developing a support infrastructure, such as robust performance monitoring, help desks, and governance committees, can help address issues as they arise and ensure the long-term success of AI initiatives (Rahimi et al. 2024; Nair et al. 2024).
   - Additionally ensuring that there are **adequate resources** and **support from hospital management** facilitates AI adoption; both **high-quality and secure data** along with ensuring that AI systems are **capable of learning and adapting over time,** helps overcome barriers to AI adoption (Rahimi et al. 2024).

While AI has the potential to transform healthcare delivery in hospitals, its successful implementation is contingent upon addressing various technical, legal, ethical, and organizational challenges. By adopting a structured approach that includes strong leadership, stakeholder engagement, and a focus on data quality and workflow integration, hospitals can effectively leverage AI to improve patient outcomes and operational efficiency to realize full digital maturity (Nair et al. 2024). These barrier overcoming strategies provide actionable recommendations for HSOs to evolve into learning or intelligent healthcare systems (Rahimi et al. 2024).

4 Sustainability of AI in Supporting Global Health: European Union (EU)

The European Union (EU) has also recognized the transformative potential of AI in healthcare, emphasizing its role in enhancing healthcare delivery, diagnostics, drug discovery, and hospital administration. Key to this recognition is the importance of maintaining trust and adhering to privacy and ethical standards. As a consequence, the EU has passed the Artificial Intelligence Act—AI Act (AIA) and European Health Data Space (EHDS) which has a significant impact on healthcare both within and outside the EU (Sheth et al. 2023).

The EU and Competitive Landscape: The EU's regulatory framework aims to foster innovation while ensuring safety and trust. The "Brussels Effect," where EU regulations influence global standards, making Europe a competitive hub for biomedical research and AI development, has wide-reaching implications outside the EU. It highlights the importance of global collaboration in AI-driven healthcare. Multinational corporations and research institutions must navigate diverse legal and ethical landscapes to develop and deploy AI solutions that benefit global health (Sheth et al. 2023).

The Triangle of Trust: A critical component of successful AI integration in healthcare is the "triangle of trust" between AI developers, healthcare providers, and patients. Trust in AI systems depends on transparency, explainability, and effective communication among all stakeholders.

State of AI in Healthcare: AI's potential in healthcare is vast, encompassing data analysis, patient monitoring, predictive analytics, and operational efficiencies. There is a strong need for robust data governance and interoperability standards to fully realize AI's benefits in healthcare.

Trust for Data and AI in Healthcare: Building and maintaining trust in AI systems is essential for their successful adoption. Various strategies can be applied to ensure data privacy, ethical AI development, and transparent communication with patients and the public.

Upcoming European Legislations: The EHDS aims to create a unified health data space across the EU, facilitating data sharing for research and clinical purposes. The EU Artificial Intelligence Act (AIA) seeks to establish ethical guidelines and regulatory standards for AI systems, promoting trust and accountability (Sheth et al. 2023).

5 How AI Is Catalyzing Digital Maturity from Traditional and Smart Hospitals to the Intelligent Hospital

Intelligent hospitals represent an evolution from traditional and smart hospitals, characterized by the integration of AI to enhance healthcare delivery. These hospitals combine advanced clinical processes, highly digitized infrastructure, and innovative management systems to provide seamless patient care and improve operational efficiency (Zulaikha et al. 2021).

Clinical Components of Intelligent Hospitals

1. **Electronic Health Records (EHRs)**: Traditionally paper-based, EHRs have evolved to digital formats, facilitating faster and more efficient data retrieval. AI enhances EHRs by analyzing patient data to predict disease risks and tailor medical care, thereby improving clinical outcomes and reducing healthcare costs.
2. **Patient Triaging and Management**: AI tools in emergency departments assist in risk stratification, diagnosis, and treatment choice, accelerating patient triage and reducing errors. AI-based systems also predict no-show rates for appointments, optimizing scheduling and resource utilization.
3. **Operation Theater Management**: AI integration in operation rooms supports advanced surgical techniques and enhances patient safety through real-time tracking and documentation of surgical instruments and procedures.
4. **Radiology and Pathology**: AI algorithms analyze complex radiological images quickly, aiding in accurate diagnosis and disease pattern recognition. In pathology, AI systems examine biospecimens with high accuracy, reducing human error and fatigue.

Non-Clinical Components of Intelligent Hospitals

1. **Data Management**: AI aids in managing "Big Data" from various healthcare sources, using machine learning and deep learning techniques to derive meaningful insights and support decision-making processes.
2. **Smart Asset and Building Management**: AI, combined with IoT and RFID, enhances hospital asset tracking, preventing loss and theft while optimizing resource utilization. Intelligent building systems use AI for energy optimization and efficient facility management.
3. **Billing and Financial Management**: AI improves the accuracy and efficiency of the billing process by integrating with EMRs and automating the coding and charging system, reducing human error (Zulaikha et al. 2021).

Advantages:

- **Enhanced Patient Care**: AI improves diagnostic accuracy, patient monitoring, and personalized treatment plans, leading to better health outcomes.

- **Operational Efficiency**: AI streamlines hospital operations, from scheduling and resource management to automated billing and smart infrastructure, increasing overall efficiency and reducing costs.
- **Public Health Benefits**: AI aids in disease surveillance and public health measures by analyzing large patient datasets.

Challenges:

- **Data Management and Privacy**: Ensuring the security and privacy of health data remains a significant challenge. Integrating AI systems with existing hospital infrastructure requires careful planning and collaboration among various stakeholders.
- **Cost and Maintenance**: High initial investment and ongoing maintenance costs for AI systems and smart infrastructure can be prohibitive for some institutions (Zulaikha et al. 2021).

The integration of AI into healthcare services is pivotal in creating intelligent hospitals that offer superior patient care, operational efficiency, and public health benefits. Despite challenges, the advantages of AI in reducing human error, enhancing patient satisfaction, and improving health outcomes make it a crucial component of future healthcare systems. Healthcare professionals must stay informed and engaged with AI developments to harness its full potential in creating and managing intelligent hospitals (Zulaikha et al. 2021).

6 Acceptance of Artificial Intelligence Among HCPs in Hospitals

To grasp how AI is effectively implemented in HSOs and matures in its adoption it is important to understand the factors influencing the acceptance of AI among healthcare professionals (HCPs) in hospital settings. The Unified Theory of Acceptance and Use of Technology (UTAUT) model has played an important role in structuring these findings.

The key findings included the following:

1. **Types of AI Systems:**
   - The review focused on clinical decision support systems (CDSS), which were the most frequently studied AI applications (21 out of 42 studies). These systems assist healthcare professionals in making clinical decisions by providing evidence-based recommendations.
2. **Main barriers to AI Acceptance:**
   - **Fear of Loss of Autonomy:** Healthcare professionals expressed concerns about AI reducing their control over clinical decisions and potentially replacing them in their roles.

- **Integration Challenges:** Difficulties in integrating AI systems into existing clinical workflows and the additional time required to learn and use these systems were significant barriers.
- **Lack of Trust and Transparency:** Concerns about the reliability and explainability of AI systems contributed to hesitancy. Professionals were wary of the "black box" nature of some AI technologies, where the decision-making process is not transparent.
- **Technical and Infrastructure Issues:** Insufficient technical infrastructure, such as inadequate Internet connectivity and lack of user-friendly interfaces, hindered the adoption of AI.

3. **Main facilitators of AI Acceptance:**
    - **Training and Education:** Providing adequate training and education on AI systems was a crucial facilitator. Familiarity with AI tools increased confidence and willingness to use them.
    - **Early Involvement of End Users:** Involving healthcare professionals in the development and implementation stages of AI systems ensured that the tools met their needs and fit into their workflows, enhancing acceptance.
    - **Perceived Benefits:** When healthcare professionals recognized the potential benefits of AI, such as improving diagnostic accuracy and reducing workload, they were more likely to accept and use these systems.
4. **Differences Among Professional Groups:**
    - Acceptance levels varied among different professional groups and specialties. For example, radiologists showed higher acceptance of AI in diagnostic imaging compared to other specialties, potentially due to the technical nature of their work and early exposure to AI tools.

It is therefore critical that HSOs take into account both technical and human factors to enhance AI acceptance in hospitals. Tailored training programs, transparent AI systems, and the involvement of healthcare professionals in the development process are key strategies for fostering acceptance. Additionally, further research is required to explore the long-term impacts of AI on clinical practice and healthcare outcomes. While AI is transforming healthcare delivery its successful integration depends on overcoming barriers related to trust, technical infrastructure, and workflow integration. By understanding and addressing the concerns of healthcare professionals, stakeholders can develop more effective strategies to promote the acceptance and use of AI in clinical settings (Lambert et al. 2023).

7 AI Powered Telehealth

Just like AI is transforming the clinical setting, it is also transforming the different digital means by which care is delivered. This expanding role of AI in telehealth is especially evident within hospital settings. Advancements in AI have significantly contributed to the growth and expansion of telehealth by enabling more efficient, accessible, and personalized healthcare services. Telehealth encompasses a broad range of digital information and communication technologies that facilitate the delivery of health-related services remotely. It is categorized into four main subtypes: synchronous (real-time interactions between patients and providers), asynchronous (store-and-forward exchanges), remote monitoring (continuous patient data collection and transmission), and mobile health (Andrikopoulou 2023).

The primary objective of telehealth is to overcome barriers in delivering high-quality care, particularly those arising from geographic and temporal constraints. Rather than replacing traditional in-person care, telehealth aims to democratize and commoditize access to healthcare, ensuring that high-value care is available to a broader population.

However, the rapid expansion of telehealth has brought about several challenges, particularly in the realms of regulation and technology. These include the need for updated best practice guidelines, adaptation of standards for data exchange, expansion of reimbursement policies, and addressing ethical concerns. Ethical challenges are particularly significant, encompassing issues such as data privacy, security, governance, AI-introduced biases, the opacity of some AI/machine learning algorithms (often referred to as the "black box" problem), and the potential for AI to exacerbate health disparities and inequities (Andrikopoulou 2023).

The disparities in access to and use of telehealth were underscored during the COVID-19 pandemic, which highlighted existing inequities. This recognition has given rise to the field of digital determinants of health (DDOH), which explores factors like digital literacy, access to technology, and community infrastructure (e.g., availability of WiFi and broadband internet) that influence individuals' ability to benefit from telehealth services (Andrikopoulou 2023).

This discussion underscores the need for ongoing research, policy development, and technological advancements to address these challenges and ensure that telehealth, empowered by AI, can achieve its full potential in improving healthcare access and outcomes.

The applications of AI in telehealth can be identified in several key areas: remote patient monitoring, healthcare information technology, intelligent assistance diagnosis, and information analysis collaboration (Pacis et al. 2018).

1. **Patient Monitoring:** AI enhances telemedicine by enabling remote patient monitoring, which allows for continuous and cost-effective patient care. Technologies like telepresence robots, equipped with AI and vision systems, facilitate real-time consultations and monitoring, especially in hospital settings. AI also supports the development of smart wearable devices and telemedicine systems for chronic disease management and rehabilitation, providing real-time data to healthcare providers.
2. **Healthcare Information Technology:** AI helps manage and analyze the vast amounts of data generated in telemedicine. Through big data analytics and cloud computing, AI streamlines the retrieval, storage, and analysis of electronic health records (EHRs), improving the efficiency of telemedicine services. AI also addresses issues related to data security and confidentiality, ensuring the safe transmission of medical information.

3. **Intelligent Assistance and Diagnosis:** AI-driven tools are used for intelligent diagnosis in telemedicine, assisting healthcare providers by analyzing patient data and providing insights for treatment. These tools can operate autonomously, supporting self-diagnosis and enabling faster, more accurate medical decisions, even in emergency situations.
4. **Information Analysis and Collaboration:** AI facilitates medical research and training by enabling the analysis of large datasets and fostering collaboration among healthcare professionals across different locations. This enhances the quality of telemedicine services and supports the development of new treatment protocols and healthcare strategies.

Despite its benefits, there are challenges that come with implementing AI-based telemedicine, including the high cost of technology, the need for infrastructure improvements, and concerns over data security and patient privacy. Overall, however, the critical aspect to grasp is that AI will continue to play a pivotal role in advancing telemedicine, particularly in enhancing patient monitoring, data management, and intelligent diagnosis (Pacis et al. 2018).

7.1 The Role of AI in Different Telemedicine Specialties

1. **Tele-ICU and the Role of AI**
   One key area we see AI play an important role is tele-ICUs. The integration of AI in tele-intensive care units (tele-ICUs) enhances patient monitoring by analyzing real-time data to predict and prevent adverse events. AI-driven tele-ICUs have been shown to reduce mortality rates and improve patient outcomes, particularly in critical care settings (Sharma et al. 2023).
2. **Telecardiology and the Role of AI**
   In telecardiology, AI plays a pivotal role by enhancing the accuracy and speed of cardiovascular diagnostics and monitoring. AI algorithms are utilized to analyze large datasets from remote monitoring devices, such as electrocardiograms (ECGs), enabling the early detection of cardiac anomalies. This facilitates timely intervention, potentially reducing the incidence of adverse cardiac events. Moreover, AI supports the stratification of patients based on risk, aiding clinicians in making informed decisions regarding treatment priorities and resource allocation. The integration of AI in telecardiology thus optimizes patient care by enabling precise, data-driven management of cardiovascular conditions in a remote setting (Sharma et al. 2023).
3. **Teleoncology and the Role of AI**
   AI in teleoncology significantly advances the precision and personalization of cancer care delivered remotely. By analyzing complex datasets, including genomic data, imaging, and patient histories, AI systems can assist in diagnosing various cancers with high accuracy. Additionally, AI contributes to treatment planning by predicting patient responses to therapies and identifying the most effective treatment protocols. In a teleoncology framework, AI also enhances the

ability to monitor patient progress and adapt treatment plans dynamically, thereby improving outcomes in cancer management. This tailored approach to cancer treatment underscores the critical role of AI in overcoming the challenges of providing oncology care at a distance (Sharma et al. 2023).

4. **Telepsychiatry and the Role of AI**
   AI's role in telepsychiatry is centered around improving the accessibility and effectiveness of mental healthcare. Through the analysis of patient interactions, AI can detect subtle cues in speech, text, and behavior that may indicate mental health conditions, providing clinicians with valuable insights that might not be apparent during traditional consultations. AI also supports the development of personalized treatment plans by identifying patterns in patient data that correlate with treatment outcomes. Furthermore, AI-driven tools in telepsychiatry can offer continuous monitoring and support, helping patients manage their conditions more effectively between clinical encounters. This application of AI in telepsychiatry enhances the quality of mental healthcare, particularly in underserved populations (Sharma et al. 2023).
5. **Telestroke and the Role of AI**
   In telestroke care, AI is instrumental in improving the speed and accuracy of stroke diagnosis and treatment. AI algorithms assist in the rapid analysis of neuroimaging data, helping to identify stroke types and locations with greater precision. This expedited analysis is crucial in stroke care, where time-sensitive decisions are critical to patient outcomes. AI also supports the prioritization of patients for thrombolytic therapy or surgical intervention, enhancing the overall efficiency of telestroke networks. By enabling more accurate and faster decision-making, AI significantly contributes to improving survival rates and reducing long-term disabilities associated with stroke (Sharma et al. 2023).
6. **Teleurology and the Role of AI**
   AI's integration into teleurology facilitates more accurate diagnostics and personalized treatment plans for urological conditions managed remotely. AI models are used to analyze data from imaging studies, lab results, and patient histories to diagnose conditions such as kidney stones, urinary tract infections, and prostate disorders. Additionally, AI supports the ongoing monitoring of patients with chronic urological conditions, allowing for timely adjustments to treatment plans. By enhancing the diagnostic and monitoring capabilities of teleurology, AI plays a critical role in improving patient outcomes and the efficiency of urological care delivered through telemedicine (Sharma et al. 2023).
7. **Teleophthalmology and the Role of AI**
   AI, particularly deep learning (DL), has significantly improved image recognition and classification in ophthalmology. AI has been instrumental in enhancing the diagnosis and monitoring of retinal diseases such as diabetic retinopathy (DR), age-related macular degeneration (AMD), and retinopathy of prematurity (ROP). Key applications include the development of AI algorithms for automated screening, prognostic predictions, and treatment recommendations, which are crucial in early detection and management of retinal conditions. In the near future AI and

telemedicine will be central to retinal disease management, suggesting that this model could improve access to care, reduce healthcare costs, and enhance patient outcomes (Wu et al. 2023).

The integration of AI across various telemedicine specialties—tele-ICU, telecardiology, teleoncology, telepsychiatry, telestroke, teleurology, and teleophthalmology—demonstrates its transformative potential in enhancing the accuracy, efficiency, and personalization of remote healthcare services. AI not only improves diagnostic precision but also facilitates dynamic treatment planning and patient monitoring, addressing the unique challenges of delivering specialized care in a telemedicine environment (Sharma et al. 2023).

7.2 Integration of AI-Powered Telehealth in HSOs

The integration of AI into telehealth within hospital settings can be conceptualized around five foundational pillars (Fig. 5).

1. **Data Quality**
   Ensuring the reliability and accuracy of AI-driven telehealth solutions fundamentally depends on the quality of the data used. High-quality data is critical for generating actionable insights that can support personalized patient care. Without robust data, AI systems cannot deliver the precision necessary to effectively guide clinical decisions or predict patient outcomes.
2. **Health Equity**
   The role of AI in promoting health equity is multifaceted. Developers, clinicians, investors, and regulators must prioritize the creation of AI-based telehealth solutions that are accessible and applicable across diverse socioeconomic groups. Simultaneously, AI has the potential to advance health equity by enabling broader access to high-quality healthcare services, particularly for underserved populations, thus narrowing the gap in healthcare disparities.
3. **Research and Innovation**
   Research occupies a central role in the evolution of AI-enabled telehealth. Not only is research critical during the development and ideation phases of AI solutions, but the insights generated through AI applications can also fuel further research, leading to new discoveries and innovations in healthcare. This cyclical relationship underscores the importance of continued investment in AI research to drive progress in telehealth.

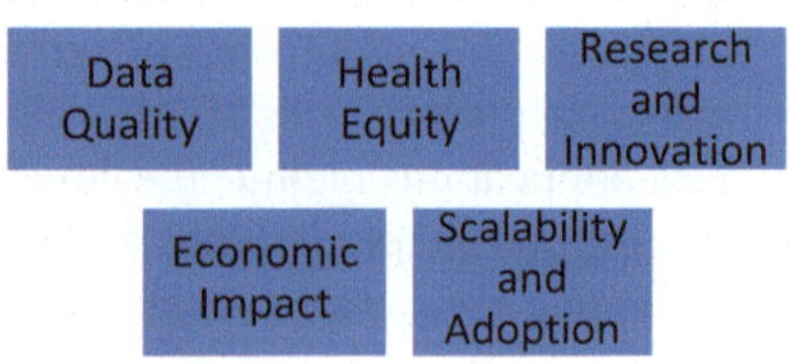

Fig. 5 Five Foundational Pillars for Integrating AI into Telehealth in Hospital Settings

4. **Economic Impact**
   For AI solutions in telehealth to gain widespread adoption in hospitals, they must demonstrate a positive impact on economic growth. There is a strong hypothesis that AI-driven telehealth can lead to more cost-effective care by optimizing resource utilization and reducing unnecessary interventions. However, existing studies offer mixed results, and larger, more comprehensive research is needed to definitively assess the economic benefits of AI in telehealth.
5. **Scalability and Adoption**
   Finally, the adoption of AI-based telehealth solutions is contingent on their ability to scale effectively within different healthcare environments. Demonstrating scalability, coupled with clear evidence of economic and clinical benefits, is essential for achieving broad acceptance among healthcare providers and institutions.

   This framework highlights the key considerations for successfully integrating AI coupled with telehealth in hospital settings, emphasizing the importance of data quality, equity, research, economic impact, and scalability (Andrikopoulou 2023).

7.3 The Future of AI in Telehealth

Over the past 5 years applications of AI in telehealth have improved healthcare delivery and created new models of care in hospitals and other settings. Broadly speaking two primary areas where AI has made significant contributions include enhancing the quality of existing clinical practices and supporting the development of new healthcare delivery models.

1. **Quality Improvement in Clinical Practice:** AI has been used to analyze vast amounts of healthcare data, allowing for more precise clinical decision-making and better patient outcomes. In telehealth, AI supports activities such as tele-diagnosis, tele-monitoring, and remote patient assessment, making it possible to provide high-quality care even at a distance. AI has also assisted in the process of history taking and clinical evaluation by automating parts of the diagnostic process, thereby saving time and reducing the workload for HCPs (Kuziemsky et al. 2019).
2. **Development of New Models of Care:** AI has facilitated the creation of new telehealth models that can deliver care more efficiently and effectively. For instance, AI-powered virtual assistants and chatbots have been developed to interact with patients, provide health information, and even assist in the management of chronic conditions. Remote patient monitoring systems, supported by AI, can continuously track patients' health data and provide timely interventions, particularly in managing chronic diseases like COPD and diabetes (Kuziemsky et al. 2019).

However, as AI is increasingly integrated into telemedicine, one may say that it has cemented its role in telemedicine more specifically by enhancing clinical efficiency, improving diagnostic accuracy, and expanding the reach of healthcare services.

1. **Enhanced Clinical Efficiency:** AI helps streamline administrative tasks, allowing healthcare providers to focus more on patient care. For example, AI-driven tools can handle scheduling, patient engagement, and data analysis, reducing the burden on medical staff and improving overall workflow (Pratt 2023).
2. **Improved Diagnostic Accuracy:** AI technologies, particularly in medical imaging and patient monitoring, enable more accurate and timely diagnoses. AI algorithms analyze vast datasets, including patient histories and imaging results, to assist clinicians in making informed decisions quickly. This capability is especially valuable in emergency settings and for managing chronic conditions (Pratt 2023).
3. **Expansion of Healthcare Access:** AI-powered telemedicine extends the reach of healthcare services, particularly in rural and underserved areas. Remote patient monitoring tools allow clinicians to oversee patients' health in real time, regardless of their physical location. This technology is critical for managing chronic diseases and ensuring continuous care (Pratt 2023).

8 Risk Mitigation Strategies Relative to The Future of AI's Impact on Digital Maturity in Hospitals

Although AI has already made significant inroads into telemedicine, its future development must be carefully managed to maximize benefits and mitigate risks. AI-enabled telehealth also brings its own set of challenges such as ensuring equity in access to these technologies, preventing the digital divide from widening, difficulties in system integration, high costs, and concerns about data privacy and algorithmic bias, and addressing the ethical implications of AI in healthcare. The "black box" nature of AI where decision-making processes are not fully transparent can also lead to mistrust among healthcare providers. There continues to be a need for ongoing validation of AI algorithms, and the importance of integrating AI into healthcare systems in a way that prioritizes patient empowerment and provider support. Best practices emphasize the need for HCPs to play a central role in guiding the implementation of AI in telemedicine. Physicians' expertise is crucial in ensuring that AI tools are used effectively and ethically, enhancing rather than replacing clinical judgment (Sharma et al. 2023). As AI continues to evolve, it will likely play an even more integral role in healthcare delivery, which must be supported by strong governance frameworks (Kuziemsky et al. 2019; Pratt 2023).

One cannot lose sight of the fact that coupled with advancements in the Internet of Things (IoT) and big data, AI is poised to revolutionize healthcare delivery by enabling more personalized and efficient telemedicine services. In medical imaging

AI, particularly deep learning (DL) algorithms like convolutional neural networks (CNNs) have proven effective in processing and analyzing medical images, such as those used in ophthalmology. These technologies enable rapid and accurate diagnosis, which is critical in telemedicine, where real-time analysis can significantly impact patient outcomes. In the area of personalized telemedicine, AI is tailoring telemedicine services to individual patients by analyzing large datasets and providing precise, personalized medical advice. This approach reduces the need for in-person visits and allows for continuous monitoring and management of chronic conditions, thus improving patient care while minimizing the burden on healthcare systems. In spite of the challenges (ethical concerns, data privacy issues, and the need for robust AI algorithms) that can be overcome by validation of healthcare AI models the future of AI in telemedicine will involve the development of cloud-based AI platforms that can be accessed globally, improving accessibility and the quality of healthcare services, especially in underserved regions (Ying-Chuna et al. 2020).

The Massachusetts Institute of Technology (MIT) recently developed the AI Risk Repository report consisting of a comprehensive synthesis of AI risks, compiling 777 identified risks from 43 taxonomies into a living database. This effort addresses the challenges posed by the diverse and often conflicting risk classifications found in existing literature. The report introduces two primary taxonomies: a high-level Causal Taxonomy, which categorizes risks based on the entity responsible (human or AI), the intentionality (intentional or unintentional), and the timing (pre- or post-deployment), and a mid-level Domain Taxonomy, which organizes risks into seven domains, including discrimination, privacy, misinformation, malicious use, human–computer interaction, socioeconomic and environmental harms, and AI system safety. The report underscores that most identified risks are attributed to AI actions post-deployment and are often unintentional. Key areas of concern include AI system safety, socioeconomic impacts, and discrimination. The Repository serves as a foundational tool for researchers, policymakers, and industry professionals to better understand, audit, and mitigate AI risks. This work is positioned as a crucial step toward creating a coordinated and comprehensive approach to AI risk management, with implications for the development of future regulations, standards, and best practices in AI deployment, particularly in critical sectors like healthcare and hospitals. The report highlights the necessity of continuous updates to the Repository and the importance of interdisciplinary collaboration in addressing the evolving landscape of AI risks (Slattery et al. 2024).

Stanford University's Human-Centered Artificial Intelligence (HAI) department explored the complex legal landscape surrounding the use of AI in healthcare, particularly focusing on liability issues when AI tools contribute to patient harm. Their research highlighted that despite the tremendous potential of AI to transform healthcare, significant concerns persist regarding legal accountability. They identified a critical gap in case law, with limited precedents on physical injuries caused by AI or software systems in healthcare. The analysis of 51 relevant cases showed that liability claims typically arise from software defects in medical devices, errors in AI-driven decision-making tools, or the improper use or maintenance of AI systems by healthcare providers (Mello and Guha 2024).

One of the key challenges identified was the difficulty in applying traditional product liability doctrines to AI and software, due to the intangible and opaque nature of these technologies. Plaintiffs often struggle to meet the legal requirements for proving defects, particularly when they must demonstrate a safer alternative design or foreseeability of harm, given the "black box" nature of AI models. The HAI research noted that while tort law has historically adapted to technological changes, the evolution of legal principles in response to AI in healthcare remains uncertain (Mello and Guha 2024).

To address these challenges, the HAI proposed a risk assessment framework for healthcare organizations, emphasizing the importance of carefully evaluating the liability risks associated with different AI tools before implementation. The HAI also recommends negotiating robust licensing agreements with AI developers to clarify responsibilities and include indemnification clauses that allocate liability appropriately. The findings suggest that as AI continues to integrate into healthcare, there will be a growing need for legal frameworks that can accommodate the unique risks posed by AI. Healthcare organizations, policymakers, and legal professionals must collaborate to ensure that AI tools are deployed safely and that liability risks are managed effectively, thereby supporting the responsible adoption of AI in hospitals and other healthcare settings (Mello and Guha 2024).

9 A Global Regulatory Perspective on AI

In August of 2024, UNESCO[1] released a paper on AI regulation that outlines several key findings relevant to the impact of AI on digital maturity in hospitals, particularly through its discussion on the global landscape and emerging regulatory approaches for AI. The paper identifies the rapid increase in AI regulations worldwide, highlighting that since 2016, over 30 countries have enacted laws specifically addressing AI, with further legislative discussions ongoing. It describes nine distinct regulatory approaches to AI, ranging from principles-based to liability approaches, emphasizing a flexible and tailored strategy to managing AI's diverse impacts (Gutiérrez 2024). These regulatory frameworks are designed to address the ethical, social, and technical challenges posed by AI.

- The *Principles-Based Approach* sets out ethical guidelines and foundational principles to direct the development and deployment of AI systems, focusing on ensuring that AI operations are responsible, human-centric, and respectful of human rights.

[1] The United Nations Educational, Scientific and Cultural Organization (UNESCO) is a specialized agency of the United Nations with the aim of promoting world peace and security through international cooperation in education, arts, sciences and culture.

- The *Standards-Based Approach* delegates regulatory authority to organizations that establish technical standards, ensuring AI systems meet predefined safety and performance benchmarks.
- The *Agile and Experimentalist Approach* involves creating flexible regulatory environments, such as regulatory sandboxes, which allow AI technologies to be tested and developed in controlled settings to foster innovation while maintaining oversight.
- The Facilitating and Enabling Approach promotes an environment conducive to AI innovation, encouraging the responsible development and use of AI through supportive policies and infrastructure.
- The *Adapting Existing Laws Approach*, instead of creating new laws, modifies existing regulations across various sectors to incorporate AI-specific considerations, providing a gradual adaptation to AI's integration.
- The *Access to Information and Transparency Mandates* requires the disclosure of AI systems' operations to ensure public transparency and accountability, helping users understand how AI decisions are made.
- The *Risk-Based Approach* is tailored based on the risk level of different AI applications, ensuring that higher-risk AI systems undergo stricter regulatory scrutiny.
- The *Rights-Based Approach* focuses on protecting individual rights, emphasizing the importance of safeguarding human rights through stringent regulatory measures.
- The *Liability Approach* assigns legal responsibility and penalties for the misuse or harmful effects of AI systems, ensuring accountability for AI's societal impacts.

Combining these approaches into one can effectively address the multifaceted challenges AI poses to digital maturity in hospitals, particularly regarding ethical considerations, patient safety, and regulatory compliance. The adaptive and rights-based approaches, for instance, are critical in ensuring that AI integration in healthcare respects patient rights and adapts to evolving technological landscapes, promoting safe and ethical digital maturity. These findings provide a comprehensive framework that can inform hospital administrators and policymakers about the diverse strategies available for integrating AI into healthcare systems, ensuring that its adoption enhances digital maturity while mitigating potential risks (Gutiérrez 2024).

10 Ethical Considerations in the Current Landscape of AI Applications in Hospitals

AI technologies in hospitals are utilized to enhance patient care by improving the accuracy and timeliness of medical information, tailoring treatments to individual patients, and integrating with hospital information systems. AI supports various tasks such as exercise promotion, medication adherence, chronic disease management, and disease diagnostics, providing superior results compared to traditional methods. The five core principles of biomedical ethics which apply to the digital maturity catalyzed by AI in hospitals include beneficence, non-maleficence, autonomy, justice, and by nature of being AI explicability (Floridi et al. 2018; Beauchamp and Childress 2019). Unfortunately, the current ethical discourse on AI ethics or responsible AI is fragmented and often lacks a systematic approach. However, an interdisciplinary approach and attention to key ethical considerations can proactively alleviate any anticipated challenges. Some of the challenges that can be anticipated include bias and discrimination; opacity and trust; responsibility and liability; privacy and confidentiality; and transparency and explainability. In spite of these obvious challenges further research is necessary into a more detailed examination of ethical principles in AI applications, interdisciplinary research to address ethical challenges, data interoperability, and the development of frameworks and guidelines to ensure responsible AI use in healthcare. Digital maturity catalyzed by AI should ensure the technology benefits patients while maintaining high ethical standards to fully realize the potential of AI in hospitals and healthcare at large (Mirbabaie et al. 2022).

11 Conclusion

AI has the potential to significantly transform healthcare by improving diagnosis, treatment, and patient care. However, it is crucial to address ethical issues and ensure that AI technologies are implemented responsibly, prioritizing patient-centered approaches and equity in healthcare (Varnosfaderani and Foruzanafar 2024). Regulators, policymakers, researchers, and the public need to ensure the responsible and effective implementation of AI in healthcare. This includes fostering trust, enhancing data governance, promoting education, and encouraging global collaboration (Sheth et al. 2023). AI's full potential to revolutionize healthcare in hospitals by improving diagnostics, personalized treatment, and patient care can only be realized by addressing ethical issues, ensuring data security, and fostering interdisciplinary collaboration. The latter are key to maximizing AI's benefits in clinical practice and realizing full digital maturity. Continued research, innovation, and education are also vital for the responsible and effective implementation of AI in healthcare (Alowais et al. 2023). The development of AI and its responsible implementation and adoption present an exciting opportunity for modern medicine in the hospital setting. The ADAM framework supports the integration of AI by delineating different forms of AI and setting standards for their use in clinical practice (Kazzazi 2021). The proposed five-level framework also offers a scaffold for classifying and implementing AI in healthcare and can transform the adoption and regulation of AI in healthcare, ensuring its benefits are realized while minimizing risks (Yasin et al. 2024). In spite of the challenges that come with AI which could vary from validating datasets to integrating AI into live clinical workflows, this chapter provides actionable recommendations for hospitals to overcome these barriers and leverage AI for achieving an intelligent health system (HIS). By aligning the identified enablers and barriers with the 3-horizon framework, HSOs can develop a comprehensive roadmap for the successful digital transformation (Rahimi et al. 2024).

Typically, when the impact of AI is assessed relative to digital maturity in hospitals, the focus is on model performance. Minimal attention is paid to the interaction between clinicians and AI systems. Seven categories of this interaction have however been identified and they play an important role in the digital maturity of hospitals. They include usability and user experience, system performance, clinician trust and acceptance, impact on patient care, communication and collaboration, ethical and professional concerns, and clinician engagement and workflow. Each category includes various traits that define how clinicians perceive and interact with AI tools in clinical practice. Usability and user experience are crucial in determining how easily clinicians can learn and use AI systems. Key traits include ease of use, system learnability, and user satisfaction. System performance, including the accuracy and reliability of AI outputs, is fundamental in gaining clinician trust. Trust and acceptance of AI systems are critical for their integration into clinical workflows. Positive impacts on patient care, such as improved decision-making and patient outcomes, are essential for clinicians to value AI integration. Effective communication and collaboration among clinicians, facilitated by AI systems, are also essential for improving team dynamics and patient care coordination. Consequently, developing a framework based on these seven categories can have a significant impact on digital maturity relative to AI integration in hospitals (Perivolaris et al. 2024). The UNESCO description of nine distinct regulatory approaches to AI that range from principles-based to liability approaches, emphasizing a flexible and tailored strategy to managing AI's diverse impacts, can be combined into one standard regulatory framework to address ethical, social, and technical challenges posed by AI (Gutiérrez 2024).

For the increasing integration of artificial intelligence (AI) tools, particularly large language models (LLMs) and Large Concept Models (LCMs), into healthcare organizations, scientific research, and education, there is an ongoing demand for improved AI literacy among HCPs, scientists, and students. With regard to AI in education and research universities like Stanford and Oxford are adopting flexible policies to incorporate AI tools in teaching and research, encouraging AI literacy while ensuring ethical and appropriate use. The importance of critical thinking in AI use is underscored, with some educators restricting AI to develop students' analytical skills. At the same time, challenges with AI Tools will have to be addressed. AI, especially LLMs, often functions as a "black box," which can be problematic in scientific research where transparency and understanding are crucial. Researchers must balance the efficiency AI offers with the need to verify and understand AI-generated outputs, particularly given the risks of bias, inaccuracies, and "hallucinations" (false outputs). In biomedical research AI's role in data analysis and hypothesis generation is increasingly recognized, but its integration into biology and medicine requires careful validation to ensure that AI predictions align with scientific observations. Collaborative efforts are essential to develop AI models that are tailored to specific scientific domains, such as single-cell data analysis in plant biology. In education, there is a significant demand for structured training and guidelines on using AI tools in research. Many scientists express a need for better resources to understand and apply AI effectively in their work. Initiatives like deep learning courses for microscopy image analysis are examples of efforts to bridge the educational gap in AI literacy. The never-ending battle with responsible AI demands attention to ethical implications of AI in science, particularly the importance of truthfulness and the risks of AI producing misleading or harmful outputs. Researchers are urged to maintain a critical approach when using AI, recognizing its limitations and ensuring that it supports rather than undermines the scientific process (Marx 2024).

References

Ahmed MI, Spooner B, Isherwood J et al (2023) A systematic review of the barriers to the implementation of artificial intelligence in healthcare. Cureus 15(10):e46454. https://doi.org/10.7759/cureus.46454

Alowais SA, Alghamdi SS, Alsuhe N et al (2023) Revolutionizing healthcare: the role of artificial intelligence in clinical practice. BMC Med Educ 23(1):1–15. https://doi.org/10.1186/s12909-023-04698-z

Andrikopoulou E (2023) The rise of AI in telehealth. In: Freeman AM, Bhatt AB (eds) Emerging practices in Telehealth: best practices in a rapidly changing field. Elsevier, San Diego, pp 183–207. https://doi.org/10.1016/B978-0-443-15980-0.00011-9

Beauchamp T, Childress J (2019) Principles of biomedical ethics: marking its fortieth anniversary. Am J Bioeth 19(11):9–12. https://doi.org/10.1080/15265161.2019.1665402

Dagan N, Devons-Sberro S, Paz Z et al (2024) Evaluation of AI solutions in health care organizations-the OPTICA tool. NEJM AI 1(9). https://doi.org/10.1056/AIcs2300269

Floridi L, Cowls J, Beltrametti M et al (2018) AI4People - an ethical framework for a good AI society: opportunities, risks, principles, and recommendations. Mind Mach 28:689–707. https://doi.org/10.1007/s11023-018-9482-5

Gutiérrez JD (2024) Chapter 24: Critical appraisal of large language models in judicial decision-making. In: Handbook on public policy and artificial intelligence. Edward Elgar Publishing, Cheltenham. https://doi.org/10.4337/9781803922171.00033

Kazzazi F (2021) The automation of doctors and machines: a classification for AI in medicine (ADAM framework). Future Healthc J 8(2):e257–e262. https://doi.org/10.7861/fhj.2020-0189

Kuziemsky C, Maeder AJ, John O et al (2019) Role of artificial intelligence within the Telehealth domain. IMIA Yearb Med Inform 28(1):35–40. https://doi.org/10.1055/s-0039-1677897

Lalli M, Amutha S (2020) Applications of machine learning and deep learning in healthcare domain - a literature review. Int J Electr Eng Technol 11(8):113–126. https://doi.org/10.34218/IJEET.11.8.2020.011

Lambert SI, Murielle M, Sopka S et al (2023) An integrative review on the acceptance of artificial intelligence among healthcare professionals in hospitals. NPJ Digit Med 6:111. https://doi.org/10.1038/s41746-023-00852-5

Marx V (2024) Quest for AI literacy. Nat Methods 21:1412–1415. https://doi.org/10.1038/s41592-024-02369-5

Mello MM, Guha N (2024) Understanding liability risk from healthcare AI. Stanford: Stanford University Human-Centered Artificial Intelligence [Policy brief]. https://hai.stanford.edu/sites/default/files/2024-02/Liability-Risk-Healthcare-AI.pdf. Accessed 1 Aug 2024

Mirbabaie M, Hofeditz L, Frick NR et al (2022) Artificial intelligence in hospitals: providing a status quo of ethical considerations in academia to guide future research. AI & Soc 37:1361–1382. https://doi.org/10.1007/s00146-021-01239-4

Nair M, Svedberg P, Larsson I, Nygren JM (2024) A comprehensive overview of barriers and strategies for AI implementation in healthcare: Mixed-method design. PLoS One 19(8):1–27. https://doi.org/10.1371/journal.pone.0305949

Owoyemi A, Osuchukwu J, Salwei M, Boyd A (2024) Checklist to support the development and implementation of AI in clinical settings. Preprint at doi:https://doi.org/10.1101/2024.08.08.24311701

Pacis DM, Subido ED Jr, Bugtai NT (2018) Trends in telemedicine utilizing artificial intelligence. AIP Conf Proc 1933:040009. https://doi.org/10.1063/1.5023979

Pallavi T, Rao AL, Bhanuteja T, Murthy TEGK (2022) Artificial intelligence: applications in healthcare industry. Int J Res AYUSH Pharm Sci 6(5):622–627. https://doi.org/10.47070/ijraps.v6i5.136

Perivolaris A, Adams-McGavin C, Madan Y et al (2024) Quality of interaction between clinicians and artificial intelligence systems: a systematic review. Future Healthc J 11(3). https://doi.org/10.1016/j.fhj.2024.100172

Pratt MK (2023) How AI has cemented its role in telemedicine. https://www.techtarget.com/searchenterpriseai/feature/How-AI-has-cemented-its-role-in-telemedicine. Accessed 1 Aug 2024

Rahimi AK, Pienaar O, Ghadimi M et al (2024) Implementing AI in hospitals to achieve a learning health system: systematic review of current enablers and barriers. J Med Internet Res 26:e49655. https://www.jmir.org/2024/1/e49655

Sahni N, Stein G, Zemmel R, Cutler DM (2023) The potential impact of artificial intelligence on healthcare spending, vol No. 30857. National Bureau of Economic Research., Working Paper Series,, Cambridge. https://doi.org/10.4103/jehp.jehp_402_23

Sharma S, Rawal R, Shah D (2023) Addressing the challenges of AI-based telemedicine: best practices and lessons learned. J Educ Health Promot 1(338):1–9. https://doi.org/10.4103/jehp.jehp_402_23

Sheth M, Sundler M, Sandstrom A et al (2023) Sustainable AI to drive global health. Sustainable AI to drive global health [White paper]. https://www.feam.eu/wp-content/uploads/Sustainable-AI-to-Drive-Global-Health-white-paper-11-Sep.pdf. Accessed 1 Aug 2024

Slattery P, Saeri AK, Grundy EA et al (2024) The AI risk repository: a comprehensive meta-review, database, and taxonomy of risks from artificial intelligence. The Massachusetts Institute of Technology. https://airisk.mit.edu/#Repository-Overview. Accessed 1 Aug 2024

Varnosfaderani SM, Forouzanfar M (2024) The role of AI in hospitals and clinics: transforming healthcare in the 21st century. Bioengineering 11(4), 337, pp. 1–38. doi:https://doi.org/10.3390/bioengineering11040337

Wu CT, Lin TY, Lin CJ, Hwang DK (2023) The future application of artificial intelligence and telemedicine in the retina: a perspective. Taiwan J Ophthalmol 13(2):133–141. https://doi.org/10.4103/tjo.TJO-D-23-00028

Yasin ZM, Bates DW, Shah N (2024) A five-level path to AI automation in health care doi:https://doi.org/10.1377/forefront.20240617.335248. Accessed 1 Aug 2024

Ying-Chuna J, Chung-Lan K, Aliaksandr AY et al (2020) The era of artificial intelligence–based individualized telemedicine is coming. J Chin Med Assoc 83(11):980–983. https://doi.org/10.1097/JCMA.0000000000000374

Zulaikha MRNS, Adibah SNS, AB N, Mahmud A (2021) The application of artificial intelligence (AI) in highly intelligent hospitals. Int J Public Health Clin Sci 8(6):1–18

Social Determinants of Health: Expanding our Understanding of Digital Maturity

Armin Scheuer, Nina Haffer, Valerie Kirchberger, and Sylvia Thun

Abstract

The role of hospitals is evolving beyond traditional medical treatment to encompass the social determinants of health (SDH) that significantly impact patient outcomes. This transformation is driven by the growing recognition of the interconnectedness between health and social factors and the shift toward value-based healthcare (VBHC). Digital maturity plays a crucial role in empowering this evolution, as advanced digital infrastructures enable the seamless collection, analysis, and sharing of comprehensive health and social data. Key to this process is the advancement of interoperable electronic health records (EHR) through health information exchange platforms, which support the seamless exchange of information across diverse platforms and stakeholders. By focusing on VBHC and leveraging digital technologies, hospitals can prioritize treatments that offer the greatest benefit to patients, reduce health disparities, and promote overall population health. This approach enhances healthcare outcomes and contributes to a more inclusive and equitable healthcare system.

Keywords

Social determinants of health · Interoperability · Value-based care · Digital health · Health informatics

A. Scheuer (✉)
Lemonmint HealthTech Communications GmbH, Berlin, Germany
e-mail: armin.scheuer@lemonmint.org

N. Haffer · S. Thun
Charite Universitätsmedizin, Berlin, Germany
e-mail: nina.haffer@charite.de; sylvia.thun@bih-charite.de

V. Kirchberger
Evela Health GmbH, Berlin, Germany
e-mail: valerie.kirchberger@evela.health

A. Scheuer, J. Studzinski (eds.), *Digital Maturity in Hospitals*,
https://doi.org/10.1007/978-3-031-80704-6_17

Abbreviations

CCA	Core competency area
CDR	Clinical data repositories
EHR/EPA	Electronic health record
EMRAM	Electronic Medical Record Adoption Model
HIS	Health information system
IHE	Integrating the Healthcare Enterprise
KSCs	Knowledge, skills, and competencies
PROM	Patient-reported outcome measures
REST	Representational state transfer
SDH	Social determinants of health
SOA	Service-oriented architecture
VBHC	Value-based healthcare

1 Introduction

Social determinants of health (SDH)—conditions tied to our birth, growth, living, working, and aging—are fundamental to our well-being. Recognizing their role in health equity and disparity reduction is critical (WHO n.d.). As hospitals expand their influence into patients' homes and communities, leveraging digital technologies and data becomes essential for understanding and addressing the SDH affecting their patient groups.

The WHO views social determinants of health as interconnected and multidimensional factors that impact health outcomes at individual, community, and societal levels. These determinants include socioeconomic status, education, employment opportunities, social support systems, access to healthcare, housing conditions, and environmental factors. Collectively, they influence access to resources, opportunities, and power, thereby shaping overall health and quality of life (WHO n.d.).

In the USA, social determinants of health have been a major topic of concern and research. The country's healthcare system has been grappling with health disparities across various racial, ethnic, and socioeconomic groups, highlighting the significant impact of these non-medical factors. According to the Robert Wood Johnson Foundation, up to 80% of a person's health in the USA is determined by SDH (Manatt et al. 2019). Efforts have been made to address these disparities through policy interventions and targeted public health initiatives, such as the Healthy People 2030[1] national objectives developed under the US Department of Health and Human Services.

[1] https://health.gov/healthypeople/objectives-and-data/social-determinants-health

In Europe, social determinants of health also play a crucial role in shaping health outcomes. However, due to more evenly distributed wealth and relatively equitable health systems, Europe often sees fewer health disparities among its population compared to the USA. While wealth distribution in the UK and the USA can create "poor societies with very rich people" (Burn-Murdoch 2022), Europe tends to maintain higher living standards for its less affluent populations. This equity, paired with comprehensive healthcare coverage and accessible social services, leads to superior health outcomes overall. A 2021 study by the Commonwealth Fund affirmed Europe's universal health coverage as among the most equitable in high-income countries, placing the USA at the lower end of the spectrum (Schneider et al. 2021).

1.1 Digital Maturity in the Context of SDH

In recent years, health systems globally have witnessed a significant push toward digital maturity, primarily driven by regulatory mandates. In Germany, a baseline assessment focused on procedural compliance, evaluating the availability and use of digital technologies in hospitals. While the digital maturity assessment evaluated hospitals' capabilities to capture and share data, it did not specifically address the integration of SDH data into medical practice. However, there is a significant opportunity regarding the integration of SDH data, and it is suggested that the next step in digital maturity should focus on incorporating SDH data into digital maturity frameworks. Ultimately, the beneficiaries of digitally mature healthcare systems must be the people, especially as the sector begins to emphasize prevention over treatment.

1.2 Strengthening Prevention over Treatment

The burden of disease underscores the urgency of the shift from treatment to prevention. Preventable non-communicable diseases (NCDs), particularly those affecting the cardiovascular and metabolic systems, continue to be the leading cause of death globally. According to the World Health Organization (WHO), cardiovascular diseases (CVDs) are accounting for 17.9 million deaths annually, which represents 32% of all global deaths.[2] Most of these deaths are due to heart attacks and strokes, which are largely preventable through lifestyle changes and addressing key risk factors such as unhealthy diets, physical inactivity, and tobacco use.

Other NCDs, including diabetes and chronic respiratory diseases, also contribute significantly to global mortality. These diseases are driven by modifiable behavioral risk factors and metabolic risk factors such as high blood pressure, raised blood

[2] https://www.who.int/news-room/fact-sheets/detail/cardiovascular-diseases-(cvds)

glucose, and obesity. Together, these conditions account for a substantial portion of premature deaths, emphasizing the critical need for preventive measures and early interventions.[3,4]

A "super" SDH in this context is health literacy, including digital health literacy, which is directly linked to health outcomes (van Kessel et al. 2022). Health literacy is a measure of one's ability to understand healthcare information. Health literacy is strongly connected to socioeconomic status and influences health outcomes (Braveman and Gottlieb 2014). Childhood obesity and cardiovascular disease in adults are prime examples of conditions influenced by these determinants. To effectively address these and other SDHs, it is imperative to reach individuals who are often removed from the healthcare discourse. This requires two key advancements in healthcare:

1. **Advancement of Value-Based Healthcare (VBHC):** Embracing value-based healthcare (VBHC) is essential to align healthcare outcomes with patient well-being, ensuring that services deliver the most value to those who need them. By focusing on VBHC, the healthcare system can prioritize treatments and interventions that provide the greatest benefit to patients. Viewing individuals through the lens of their backgrounds and varying outcomes allows for a more nuanced and ethical approach to healthcare.
2. **Access to Health Data:** VBHC requires a high volume of patient data across the entire continuum of care—not only health data but also SDH data, i.e., information related to education, social circumstances, and disease incidence. This data must be captured, accessed, and shared by patients and their healthcare providers, which necessitates a high level of digital maturity, particularly concerning electronic health records (EHR) and interoperability.

While the transition to VBC is slower in Europe, there are successful VBC-based health systems such as Kaiser Permanente in the USA, also known for their determined use of health data and technology to enhance patient engagement and improve outcomes.

Advancing interoperability and electronic health records through health information exchange platforms is at the core of every digital maturity framework. Interoperability supports the seamless exchange of information across diverse platforms and stakeholders, ensuring that every piece of data contributes to a holistic understanding of patient needs and societal health trends. This interconnectedness, driven by digital maturity, enables healthcare providers to make more informed decisions, leading to improved health outcomes and a more inclusive and effective healthcare system.

---

[3] https://www.paho.org/en/topics/noncommunicable-diseases

[4] https://www.who.int/data/gho/data/themes/topics/topic-details/GHO/ncd-mortality

2 What Is Value-Based Healthcare

VBHC is a healthcare delivery model that prioritizes patient outcomes relative to the costs of achieving those outcomes. The primary aim of VBHC is to improve the quality of care provided to patients by focusing on their overall health and well-being rather than the volume of services delivered. This model involves structuring healthcare around patients' needs and creating a system where providers are rewarded for helping patients achieve better health results in a cost-effective manner.

VBHC emphasizes the importance of measuring patient outcomes (patient-reported outcome measures = PROMs) and using these metrics to inform care practices and policies. By collecting and analyzing data on patient health outcomes, healthcare providers can identify the most effective treatments and practices, thereby continuously improving care quality. Additionally, VBHC encourages the integration of care across different providers and settings, ensuring that patients receive coordinated and comprehensive care tailored to their specific needs.

One of the core principles of VBHC is the alignment of incentives. This means that financial incentives for healthcare providers are linked to the quality and efficiency of the care they deliver. For example, providers might receive bonuses for achieving certain health outcomes or reducing hospital readmissions, thereby encouraging them to focus on long-term patient health rather than short-term interventions.

Moreover, VBHC supports patient-centered care, where patients are actively involved in their own health decisions. This approach fosters better patient–provider communication, enhances patient satisfaction, and empowers patients to manage their own health more effectively. By involving patients in their care plans and decisions, healthcare providers can ensure that treatments are aligned with the patients' preferences and goals.

Overall, VBHC aims to create a more sustainable healthcare system that delivers high-quality care while controlling costs. By focusing on outcomes, fostering care coordination, and aligning financial incentives with patient health, VBHC seeks to enhance the overall health of populations and improve the efficiency of healthcare delivery.

3 What Is Interoperability?

Increasing digitalization is generating a lot of data. The lack of standards prevents this data from being exchanged across system boundaries. This, in turn, results in so-called data silos.

Interoperability is defined as the ability of systems to exchange and interpret data in a compatible manner (Lehne et al. 2019). The aim is to integrate multiple systems from multiple stakeholders without obstacles (Mertens and Wieczorrek 2000). In the medical sector, interoperability enables seamless communication between healthcare providers across interfaces. Data from different institutions is interoperable if it can

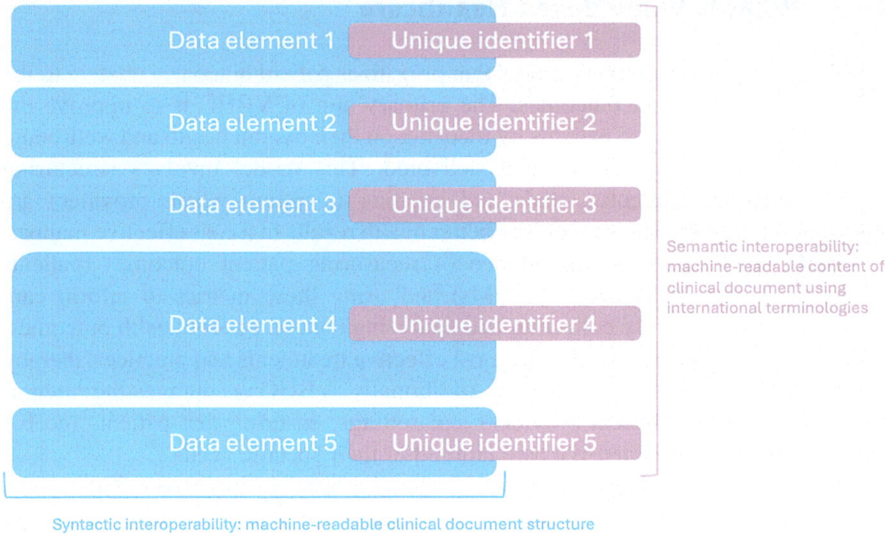

Fig. 1 Scheme of a machine-readable clinical document structure

be used seamlessly in the respective institution's patient care and decision support programs without having to be translated or converted.

Digital health information is exchanged between two related information processing systems or system components via interfaces, so-called connection points. Error-free translation at an interface used to network systems and technologies can be achieved by using national and international standards. In other words, defining common standards makes a smooth data flow possible. The generic term interoperability refers to a level model in which four mutually influencing levels are distinguished: technical, syntactic, semantic, and organizational interoperability (Johner and Haas 2009).

Interoperability in healthcare is crucial for improving the quality of care and effectively sharing health information (Lehne et al. 2019). Once data interoperability becomes the new normal, the way is paved to pursue artificial intelligence approaches and use machine learning and natural language processing to gain actionable insights. These can guide professionals and improve the quality of care. Predictive data analysis leads to a more robust identification of patient risks and the development of new clinical treatment pathways (Ghosh 2020).

The increase in data quality through data interoperability enables, on the one hand, a higher availability of evaluable data and, on the other hand, an increased variety of usable data which can also reduce data gaps.

In order to increase the usability of digital clinical data, it must be transferred across system boundaries in a format that is not only human-readable but also machine-readable. Furthermore, the content of the digitized document must be identified in terms of its meaning automatically using international terminology standards. Only such factors allow automatic evaluations (Fig. 1).

3.1 Syntactic Interoperability

Syntactic interoperability is defined as the definition of data formats and structures (Lopez and Blobel 2009). This enables data to be received from different systems and identified as such. The definition of such formats and structures for data exchange in a machine-readable and human-understandable form is essentially ensured in the healthcare sector by standardization organizations such as Health Level 7 (HL7) and Integrating the Healthcare Enterprise (IHE). The HL7 standard Fast Healthcare Interoperability Resources (FHIR) is particularly important in the context of patient-centered interoperability and web-based technologies (Thun 2017). In addition to its increasing use in systems and processes worldwide and the growing developer community, the standard is particularly suitable due to its lightweight and expandable architecture based on web standards and protocols such as representational state transfer (REST). FHIR focuses on structured data exchange in the form of a REST programming interface (REST application programming interface and search), message transfer, document exchange, a service-oriented architecture (SOA), and persistent storage. Health concepts are defined within FHIR, so-called resources, which can be accessed and exchanged via web technologies. Resources are modular, expandable information units. Relationships are created from links between resources via Uniform Resource Locators (URLs). For FHIR profiling in Germany, the Extensible Markup Language (XML) and JavaScript Object Notation (JSON) data format can be used, a human- and machine-readable language for text files for data exchange across system boundaries, regardless of the respective underlying system (HL7 International 2023).

3.2 Semantic Interoperability

While standards such as HL7 FHIR already describe the basic syntax of the data, semantic interoperability refers to the entirety of the definitions of the meaning content of messages (Lopez and Blobel 2009). The overarching goal at the semantic interoperability level is to interpret the information units of the participating systems in the same way. When data generated in one system is used in another, it is desirable that this use matches the meaning or semantics of the data in order to avoid increased workload. To achieve semantic interoperability, international and national classification systems such as nomenclatures, e.g., Logical Operation Identifiers Names and Codes (LOINC) and Unified Codes for Units of Measure (UCUM), or taxonomies and classification systems such as the International Statistical Classification of Diseases and Related Health Problems 10 (ICD-10), operation and procedure codes (OPS), or ontologies such as the Systematized Nomenclature of Medicine (SNOMED CT) are used (Lin et al. 2011).

4 How Can Digital Maturity Support Value-Based Healthcare

One of the fundamental basics of mature digital systems is interoperability. It is a necessary requirement for a data-driven and patient-centered system, in other words a value-based system. Interoperability is especially important for one of the key data points of VBHC: the collection of patient-reported outcomes. Data means that individual treatment plans can be created and patient data can be analyzed more precisely.

Hundreds of models exist worldwide to measure the digital maturity of hospitals and other healthcare organizations. (Carvalho et al. 2016; Tarhan et al. 2020)

- The NHS's digital maturity assessment "helps providers and integrated care systems across England to understand their level of digital maturity by identifying key strengths and gaps in the provision of digital services."[5] The assessment measures maturity against the seven dimensions of the What Good Looks Like (WGLL) framework.[6]
- CHIME's Digital Health Most Wired[7] comprises 10 stages; from stage 7 onward health organizations meet the criteria for being designated as "most wired"—which includes deployed technologies and strategies for health information exchange to help them analyze their data.
- The HIMSS Electronic Medical Record Adoption Model (EMRAM)[8] comprises eight levels (0 to 7) that map the maturity of digital applications around the hospital information system (HIS) and, in particular, the degree of implementation and use of digital functions to support patient care. Interoperability is an integral part of the EMRAM model. A fully digitally mature hospital (level 7) would provide complete interoperability enabling seamless data exchange and use across the healthcare ecosystem.

The traceability of patients' journeys through the healthcare system enables a more precise analysis of their state of health and the success of their treatment. As a result, the integration of PROMs is becoming increasingly important. The standardized and electronic recording of patient-reported outcomes can significantly influence the quality of treatment.

Processes can be automated and improved through automatic evaluations with the help of AI. The standardized evaluation of large amounts of data also makes it possible to identify deeply rooted human biases. One particular subject is ethical concerns with automation of processes, since the same algorithms are used on large collections of data.

---

[5] https://www.england.nhs.uk/digitaltechnology/connecteddigitalsystems/digital-maturity-assessment/

[6] https://transform.england.nhs.uk/digitise-connect-transform/what-good-looks-like/what-good-looks-like-publication/

[7] https://www.dhinsights.org/most-wired

[8] https://www.himss.org/what-we-do-solutions/maturity-models-emram

Racial biases, for example, resulted in an underrepresentation of diverse skin types in machine learning research for skin cancer detection. The analysis of large amounts of data in research makes it easier to identify unethically selected study populations. This will make it easier to uncover social prejudices in the future (Guo et al. 2022).

5 Digital Maturity Assessment Identifies Gaps in Information Exchange

The German Ministry of Health has completed the world's largest assessment of digital maturity to date. In the period from October 5, 2021, to December 17, 2021, a total of 1624 hospitals, representing 91% of all hospitals in the country, participated in the DigitalRadar[9] online self-assessment of their digital maturity as part of a national innovation funding program for hospitals.

The assessment focused on seven dimensions to identify critical barriers and gaps in digital maturity. The dimension "information exchange" specifically looked into the hospital's ability to securely communicate clinical data and hospital indicators both internally and externally using recognized technical, syntactic, and semantic standards and terminologies (e.g., IHE, HL7, SNOMED CT, or the Interoperability Directory of gematik[10]). This dimension aimed at understanding the extent to which seamless interoperability of digital services is achieved, eliminating media discontinuities for users, and ensuring efficient integration of third-party products, including data outputs from medical devices (Table 1).

5.1 Findings About the State of Information Exchange in German Hospitals

The average DigitalRadar Score (DR Score) was 33.3 out of 100 possible points, with the majority of hospitals (70%) scoring between 23 and 44 points (Amelung et al. 2022). The information exchange dimension stands out as one of the three lowest-performing categories, with an average score of 25.1%. This is significantly lower than other dimensions, such as structures and systems (55.3%) and resilience management and performance (45.2%). The low performance of information exchange indicates substantial challenges in the effective sharing of clinical data both within and between hospitals (Table 2).

Drilling down into the sub-dimensions and mapping the hospital performance to the respective dimensions in EMRAM—one of the maturity models applied in DigitalRadar—reveals that the infrastructure for data exchange in German hospitals is relatively well-developed, with the majority of hospitals having the necessary

[9] https://www.digitalradar-krankenhaus.de/

[10] Gematik is Germany's national digital health agency: https://www.gematik.de/

Table 1 DigitalRadar questions for dimension "information exchange/interoperability" DigitalRadar Krankenhaus Konsortium (2021)

Question	Answering options
Is there a binding strategy that ensures recognized, non-proprietary interoperability standards for the exchange of health data are applied in the procurement of IT systems, communication interfaces, and/or internal programming routines? For example, standards and terminologies such as HL7, IHE, FHIR, DICOM, GS1, CEN, and SNOMED CT. This allows for the efficient integration of third-party products	(a) Not implemented (b) Strategy exists, not yet implemented (c) Strategy exists, partially implemented (d) Strategy exists, partially implemented and success measured (e) Strategy exists, fully implemented and success measured (f) Don't know
If "Strategy exists": Which interoperability standards and terminologies are in use?	Please select all applicable options: • HL7 • IHE • FHIR • DICOM • GS1 • UCUM (according to BfArM specifications) • SNOMED CT • LOINC • ISO/IEEE 11073 • SDC • ORPHANET • OMICS terminologies • Others (please specify)
If "IHE": Which IHE profiles do you use?	
All discrete data generated from structured input forms are normalized to a standardized/controlled medical vocabulary (e.g., LOINC, SNOMED, ICD) for all clinical documentation	(a) Not implemented (b) Minimally implemented (c) Partially implemented (d) Mostly implemented (e) Fully implemented (f) Don't know
Digital admission management enables communication with existing HIS and ERP systems This means that digitally captured patient data is also automatically and interoperably available for downstream organizational processes	(a) No digital admission management available (b) Not implemented (c) Minimally implemented (d) Partially implemented (e) Mostly implemented (f) Fully implemented (g) Don't know
If "HIS in operation": Please indicate the extent to which the HIS is integrated at the application level The goal is for clinical users to access clinical information from different departments through a unified user interface. This can be achieved either through a single vendor's unified solution or through various integrated software modules that function as a common repository	(a) Based on a central database/repository (b) Based on multiple databases/repositories that appear seamless to the user (c) Based on multiple databases/repositories that are not fully integrated (d) Don't know

(continued)

Table 1 (continued)

Question	Answering options
For nursing and treatment documentation, text modules can be used that are based on interdisciplinary and uniformly stored terminologies (e.g., international standards such as SNOMED, ICD, LOINC, or ORPHANET)	(a) Not implemented (b) Minimally implemented (c) Partially implemented (d) Mostly implemented (e) Fully implemented (f) Don't know

Table 2 Overall DigitalRadar performance by dimension (Adapted from Amelung et al. 2022)

Dimension	Average achievement (%)
Structures and systems	55.3
Resilience management and performance	45.2
Organizational management and data management	40.6
Clinical processes	38.5
Information exchange	*25.1*
Telehealth	18.0
Patient participation	5.3

systems in place for secure data capture, storage, and internal exchange of health information. HIS and clinical data repositories (CDR) are widely implemented, which is a crucial foundation for achieving interoperability. Impressively, nearly 50% of hospitals meet over 90% of the criteria for health information exchange in EMRAM, indicating a robust internal data exchange capability. Even hospitals at the lowest maturity level fulfill at least 80% of the requirements for HIS/CDR, highlighting a strong base for digital data management within individual hospitals.

However, significant barriers remain in the exchange of data between hospitals. While internal interoperability within hospitals is relatively strong, the ability to share data seamlessly between different hospitals and healthcare institutions is still limited. This lack of external interoperability hinders the overall efficiency and effectiveness of the healthcare system, preventing a cohesive and integrated approach to patient care. To address this, there is a need for standardized data formats, unified communication protocols, and enhanced regulatory frameworks that facilitate secure and efficient data exchange across the entire healthcare network. Without these improvements, the full potential of digital maturity and the benefits of interconnected healthcare systems cannot be realized (Table 3).

Digital maturity assessments like DigitalRadar can identify specific gaps and guide targeted investments in technology, infrastructure, and training to enable VBHC. This strategic approach—especially if implemented at a national health system level—ensures efficient resource allocation to areas that will put the necessary digital infrastructure in place to gather, use, and exchange data effectively. The seamless exchange of health information across different systems and providers is essential for coordinating care, improving patient outcomes, and reducing healthcare costs.

Table 3 Detailed analysis of the information exchange dimension

Sub-dimension in "information exchange"	Possible points	Achievement	Key insights
Information exchange with external actors	6.0	23.4%	Low score indicates significant room for improvement in sharing data with external entities
Information exchange with patients	2.5	5.0%	Very low score shows a critical need for better mechanisms to facilitate patient access to their health data
Information exchange among clinical staff	2.0	11%	Indicates challenges in internal communication and data sharing among healthcare professionals
Integration of medical devices	9.6	16.2%	Low achievement suggests difficulties in effectively integrating data from medical devices into the hospital information systems
Interoperability	4.0	48.5%	While higher relative to other sub-dimensions, it still reflects substantial challenges in achieving seamless data integration across different systems and platforms

6 The Increasing Impact of Hospitals on Their Communities

While hospitals primarily focus on providing medical treatment and care, they are increasingly recognizing the importance of addressing SDHs that influence health outcomes. This expanded role of hospitals is driven by several global trends, of which two are particularly relevant in the context of SDH:

- **Anchor Hospitals:** The consolidation in the hospital sector forms larger organizations and health networks that can serve as anchor institutions in their communities or regions. These large, stable, and economically impactful organizations are well-equipped with medical expertise, data, local insight, and influence to understand and manage the SDHs of their population cohorts (Ubhayakar et al. 2017). Large European health systems, such as Greater Paris University Hospitals (AP-HP) and Vivantes—Netzwerk für Gesundheit GmbH in Berlin, exemplify efforts to pursue an anchor mission. They play a vital role in facilitating "meaningful action to address the groundwater conditions that have the most significant impact on residents' health and community well-being" (Dave et al. 2021). By acting as anchor institutions, these hospitals can leverage their resources to implement community-wide health initiatives, collaborate with local organizations, and address broader social determinants that affect health outcomes. This holistic approach ensures that healthcare extends beyond the hospital walls, promoting overall community well-being and health equity (Franz et al. 2019).

- **Integration of health and social care to improve prevention:** The integration of health and social care systems aims to provide comprehensive services that address both medical and social needs and is a policy focus in many countries (Exley et al. 2024). This approach acknowledges that health outcomes are influenced by a variety of SDHs beyond clinical care. By integrating these services, hospitals in cooperation with a variety of local health services can offer more holistic care, leading to improved patient outcomes and a shift from treatment to prevention. This shift involves identifying and mitigating risk factors associated with SDHs. By prioritizing prevention, hospitals can help reduce the incidence of chronic diseases, improve population health, and decrease healthcare costs in the long run.

For hospitals to record and use SDH data, they will require enhanced systems and tools, advanced data analytics, and the ability to share that data among the multiple stakeholders involved in delivering care. According to HIMSS, there is a lack of standardization in capturing attributes of SDH, making it challenging to identify social needs, implement interventions, and measure their impact. By structuring and encoding the information based on standard code systems, it becomes interpretable and can be used by decision support tools to alert service providers to social risks that may influence treatment decisions. Efforts are being made to incorporate SDH elements into existing health-related terminologies, but further development is needed to capture the full breadth of observations related to social needs and interventions (HIMSS 2021).

Once hospitals have standardized processes and data on SDH, they can use it to improve patient outcomes and overall community health:

- Hospitals can implement screening protocols to identify patients' social needs and risk factors, leading to targeted interventions.
- By collaborating and connecting patients with appropriate social services and community health workers, hospitals can reduce readmissions.
- Partnering with local organizations and advocating for policies addressing SDH allows hospitals to engage in community-wide initiatives.
- Additionally, by promoting health education and preventive care, hospitals empower patients to take control of their health, ultimately contributing to better health outcomes.
- Through research and data analysis, hospitals can identify patterns, tailor interventions, and contribute to understanding the links between SDH and health outcomes, enhancing their ability to address the specific needs of their patient populations.

By focusing on SDH, hospitals will move beyond "the internal–external dichotomy" and evolve into complex institutions that are "increasingly entwined" with the communities they serve and the health policies of their countries. Equipped with adequate digital infrastructure, hospitals can thus support new approaches in public health—in particular in the fight against non-communicable diseases (NCD), such as diabetes.

6.1 The Case of Diabetes

Diabetes is one of the most prevalent NCDs, impacting millions of people worldwide. There is a strong association between poor diabetes control and individual adverse SDH, and studies found that in particular low income, lack of health insurance, food insecurity, and limited language proficiency in particular are significantly associated with uncontrolled diabetes (Kim et al. 2021).

Digital health technologies offer promising solutions to address the challenges posed by social determinants of health in diabetes management. These technologies leverage the widespread use of smartphones and Internet connectivity to deliver personalized interventions that encourage behavior change and enhance patient engagement. Mobile Health Apps: Diabetes management apps provide users with tools to track blood glucose levels, monitor physical activity, and manage dietary habits. They offer personalized recommendations and reminders, promoting healthier lifestyle choices and medication adherence.

1. Telemedicine: Virtual healthcare consultations enable patients to access healthcare professionals regardless of their location, overcoming geographical barriers. This is especially beneficial for individuals living in remote or underserved areas with limited access to diabetes specialists.
2. Wearable Devices: Devices like fitness trackers and continuous glucose monitors provide real-time health data, empowering individuals to take proactive steps toward managing their diabetes through increased physical activity and lifestyle modifications.
3. Social Support Platforms: Online communities and support groups foster social connections and emotional support for individuals living with diabetes. These platforms help combat feelings of isolation and provide a space for sharing experiences and knowledge.

By addressing the social determinants of health and promoting behavior change, digital health technologies can positively impact diabetes prevalence and progression. Increased access to education, nutritional resources, and remote healthcare consultations can lead to improved diabetes management outcomes, reduced risk of complications, and enhanced overall quality of life for individuals living with diabetes.

In a digitally mature health system that is interoperable and able to exchange health information, all these technologies can deliver data into a central data repository—such as a patient's electronic health records—and thus deliver care providers a holistic view of their patients that also includes data and information on SDHs.

7 Discussion

The discussion on digital maturity in healthcare, particularly through the lens of VBHC and interoperability, underscores the transformative potential of focusing on patient and user needs. While regulatory mandates have laid the groundwork for digital advancement in countries around the world, it is imperative to center these efforts on improving patient outcomes and experiences. This approach not only enhances the effectiveness of healthcare delivery but also aligns with the ethical imperative of prioritizing the well-being of individuals.

A key insight is the connection between health literacy, SDHs, and digital maturity. We have aimed to highlight that health literacy extends beyond mere understanding of healthcare information. Our broader definition of health literacy is crucial, given its strong correlation with social status and health outcomes. The recognition of SDHs as significant contributors to health disparities challenges the conventional underestimation of these factors and calls for a more integrated approach to healthcare.

Another critical aspect is the integration of VBHC principles with digital maturity. VBHC emphasizes patient outcomes relative to the costs of achieving those outcomes, promoting a patient-centered approach that aligns financial incentives with quality care. This model's reliance on PROMs illustrates the importance of incorporating patient feedback into care practices. However, achieving this integration requires robust interoperability, ensuring seamless data exchange across platforms and stakeholders.

As discussed, interoperability is fundamental to realizing the full potential of digital maturity. By facilitating seamless communication and data exchange, interoperability supports a comprehensive understanding of patient needs and societal health trends. Using standards like HL7's FHIR is a step toward achieving this goal, but implementing these standards must be consistent and widespread to be truly effective.

Ethical considerations, particularly around algorithmic biases and the automation of healthcare processes, also warrant critical discussion. The underrepresentation of diverse populations in machine learning research highlights the risk of perpetuating health disparities through digital solutions. Addressing these biases is crucial to ensuring equitable healthcare outcomes.

Looking ahead, several areas require further research to build on the foundations discussed in the chapter. First, there is a need for more comprehensive studies on the impact of SDHs on health outcomes. These studies should explore the mechanisms through which socioeconomic factors influence health and identify strategies to mitigate these effects. Further research is also needed to refine and expand the implementation of VBHC. This includes developing more sophisticated methods for collecting and analyzing PROMs, ensuring that these measures are integrated into everyday clinical practice. Additionally, exploring different models of care coordination and their effectiveness in various healthcare settings can provide valuable insights for optimizing VBHC implementation. Interoperability remains a critical area for future research. Studies should focus on the practical challenges of achieving interoperability, including adopting standards like FHIR across diverse healthcare systems. Investigating the impact of interoperability on healthcare outcomes and identifying best practices for its implementation will be crucial for advancing digital maturity.

Lastly, ethical concerns related to digital healthcare demand ongoing attention. Research should aim to develop frameworks for identifying and mitigating biases in healthcare algorithms. This includes not only technical solutions but also policies and guidelines to ensure that digital tools are used ethically and equitably.

In conclusion, while significant strides have been made toward digital maturity in healthcare, centering these efforts on patient and user needs remains paramount. By integrating VBHC principles and enhancing interoperability, the healthcare system can become more inclusive, effective, and patient-centered. Addressing the social determinants of health and improving health literacy are essential steps in this direction. Future research should continue to explore these areas, ensuring that digital advancements translate into tangible benefits for all individuals.

References

Amelung V, Angelkorte M, Augurzky B et al (2022) Ergebnisse der ersten nationalen Reifegradmessung deutscher Krankenhäuser [Zwischenbericht]. https://www.digitalradar-krankenhaus.de/download/220914_Zwischenbericht_DigitalRadar_Krankenhaus.pdf. Accessed 16 July 2024

Braveman P, Gottlieb L (2014) The social determinants of health: it's time to consider the causes of the causes. Public Health Rep 129(Suppl 2):19–31. https://doi.org/10.1177/00333549141291S206

Burn-Murdoch J (2022) Britain and the US are poor societies with some very rich people. https://www.ft.com/content/ef265420-45e8-497b-b308-c951baa68945. Accessed 6 Aug 2023

Carvalho JV, Rocha Á, Abreu A (2016) Maturity models of healthcare information systems and technologies: a literature review. J Med Syst 40(6):1–10. https://doi.org/10.1007/s10916-016-0486-5

Dave G, Wolfe MK, Corbie-Smith G (2021) Role of hospitals in addressing social determinants of health: a groundwater approach. Prev Med Rep 21:101315. https://doi.org/10.1016/j.pmedr.2021.101315

DigitalRadar Krankenhaus Konsortium (2021) Instrument zur Evaluierung des Reifegrades der Krankenhäuser hinsichtlich der Digitalisierung. Schriftenreihe in Health Economics, Management and Policy No. 2021-01. https://www.econstor.eu/handle/10419/274666. Accessed 6 Aug 2023

Exley J, Glover R, Mccarey M et al (2024) Governing integrated health and social care: an analysis of experiences in three European countries. Int J Integr Care 24(1):9. https://doi.org/10.5334/ijic.7610

Franz B, Skinner D, Wynn RJ (2019) Urban hospitals as anchor institutions: frameworks for medical sociology. Socius: Sociol Res Dyn World 5(1). https://doi.org/10.1177/2378023118817981

Ghosh S (2020) Interoperabilität im Gesundheitswesen für das Engagement der Verbraucher [Whitepaper]. https://www.virtusa.com/de/insights/whitepaper/healthcare-interoperability. Accessed 16 July 2024

Guo LN, Lee MS, Kassamali B et al (2022) Bias in, bias out: underreporting and underrepresentation of diverse skin types in machine learning research for skin cancer detection—A scoping review. J Am Acad Dermatol 87(1):157–159. https://doi.org/10.1016/j.jaad.2021.06.884

HIMSS (2021) Social determinants of health. https://www.himss.org/resources/social-determinants-health. Accessed 6 Aug 2023

HL7 International (2023) FHIR overview - developers. https://hl7.org/fhir/overview-dev.html. Accessed 8 Aug 2024

Johner C, Haas P (2009) Praxishandbuch IT im Gesundheitswesen: Erfolgreich einführen, entwickeln, anwenden und betreiben. Carl Hanser Verlag, München

Kim EJ, Abrahams S, Marrast L et al (2021) Significance of multiple adverse social determinants of health on the diagnosis, control, and management of diabetes. J Gen Intern Med 36:2152–2154. https://doi.org/10.1007/s11606-020-05860-9

Lehne M, Sass J, Essenwanger A et al (2019) Why digital medicine depends on interoperability. npj Digit Med 2(79):1–5. https://doi.org/10.1038/s41746-019-0158-1

Lin MC, Vreeman DJ, Huff SM (2011) Investigating the semantic interoperability of laboratory data exchanged using LOINC codes in three large institutions. AMIA Ann Symp Proc 2011: 805–814

Lopez DM, Blobel BG (2009) A development framework for semantically interoperable health information systems. Int J Med Inform 78(2):83–103. https://doi.org/10.1016/j.ijmedinf.2008.05.009

Manatt, Phelps & Phillips, LLP (2019) Medicaid's role in addressing social determinants of health. Robert Wood Johnson Foundation. https://www.rwjf.org/en/insights/our-research/2019/02/medicaid-s-role-in-addressing-social-determinants-of-health.html. Accessed 21 Dec 2021

Mertens P, Wieczorrek HW (2000) Data X Strategien. Data Warehouse, Data Mining und operationale Systeme für die Praxis. Springer, Berlin

Schneider EC, Shah A, Doty MM et al (2021) Mirror, mirror 2021: reflecting poorly. Health Care in the U.S. compared to other high-income countries. https://www.commonwealthfund.org/publications/fund-reports/2021/aug/mirror-mirror-2021-reflecting-poorly. Accessed 6 Aug 2023

Tarhan AK, Garousi V, Turetken O, Söylemez M, Garossi S (2020) Maturity assessment and maturity models in health care: a multivocal literature review. Digit Health 6. https://doi.org/10.1177/2055207620914772

Thun S (2017) Neuer Verband für IT-Standards: Kampf gegen das Schneckentempo. kma - Klinik Management aktuell 22(7/8):64–65. https://doi.org/10.1055/s-0036-1594742

Ubhayakar S, Capeless M, Owens R et al (2017) The anchor mission playbook. https://democracycollaborative.org/learn/publication/anchor-mission-playbook. Accessed 6 Aug 2023

van Kessel R, Wong BLH, Clemens T et al (2022) Digital health literacy as a super determinant of health: More than simply the sum of its parts. Internet Interv 27:100500. https://doi.org/10.1016/j.invent.2022.100500

World Health Organization (n.d.) Social determinants of health. https://www.who.int/health-topics/social-determinants-of-health#tab=tab_1. Accessed 8 Aug 2024

Strengthening the Healthcare Workforce to Achieve Digital Maturity at Scale

Angela Ahrendt, Georgi Chaltikyan, Fara Aninha Fernandes, and Lisa Rice-Duek

Abstract

The advancement of digital maturity of healthcare organizations and health systems hinges on the potential of digital tools to empower healthcare professionals. This empowerment extends beyond care delivery to encompass the use of digital technologies for administrative functions, ensuring seamless operations and automated clinical workflows. To achieve a digitally empowered healthcare workforce, building healthcare professionals' capacity, confidence, and trust in digital tools remains imperative amidst resource constraints faced by healthcare organizations. Integrating data and artificial intelligence within care pathways and clinical management tools offers avenues to enhance healthcare effectiveness and overall improve the health workforce's performance. As the health workforce evolves, including diverse stakeholders in the digital transformation process becomes paramount to the nurturing of integrated and sustainable digital health ecosystems. This article delineates three user groups—Digital Health Workforce, Health Workforce, and Empowered Citizens—each with defined roles crucial for improved healthcare outcomes and accelerated adoption of digital technologies within clinical practice. Each user group necessitates both technological investment and cultural shift, underlining the need for distinct training and upskilling programs. The authors propose personalized strategies

A. Ahrendt (✉)
FTI Consulting, Berlin, Germany
e-mail: angela.ahrendt@fticonsulting.com

G. Chaltikyan · F. A. Fernandes
Deggendorf Institute of Technology, Deggendorf, Germany
e-mail: georgi.chaltikyan@th-deg.de; fara.fernandes@th-deg.de

L. Rice-Duek
Barcelona Institute for Global Health, Barcelona, Spain
e-mail: Lisa.rice@isglobal.org

A. Scheuer, J. Studzinski (eds.), *Digital Maturity in Hospitals*,
https://doi.org/10.1007/978-3-031-80704-6_18

for the further development of the user groups and outline scenarios for fostering workforce-driven digital maturity to ensure healthcare's resilience in the new demographic era.

Keywords

Workforce development · Upskilling · Digital literacy · Health informatics · Core competency area frameworks

Abbreviations

EHR	Electronic Health Record
CCA	Core Competency Area
ICT	Information and Communication Technology
KSCs	Knowledge, Skills, and Competencies
DiGA	Digital Health Applications
SHI	Statutory Health Insurance
KPI	Key Performance Indicators
EDH	Environmental Determinants of Health
CDH	Commercial Determinants of Health
DHIS	Digital Health Information Systems
CPOE	Computerized Provider Order Entry

1 Foreword

The World Health Organization (WHO) estimates a global shortfall of ten million health workers by 2030 (Boniol et al. 2022). Data from WHO's National Health Workforce Accounts (NHWA), however, shows disparities in the distribution of the health workforce shortage, two-third of which are claimed to be borne by countries on the WHO support and safeguards list,[1] predominantly covering the WHO's African and Eastern Mediterranean regions. Without the improved capacity to boost enrollment, employ and retain nursing graduates, this deficit in the health workforce will include at least 5.7 million nurses worldwide (WHO 2020). While the shortage is unevenly distributed among nations, posing a significant impact,

[1]Note that Boniol et al. (2022) refer to the 2020 WHO support and safeguard list consisting of 47 countries with (1) a density of doctors, nurses, and midwives below the global median (i.e., 49 per 10,000 population); and (2) a universal health coverage (UHC) service coverage index below 50 (index range: 0–100). Since then, the 2023 WHO support and safeguard list was published which now includes 55 countries with pressing health workforce needs according to the aforementioned criteria, albeit the UHC service coverage index threshold in the 2023 list was increased from 50 to 55 in order to account for the disruptions caused to health services by the COVID-19 pandemic: https://www.who.int/publications/i/item/9789240069787

particularly on low- and middle-income countries, the heavy dependence of American and European regions on the international nursing workforce makes the 2030 projections a matter of global concern.

Parallel to shortages in the health workforce supply, the world is grappling with a population that is aging, hence fueling an increased demand for health and care services. The increasing dependency on an aging population is expected to exert profound effects on already strained health systems, and more significantly, on the labor market. The decline in the working-age population necessitates the incorporation of more individuals into the workforce. In instances where demographic shifts hinder the influx of new workers to replace those retiring, the only viable solution to bridge the workforce gap is to enhance productivity through technological advancements and skills development. However, to truly unleash the digital transformation process one cannot simply implement technology and introduce digital training schemes in a top-down manner. In this article, the authors look at tailored, collaborative, and iterative processes that consider the knowledge, skills, and competence (KSC) needs of all participants involved in the digitalization of a health organization and the health system as a whole.

The authors start by postulating ways in which digital health technologies could improve the perceived benefit from their use by highlighting the effect these have on increasing health workforce capacities, which ultimately leads to better healthcare outcomes and increased employee retention rates. In the second part, the authors identify three distinct yet closely interlinked user groups that need tailored educational and training programs to be equipped with the suite of tools and knowledge to support the transformation processes in hospitals and health systems. Lastly, the authors examine scenarios for a workforce-driven advancement of digital maturity which is the last step to ensuring sustained and resilient health delivery in the wake of a growing and increasingly aging global population.

2 The Role of Digital Health in Transforming the Workforce

2.1 Introduction

The Covid-19 pandemic profoundly challenged health systems demonstrating the paramount need for the integration of innovative healthcare technologies into mainstream care. Evolving beyond mere conveniences, digital health technologies such as Electronic Health Records, telemedicine, wearable devices, mobile health platforms, and health applications (apps) powered by data and artificial intelligence/machine learning (AI/ML) have emerged as the key enablers to the continuity of healthcare delivery and preventive measures during periods of significant strain and scarcity in the health workforce. To better understand the value digital health technologies can add, and to trust their ability to drive real and lasting change, establishing confidence in their use is a key lever. From healthcare leadership to healthcare professionals and to the ultimate recipient of healthcare, the patient, widespread acceptance can be realized when benefits become tangible across all user groups. Additionally, recognizing the importance of accessing and sharing health data becomes imperative.

Cultivating confidence in the use of digital tools and resources is not only a pathway to future-proofing healthcare but also a means to democratize the benefits of the ongoing digital transformation. By establishing competence and understanding across the spectrum, we can ensure that the promise of digital health is realized for the benefit of all, marking a substantial leap toward a more equitable and accessible healthcare landscape.

2.2 Shifting the Paradigm: From Achieving Better Healthcare Outcomes to Improving Workforce Performance and Processes

The goal of driving significant advancements in digital health technologies has long centered on achieving better healthcare outcomes. However, it is increasingly recognized that digital health technologies can also enhance the performance of healthcare professionals and bridge gaps in care delivery. In a landmark study, WHO/Europe finds that the use of mobile technologies, telemedicine, and digital tools leads to enhanced performance, improved mental well-being, and increased skills and competencies among healthcare workers. More specifically, the study shows that digital health technologies increased decision-making efficacy during clinical practice, reduced the time needed to execute tasks, improved productivity, increased access to reliable real-time data, and led to a greater ability to provide timely technical and specialized reports on activities, progressions, and remedies.

European countries are widely adopting digital health technologies to counteract workforce shortages, address the demographic development, or help improve the personalization and accessibility of care. The United Kingdom's Weight Management Program, for example, was introduced in 2023 to enable up to 48,000 obese patients with multimorbidities to access virtual services and save up to 145,000 h of clinician time (National Institute for Health and Care Excellence [NICE] 2023). In Germany, the Digital Healthcare Act from 2019 allows digital health applications, or "DiGA," to be routinely prescribed by doctors for patient use for a variety of diagnoses and reimbursed by statutory health insurance (SHI). According to Germany's Digitalisation Strategy for Health and Care (2023), the Federal Ministry of Health of Germany foresees that by the end of 2025, 50% of all publicly funded hospitals in Germany will improve their digital maturity by at least two levels in at least two categories of the DigitalRadar[2] benchmark. This improvement will also include better interaction between Digital Health Applications (DiGA) and the patient's Electronic Patient Record (ePA). The goal is to save clinicians time in

[2]Tool to assess digital maturity of public hospitals within the context of the Future Hospital Act (KHZG, 2020) initiated by Germany's Federal Ministry of Health and developed by the DigitalRadar research consortium.

understanding the full health picture of patients prior to hospital admission and post-discharge. In the Netherlands, as of 1 July 2023, it is compulsory for patient health records to be exchanged digitally rather than through paper or compact disc, unleashing paperless access to patient information for the provision of care and in the long term, making health data available to healthcare professionals and healthcare professional societies for performing research across disease areas and medicines.

With its Vision 2030 and Health Sector Transformation Program (2021), the Kingdom of Saudi Arabia (KSA) takes a leap forward in developing a value-aware and productive healthcare workforce that will be financially rewarded for increasing its capacities and delivering value-based and preventive care. With over 80% of the Saudi population expected to be covered by integrated and holistic health services—including telehealth, remote patient monitoring, health applications, population health management programs, etc.—and the launch of around 30 health clusters,[3] KSA is decentralizing and privatizing the health service delivery by tasking each health cluster provider to define the health needs of the population it covers and set corresponding Key Performance Indicators (KPI) and budget allocation to meet those needs. **This aligns with the broader G20 & G7 countries' consensus that spending in health to improve capacity building and achieve efficiency gains is an investment in positive long-term healthcare and economic outcomes, not a cost, as outlined in the roadmap to sustainable finance in health published by the world's most powerful economies** (G20 & G7 Health Development Partnership 2023).

In conclusion, the paradigm shift from solely focusing on achieving better healthcare outcomes to a broader perspective encompassing the improvement of workforce performance and processes could mark a pivotal moment in the evolution of digital health technologies and their full uptake. As nations worldwide embrace digital health technologies to not only enhance patient care but also empower healthcare professionals, the stage is set for a transformative era where technology becomes a catalyst for comprehensive healthcare reform. This evolution could signify a strategic leap toward a future where, if the challenges including the workforce crisis are addressed and a digitally empowered workforce is nurtured, the synergy between technological innovation and healthcare delivery has the potential to propel the industry into new frontiers of efficiency, effectiveness, and ultimately, improved health outcomes for individuals and communities.

[3] An integrated network of healthcare providers under one administrative structure serving approximately one million people (similar to National Health Service's (NHS) Integrated Care Systems in England).

2.3 Leveraging Data and Artificial Intelligence

2.3.1 Data

We live in an era where data not only empowers individuals and benefits communities but also bolsters economies at large. Today, data is playing a critical role in informing decision-making, driving innovations, and shaping policies. The use of data in healthcare is no exception to this—although in many cases the sector is still catching up with other industries. With the publication of the proposal for a regulation on the European Health Data Space (EHDS) the European Commission (2022) described health data as "the blood that runs through the veins of our healthcare systems." The EHDS is the world's first legal framework for sharing health data that aims to amass a substantial volume of electronic health data, encompassing both personal and non-personal health information, for primary and secondary purposes. Data literacy and representative data cohorts, therefore, are key requirements for the EHDS to reach its full potential and fully embrace the concept of data altruism. Not only among citizens whose data needs to become available for further use but equally important among healthcare professionals to be properly equipped with the data analytics skills and knowledge to enable better provision of care and reduce workloads.

Following a nearly 2-year debate by the EU institutions, the final text of the EHDS legislation introduced in its recitals and Chapter V articles the imperative of a data-literate health workforce and citizenry for its successful update. More specifically, the Article "Training and information for health professionals" emphasizes the need for robust training schemes that are easily accessible and go hand in hand with user-friendly interfaces and interoperable digital health solutions provided by the industry. The Article "Digital health literacy and digital health access," on the other hand, requires EU member states to promote digital health literacy and relevant skills for citizens and patients. It includes awareness campaigns tailored to specific groups to inform them about the use of electronic health data. Additionally, EU member states are tasked with promoting access to infrastructure for managing individuals' electronic health data effectively, through easy-to-use and inclusive portals.

However, to transform data into value for the healthcare professional, burdensome tasks as part of the day-to-day duties such as managing sheer volumes of data, documentation processes that require manual data input into structured formats, and overall time-consuming and duplicative efforts with siloed datasets need to be removed. European medical societies representing healthcare professionals such as the Biomedical Alliance in Europe (2023) and the Standing Committee of European Doctors [CPME] (2022) have been vocal about the need for EHDS to facilitate the work of healthcare professionals, and not lead to additional workload and resources in managing health databases and preparing them for data processing. Healthcare professionals need specialized skills to ensure accurate and meaningful data gathering, which is why supporting and training healthcare professionals is essential for the collection of high-quality, usable health data.

To prepare countries and healthcare organizations for the forthcoming implementation of the EHDS, the EUs eHealth Network (2023), a voluntary network composed of EU Member States' competent authorities dealing with eHealth, published a Capacity Building Health Data manual with the aim to enhance the development of and improve the skills and abilities of staff working on national digital health services. The focus in the first instance is on the primary use of health data, i.e. the management of Electronic Health Records, ePrescription and eDispensation, patient summary, electronic reports and medical images, telehealth (incl. mHealth). The manual is complemented by a Capacity Building Catalog mapping the state of play of the Member States' digital health transformation and health data sharing infrastructures.

The next step in the process would be ensuring user engagement and participation, i.e. motivating healthcare professionals to support the effectiveness of the data collection process and use electronic health data accordingly. Beyond training and specialized data skills, challenges in integrating data from diverse sources or platforms, i.e. achieving interoperability across care settings and even more so, cross-border, will need to be overcome to ensure the seamless flow of information and cultivate trust among the healthcare professionals in the simplicity and security of health data use with the right standards and safeguards in place. Addressing these challenges is essential to ensure that data collection efforts are effective, ethical, and do not place undue strain on healthcare professionals and health data holders, in general, such as hospitals, primary care practices, caring institutions, etc.

The EHDS is the first EU common data space in a series of 14[4] (to date) common data spaces in domains of public interest as foreseen in EU's Data Strategy.[5] With the rollout of additional data spaces in the EU and enabling the cross-sector integration of data, the "Whole Person" approach (Whitehead and Dahlgren 1991) to promoting health will finally come to fruition whereby a person's health picture will not only be determined based on clinical indicators captured in the hospital but a whole set of non-clinical metrics responsible for defining 80% of individual's health condition (Hood et al. 2016).

For example, access to public transportation, housing, nutrition, employment, and social status, etc., forms part of the so-called social determinants of health which are of paramount importance for healthcare professionals to understand drivers to patients' healthcare outcomes beyond the genetic makeup and health interventions performed within the bricks of a hospital. Beyond the conditions in which people are born, grow, work, live, and age (i.e., social determinants of health as defined by WHO[6]), the forthcoming EU common data spaces would provide, to some extent, access to data which can offer insights into two additional determinants to the health of individuals and populations: environmental and commercial determinants of health. With the growing threats to human's health posed by trends in global warming, environmental determinants of health (EDH) are gaining in importance.

[4] Agriculture, cultural heritage, energy, finance, Green Deal, health, manufacturing, language, media, mobility, research and innovation, public administrations, skills, and tourism.

[5] https://digital-strategy.ec.europa.eu/en/policies/strategy-data

[6] WHO, https://www.who.int/health-topics/social-determinants-of-health#tab=tab_1

They refer to the global, regional, national, and local environmental factors that influence human health, including physical, chemical, and biological factors external to a person.[7] Related to these are the commercial determinants of health (CDH), i.e. the private sector activities that affect people's health, directly or indirectly, positively or negatively. CDH impact a wide range of risk factors, including smoking, air pollution, alcohol use, obesity and physical inactivity, and health outcomes, such as noncommunicable diseases, communicable diseases and epidemics, injuries on roads and from weapons, violence, and mental health conditions.[8]

In a nutshell, unlocking the full potential of data and the evidence it creates for the holistic understanding of the person's health would be pivotal to upskilling the healthcare workforce, informing effective healthcare interventions, and managing population health in building the resilience of our health systems. Even more so, it brings to light the responsibility of additional actors, such as global corporations, labor unions, and the nonprofit sector, in ensuring a healthy and safe workplace and supporting employees leading a healthy lifestyle.

2.3.2 Artificial Intelligence

Healthcare AI systems use big sets of data and smart processing algorithms to learn from patterns and features in the patient information they examine. As such, these systems generate predictions and enhance clinical decision-making with remarkable accuracy and precision. Nonetheless, it is important to note that the data fueling AI-based systems is not flawless or devoid of bias, such as gender or racial bias. This can lead to biases in the resulting AI output.

Same as with digital health technologies and health data, we observe the added value of the technology based on the benefit it brings to the healthcare professional, not to the healthcare outcome. Beyond diagnosis assistance, AI systems deployed in the hospital can provide healthcare professionals with real-time support in summarizing interactions with the patient, which not only optimizes doctor's efficiency but also improves doctor–patient facetime. In essence, doctors are open to AI technologies that help reduce the laborious paperwork, as well as to improve workflows on time-consuming tasks such as manual inputting of data into structured forms. Nurses can benefit from AI technologies, especially when it comes to burdensome clerical tasks such as preparing handoffs for the next shift. HCA Healthcare (2023), one of the leading healthcare providers in the USA owning a network of private hospitals, outpatient centers, and specialist clinics, is testing a generative AI tool capable of scanning electronic health records and extracting

---

[7] Pan American Health Organization, https://www.paho.org/en/topics/environmental-determinants-health

[8] WHO, https://www.who.int/news-room/fact-sheets/detail/commercial-determinants-of-health#:~:text=Social%20determinants%20of%20health%20matter,commercial%20actors%20that%20affect%20health

pertinent information for seamless documentation and transfer of notes between nurses.

AI use-cases in healthcare, and hospitals in particular, are still in a very nascent phase. To underpin trust in their safe, legal, and ethical deployment, clinicians must deepen their understanding of how AI systems work and the specific regulations governing their uses. The EU is the first region worldwide to regulate AI, including its uses in healthcare. The EU's Artificial Intelligence Act[9] became law in 2024 that aligns with a plethora of international initiatives at the United Nations (UN), G20/G7, and OECD level and it fosters an environment conducive to AI uptake by harmonizing rules on AI across EU member states and ensuring its ethical and responsible integration across sectors, including in healthcare. For example, the Act classifies risk from the use of AI systems into categories with corresponding levels of scrutiny and compliance requirements including AI deployments with unacceptable levels of risk to the safety of humans that will be prohibited. Recognizing the accelerated pace of AI advancements and the substantial benefits it offers, especially for healthcare professionals, other nations such as the United States and the United Kingdom are also working to establish sector-specific regulations. This proactive approach ensures that the advantages of AI in healthcare are not overshadowed by potential risks to patients' well-being and health while at the same time allowing for solutions that benefit the healthcare workforce.

2.4 Seizing the Momentum Created in the Digital Era

While healthcare still counts as one of the most digitally conservative sectors, businesses and citizens are increasingly becoming interconnected and, according to dominant thinking, tend to be more open to experimenting with digital technologies and immersing themselves in a new tech-savvy world (Clarke et al. 2017). This poses a challenge to the largely traditional and process-heavy, hierarchical healthcare sector with internally siloed units, demanding a more agile and collaborative digital service instead (O'Reilly 2011). As the divide between healthcare providers and healthcare users becomes more apparent, a good governance model calls for the need to shift boundaries and tap into a more participatory learning and decision-making process in line with the expectations of citizens (Tuurnas et al. 2019). Beyond top-down healthcare reforms, the change is expected to also come from within the healthcare organization based on the premise that particularly younger healthcare professionals are operating with new sets of values and practices relevant to the digital era. In this vein, upcoming generations of healthcare professionals that are more open, collaborative, engaged in peer production, and, overall, entrepreneurial in nature, are expected to exploit new technologies, harness the power of health data, and lead the next wave of digital transformation of hospitals and healthcare organizations.

[9] https://artificialintelligenceact.eu/

3 The Health Workforce of the Future

3.1 Introduction

The digital maturity of hospitals requires the use of digital health technologies, as outlined in the previous part, in order to enhance the delivery of healthcare services, improve workflows, and ensure better patient outcomes. This process of digital transformation depends on trained healthcare professionals who have the knowledge, skills, and competencies necessary to support the integration of digital tools.

3.2 Workforce Requirements to Drive Digital Maturity of Hospitals

The workforce requirements for the digitally mature hospital featuring a functional Digital Health Information Systems (DHIS) are influenced by the integration of technology into healthcare processes. Since the digital maturity of hospitals refers to the extent to which digital technologies are effectively used to improve and transform healthcare delivery, the training and employment of a capable workforce is an essential capacity building factor. These requirements continue to evolve as healthcare organizations embrace advanced technologies. In the context of "Governance and workforce" for the digital transformation of hospitals, it is necessary to "build governance to support sustainable workforce using outcomes data" (HIMSS 2023a). This is one of the four dimensions of the "Digital Health Indicator" developed by HIMSS,[10] which was designed to measure progress toward attaining a digital health ecosystem and further advancing the transformation. This digital health indicator is useful to evaluate the ability of staff to support and to provide person-enabled care that is personalized to the unique needs, circumstances, and choices of the individual.

Healthcare maturity models provide prescriptive frameworks in the form of roadmaps that assist healthcare organizations to build their digital health ecosystems. Some examples of workforce skill requirements (measured by maturity models) in a "digitally mature" healthcare organization would include the following (HIMSS 2023b):

1. Capabilities of the organization in terms of data management and data infrastructure
2. Digital maturity of care delivery in non-acute facilities in terms of strengthening the digital capacity of health workers to deliver healthcare services and

[10] HIMSS (Healthcare Information and Management Systems Society) is a global advisor, thought leader, and member-based society committed to reforming the global health ecosystem through the power of information and technology. www.himss.org

meaningful engagement with care recipients, families, caregivers, and populations through the adoption of secure digital tools
3. Communication in healthcare that evaluates capabilities needed for coordinated patient care and identifying actions from critical stakeholder groups
4. Digital imaging strategy for improved patient health outcomes
5. Adoption and utilization of electronic medical record (EMR) functions to support optimized patient care, reduce medication errors, improve operations, and achieve a near paperless environment
6. Technology infrastructure capabilities
7. A health system's ability to track processes and products used in care by mobilizing data to create real-world evidence of impact and outcomes for patient populations

The digital health indicator and the healthcare maturity models are a tool for healthcare organizations to assess their workforce needs and provide guidance for investing in training and development programs to keep their staff up to date with the latest technologies and industry best practices.

3.3 User Groups

The KSCs of the workforce who are working toward achieving digital maturity require a comprehensive approach that addresses various aspects of the healthcare organization. These aspects are related to digital health literacy, training, continuous learning, information technology, data management, data analytics, data interpretation, interdisciplinary collaboration, change management and leadership as well as patient engagement skills. For example, workers have to be adept in using electronic health records, computerized provider order entry (CPOE), telemedicine platforms, and other digital health systems. Data scientists and informaticians should be able to extract meaningful insights from health data to improve patient outcomes. Overseeing the digital transformation of a hospital will require the insight from change management leaders who will function in providing strategies and leadership. Therefore, the KSCs are dependent on the organization of all people who work with digital health. In order to reflect the specific KSCs required for the different aspects of digital health workers or users, it is useful to categorize them into distinct groups (Table 1). WHO and ITU previously proposed differentiating all health

Table 1 Categorization of User Groups in DHIS; Group 1 and Group 2 (WHO and ITU 2012); Group 3 (Chaltikyan and Fernandes 2022)

Group 1: Digital Health Workforce	Professional users specialized in the design, development, operation, and maintenance of DHT
Group 2: Health Workforce	Professional users specialized in the delivery of healthcare services with the use of DHT
Group 3: Empowered Citizens	Nonprofessional users—recipients of healthcare services with the use of DHT, or their assistants

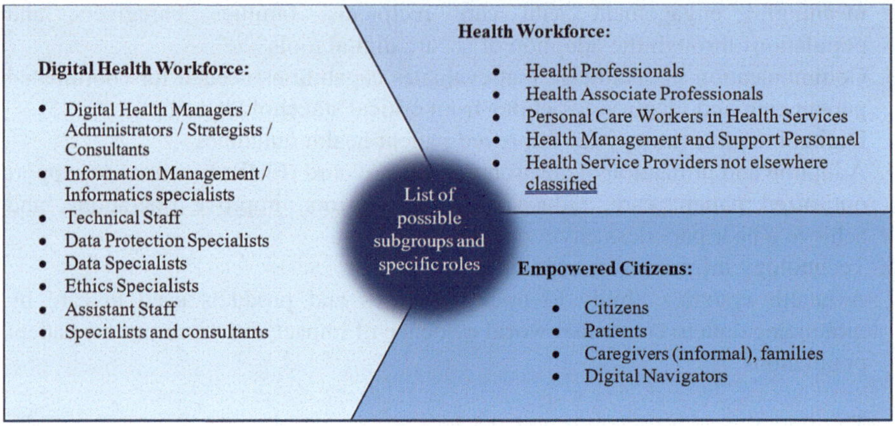

Fig. 1 List of possible subgroups in the three user groups

professionals into Health Workers vs. Health ICT Workers based on their relation with DHT (WHO and ITU 2012). Building on that, and in line with the recent developments in the domain of digital health, as well as with the growing importance of citizen engagement and citizen centeredness, we propose an updated grouping of digital health workforce and users:

By categorizing the workforce and users of the DHIS based on the scope of the users' relation to DHIS, it is possible to describe the specific roles, education pathways, and sustainability of the user groups. Figure 1 provides a view into the possible subgroups of each user group.

Group 1: Digital Health Workforce

Digital Health Workforce primarily functions as the bridge between IT staff, service improvement staff, and front-line staff in a hospital. A hospital that undergoes a digital transformation requires professionals who have knowledge of information technology, medicine, as well as skills in designing user-centered solutions. The Digital Health Workforce may consist of subgroups of personnel detailed as follows:

1. Digital Health or Digital Transformation Managers /Administrators/Strategists/ Consultants at national, regional, local, or institutional level are tasked with leadership, decision-making, and championing the adoption of digital health technologies. They ideally have a background in either medicine, nursing, pharmacy, healthcare management, or public health, or in computer/information/data science coupled with an advanced degree in digital health or health informatics. These executives oversee the work of information management/informatics specialists, technical staff, data protection specialists, ethics specialists, and assistant staff. The example roles at national/regional/local levels are Digital Health Coordinator or Strategist, and at the institutional level—Chief Information Officer (CIO), Chief Digital Officer (CDO), or Chief x Information Officer

(CxIO, where x = Medical, Nursing, and others). These professionals also coordinate development and delivery of Digital Health Educators.
2. Information Management/Informatics Specialists are engaged in designing, developing, operating, and maintaining a DHIS (HIMSS 2020).
3. Technical Staff (biomedical engineers, software engineers, network administrators, and system administrators) are IT specialists involved in developing and implementing technological solutions (Hübner et al. 2019).
4. Data Protection Specialists enforce access and security measures to protect digital health data and its secure sharing. They also develop policies for the access, collection, insertion, retrieval and storage, modification, and destruction of patient data.
5. Data Specialists consult on the intended use of the IT solution, the organization and integration of data within the EHR and other DHIS subsystems.
6. Ethicists enforce ethical principles of human respect, beneficence, and justice, including the ethics of using artificial intelligence in healthcare.
7. Various other administrative tasks are performed by technical assistants, technicians, office, and desk assistants.

3.3.1 Professional Development for User Group 1: Digital Health Workforce

The Core Competency Areas (CCAs) of Group 1—Digital Health Workforce can be grouped in eight foundational domains:

- Life sciences/biology and medicine
- Health and healthcare
- Information science and technology/computer science/data science
- Medical and health informatics/health IT
- Digital health care management
- Biomedical research and innovation
- Public and business administration and management
- Social, legal, and behavioral sciences

Education and training of these professionals can best be achieved through formal degree-seeking education as a dedicated and separate specialty in digital health or health informatics. Such degrees exist at all academic levels—Bachelor, Master, and Doctoral degrees.

This user group requires a relatively broad KSCs mix that may vary for the different types of professionals, be delivered at different levels, and often with the need for lifelong learning. Proper recognition of this user group is necessary in order to design formal education pathways in digital health. With the introduction of dedicated specialties in this area, it is required to develop programs that are tailored toward training individuals who will support the digital transformation of hospitals. In the event of the deficiency of such programs, a Digital Health Workforce may be trained from candidates with diverse backgrounds and categories (medicine, ICT, or combined) and providing in-service or professional training by recognized institutions such as the College of Healthcare Information Management Executives (CHIME), or the American Health Information Management Association (AHIMA), and others. Accreditation bodies such as the European Federation for Medical Informatics (EFMI) AC2 Accreditation Assessment Process and the International Medical Informatics Association (IMIA) ensure the delivery of high-quality educational programs in the field of health informatics and digital health.

3.3.2 Needs, Opportunities, and Challenges in User Group 1—Digital Health Workforce

Monitoring of the current state as well as the needs and requirements of Digital Health Workers is essential in order to determine workforce statistics. Obtaining data on the available trained professionals, their employment and career options, availability of capacity necessary to provide the training and upskilling, identifying the gaps, and calculating the needs is essential for supporting the digital maturity of hospitals.

Some of the opportunities of the Digital Health Workforce include:

1. Role in the improvement of patient outcomes by the development and use of digital health tools that enhance healthcare delivery.
2. Engagement in technological innovation and contributing expertise in the design of new digital health solutions for patient care, while advancing the field of digital health.
3. Interdisciplinary collaboration with various specialized health and IT fields.
4. Super-specialization in areas such as artificial intelligence, data analytics, and cybersecurity as a result of demands of the field.
5. Leadership and Entrepreneurship opportunities in managing the digital transformation of health systems.

While there are significant opportunities for professionals in this field, there are also challenges that the Digital Health Workforce encounters. Some of these challenges are related to the provision of education and career opportunities in digital health. Due to the uniqueness of the digital health field, there are still few educational courses that primarily focus on digital health as a core content of education. Other situations include the provision of digital health courses that do not completely address the needs of developing a competent Digital Health Workforce. A lack of informaticians with strong insights into healthcare further this concerning aspect. As a result, there is a challenge in achieving the required KSCs to support digital maturity of hospitals.

3.4 Group 2: Health Workforce

The Health Workforce consists of health professionals like physicians, dentists, nurses, allied health workers, pharmacists, and others. These professionals are experts skilled in treating patients and are at the center of the provision of medicine and health care. They play a key role as clinicians in decision-making at the point of care. This user group is unique since they are the individuals who are in constant contact with the ultimate beneficiaries of the DHIS. Therefore, health professionals are indispensable to the successful adoption of DHT. The subgroups and specific roles of the Health Workforce are detailed as follows (International Labor Office 2012):

1. Health Professionals: They are tasked with conducting research and obtaining scientific knowledge through the study of human and animal disorders and illnesses and ways of treating them; advising on or applying preventing and curative measures, or promoting health; preparing scientific papers and reports.
2. Health Associate Professionals: They are tasked with performing technical and practical tasks to support diagnosis and treatment of illness, disease, injuries, and impairments in humans and animals, and to support implementation of health care, treatment, and referral plans usually established by medical, veterinary, nursing, and other health professionals.
3. Personal Care Workers in Health Services: They provide personal care and assistance with mobility and activities of daily living to patients and elderly, convalescent and disabled people in healthcare and residential settings.
4. Health Management and Support Personnel: They include health systems personnel like health service managers, health economists, health policy lawyers, clinical psychologists, social workers, medical secretaries, ambulance drivers, building maintenance staff, and other general management, professional, technical, administrative, and support staff.

3.4.1 Professional Development for User Group 2: Health Workforce

The CCAs of Group 2—Health Workforce include the disciplines of medicine, dentistry, nursing, pharmacy, and allied sciences. However, from the aspect of digital health, the knowledge, skills, and competences of health workers are roughly categorized in the following basic and specialized skilled domains (Table 2):

The education and training of health professionals in digital health should reflect the current KSCs among the health professionals. It also includes the capability of acquiring quality training in using DHIS tools and the ability to empower patients. In order to provide specialized training for this user group, the first step would be to assess the digital skills within the group and further develop these skills with current educational offers. It is important to conduct surveys to assess the current state of digital competencies and identify necessary measures to develop the IT skills of the Health Workforce. Further steps include the development of formal education or other allied approaches such as built-in courses and standalone courses to supplement medical education with informatics. Additionally, digital health education for healthcare professionals can be provided through in-service and professional training scenarios, and the development of joint degrees and hybrid educational programs that combine the understanding of clinical practice, technology, change management, and health policy at the executive and postgraduate levels.

Table 2 Domains of KSCs of the Health Workforce

Digital health services (Clinical, public health, and administrative)	Informatics/ICT	Social, economic, legal, and ethics	Education, research, and innovation/data science
Health information systems	Technical knowledge and support	Privacy regulations	Artificial intelligence
Digital documentation and records management	Systems implementation	Financial management	Health data analytics
Digital and remote care delivery	Informatics concepts and processes	Patient access and engagement	Personalized medicine
Digital imaging	Data integration and interoperability	Healthcare quality and safety	Digital health education and training
Digital medication management	Telecommunications	Digital health ethics	Digital health literacy
Clinical decision support	IT advocacy	Leadership and executive management	Digital health research
Practice administration/ management	IT compliance		
Digital public health systems and services	IT risk management		
Project management	Attitudes toward IT		

3.4.2 Needs, Opportunities, and Challenges in User Group 2: Health Workforce

The assessment of the current state of digital KSCs of healthcare professionals must gauge their strength in terms of harnessing the benefits of digital health technology. The necessary data includes the number of available trained professionals, need for more capacity and resources, and ways in which it can be achieved. Some of the opportunities in the Health Workforce include:

1. Digital health technology enables healthcare professionals to deliver more personalized and efficient patient care with the potential of improving health outcomes.
2. The provision of remote monitoring, telehealth, and mobile health apps contributes to improved access to health services.
3. The use of digital health tools allows automation and improved workflows for administrative tasks thus enhancing the overall operational efficiency within healthcare organizations.
4. Use of data analytics and artificial intelligence allows for evidence-based individualized decision-making, improving clinical outcomes and resource utilization.
5. Healthcare professionals can leverage the use of digital health technology to promote patient engagement and thus empower patients.

The integration of DHT, though crucial for the provision of quality healthcare, is facing a number of challenges. Digital transformation is accompanied by complex adaptive changes that need to be accepted by the workforce. Digital Health Professionals must change the attitudes of health professionals toward the adoption of digital technology, which has been challenging in view of necessary effort, changes in daily routine brought on by technology, and doubting the value that these technologies bring about in the worklife (Socha-Dietrich 2021). Such challenges may be mitigated through the demonstration of the benefits of digital health solutions (e.g., compensating for workforce shortages) in order to increase uptake and the gradual integration of digital health solutions. Digital health is not always included in education or professional training programs, or frequently only available as elective courses. This can be addressed by encouraging health professionals to obtain additional training and certification in digital technology, provision of full-time, in-house digital health jobs with attractive job offers, sustainable career tracks, and providing lifelong learning opportunities for healthcare professionals during their working time (DigitalHealthEurope 2022). Other challenges include the existence of fragmented and incomplete patient health data and therefore, the regular use of data for improving workflows and service delivery is not yet a norm. Some health workers also report that legal, financial, and organizational aspects of work that were designed in the pre-digital era do not enable them to reap the full benefits of new technologies in healthcare. These barriers are related to safeguards against possible lack of transparency or threats to data privacy (Socha-Dietrich 2021).

3.5 Group 3: Empowered Citizens

Empowered citizens play an important role in the healthcare system. They are the ultimate beneficiaries and stakeholders of digital health. The success of digital health strategies relies on their compliance and effective and efficient adoption and utilization by this user group. Empowering a citizen, and the collective empowerment of citizens, should enable them to utilize DHT in order to manage their health by taking responsibility and an active part in informed decision-making, collaboration with physicians, digital health literacy, and reporting of outcomes. These roles can be carried out through digital health tools such as telehealth, patient portals, remote monitoring, personal health records, and personal health apps. These roles are not confined to the patient alone but also translate to informal caregivers as detailed below:

1. Patient: Is any natural person who seeks to receive or receives healthcare (DigitalHealthEurope 2022).
2. Care recipient: Is an individual who receives or seeks any care services or support considered to be integrated care, from another person or organization. In certain circumstances, a care recipient may be the term used for a person receiving a new project-based service (DigitalHealthEurope 2022).

3. Care coordinator: Is an individual who is responsible for coordinating the care and support of a care recipient. It could be a role undertaken by a healthcare professional or the care recipient or family carer (DigitalHealthEurope 2022).
4. Social care support worker: Is an employed individual who is qualified to deliver care, services, treatment, or support to a care recipient (DigitalHealthEurope 2022).
5. Informal carer: Is an individual who is registered in any organization to deliver care, services, or support to a care recipient (DigitalHealthEurope 2022).
6. Digital Navigator: Is an individual who addresses the entire digital inclusion process (home connectivity, devices, and digital skills) with community members by interacting with them, guiding and assisting them in the use of the Internet and computing devices (National Digital Inclusion Alliance n.d.).

3.5.1 Education of User Group 3: Empowered Citizens

The KSCs required by Empowered Citizens include the ability to take control of their own health and range from searching, finding, understanding, and appraising health information from electronic sources to actively participating in online and remote care. Through digital health literacy, they are able to apply the retrieved information to further their health status. Such skills are critical to harnessing all the benefits of digital health.

Norman and Skinner (2006) described this set of skills into two types, analytic (involving skills applicable to a broad range of information sources irrespective of the topic or context) and context-specific (skills contextualized within a specific problem domain or circumstance). The following are components of digital health literacy that may be used for gauging consumer digital health literacy:

1. Traditional literacy: Reading and writing skills to understand textual resources.
2. Information literacy: Knowledge of resources of information on health-related topics to search for relevant information.
3. Media literacy: Skills to understand information from media and the information delivered through this medium.
4. Health literacy: Skills to engage with the health system and apply it to self-care.
5. Computer literacy: Knowledge of using computers and other technological devices and medical based software.
6. Scientific literacy: Understanding of the nature of scientific information along with the processes and methods of application of health research findings.

The training of citizens in order to empower them consists of the KSCs of obtaining knowledge about health and in particular one's own health and the ability to retrieve valuable and precise information to address the particular health condition through an efficacious use of digital technology. Supplementing these aspects, empowered citizens should develop a positive attitude, motivation, and confidence in engaging with digital technologies to manage their health conditions. Therefore, it is imperative for the digital transformation of a health system to provide the environment for empowered citizens to act decisively and thereby benefit from the adoption of DHT.

3.5.2 Needs, Opportunities, and Challenges in User Group 3: Empowered Citizens

In order to build on the required KSCs described in Sect. 2.6.2, four components have been reported as being fundamental to the process of citizen empowerment. These are the patients' own understanding of their roles, the knowledge for engaging with the healthcare provider, the acquisition of these skills, and a facilitating environment (Affinito et al. 2020). It is critical that sufficient steps are taken by authorities to ensure an optimal level of literacy (both health and digital) and create avenues to train and sustain such education. In addition, the satisfaction levels of empowered citizens must be taken into consideration. The citizen must also be involved in collaborative programs with health professionals and engineers to co-design user-centered digital health solutions by providing their personal experiences, needs, and expectations. The Global Strategy on Digital Health provides ideas for people-centered health systems that are enabled by digital health that encourage the adoption and management of digital health technologies. Empowering citizens through the use of digital health tools offers various opportunities to enhance individual health management and healthcare outcomes.

Some of the opportunities are as follows:

1. As DHT provides insights into health conditions, treatment and preventive strategies, an empowered citizen is able to obtain such information in an effort to improve health and health outcomes.
2. With increased health literacy, empowered citizens are able to actively engage in healthcare and actively contribute toward improved health outcomes.

However, certain challenges exist which hinder the empowerment of the citizen. Challenges regarding patient attitudes in using new methods that limit patients' interaction with the technology. A common challenge in patient acceptability is the feeling of a "mental distance" created by DHT. Along with given limitations, there are doubts concerning the benefits that these technologies will provide that may discourage some users. The collection and sharing of health data through digital tools raise a host of privacy and security concerns. Citizens may be apprehensive about the protection of their sensitive health information, necessitating robust security measures. Ethical considerations, such as responsible and equitable use of DHT, need to be carefully addressed to maintain trust and transparency. All these challenges can be tackled by improving patients' attitude toward digital health, recognizing the value of person-centered DHIS, instilling digital health literacy, removing technological barriers, addressing disparities in access to digital health tools, and fostering a culture of using them.

4 Workforce-Driven Advancement of Digital Maturity

4.1 Introduction

In the last section of this article, the authors delve into how the healthcare workforce could lead the charge toward digital transformation, ensuring resilient and sustainable healthcare delivery amidst demographic shifts. This section covers scenarios showcasing the interaction of digital tools and digital literacy among healthcare professionals, illustrating the tangible benefits of such initiatives. These examples include the implementation of digital literacy programs, the adoption of innovative digital health solutions in clinical practice, and collaborative projects that include citizens in leveraging technology to improve patient outcomes. Highlighting the outcomes, such as enhanced patient care, streamlined healthcare processes, and data-driven decision-making, underscores the workforce's pivotal role in achieving digital maturity. The section concludes with recommendations for fostering a culture of digital innovation through supportive policies, continuous education, and strategic leadership, paving the way for a digitally advanced healthcare ecosystem capable of meeting future challenges.

4.2 Key Requirements for a Workforce-Driven Advancement of Digital Maturity

Advancing digital maturity within healthcare necessitates a digitally literate workforce. This foundational requirement enables healthcare professionals to seamlessly adopt and integrate digital technologies into their practice, thereby enhancing care delivery and operational efficiencies. Matthews (2021) highlights the correlation between digital literacy and a positive disposition toward new technologies among health workers, underscoring the benefits of a digitally empowered workforce. These benefits include the ability to customize digital solutions to patient needs, critically assess technological impacts, and make informed decisions based on health data insights.

Addressing the digital health literacy gap is paramount for fostering trust, facilitating technology adoption, and mitigating resistance to new innovations. Essential actions include:

- **Integrating data analytics and digital health modules** into all healthcare and allied health professional training programs, ensuring all new entrants are well-prepared to navigate, apply, and advocate for digital health solutions. For example, University of California, San Francisco (UCSF) School of Medicine has been proactive in incorporating digital health education into UCSF curriculum, offering courses and workshops on topics such as telemedicine, health informatics, and data analytics to ensure that medical students are prepared to leverage technology in healthcare delivery. Additionally, UCSF has established partnerships with industry leaders to provide students with hands-on experience

in utilizing digital health solutions. Within UCSF's campus, the Centre for Digital Health Innovation (CDHI)[11] equips future healthcare professionals with the skills needed to navigate and advocate for digital health innovations in their practice.
- **Establishing standardized competency frameworks for digital health literacy** to identify skill gaps and support targeted professional development, promoting a unified approach to upskilling across Europe. For example, Health Education England[12] has developed numerous profession specific digital health capability frameworks and is currently working on a digital skills self-assessment tool, which would allow any healthcare professional to assess their current skill gaps and more effectively target professional development needs. The authors of this article have also offered core competency frameworks for the Digital Health Workforce and Health Workforce User Groups outlined in Sects. 2.3–2.5, respectively.

In summary, addressing the digital health literacy gap through initiatives like integrating digital health modules into training programs and establishing standardized competency frameworks is essential for empowering healthcare professionals to effectively navigate and embrace technological advancements, ultimately enhancing patient care and driving innovation in healthcare delivery worldwide.

4.2.1 Creation of a Stimulating Environment for Advancing Digital Transformation

Envisaging the ideal scenarios of a digitally empowered and motivated workforce, ideally equipped with the latest knowledge on key digital and data topics, and innovative advances, would lead to a future where:

- **Scenario of agile and accessible solution creation by healthcare professionals (HCPs):**
  Clinicians are able to identify the gaps in clinical insights and devise solutions utilizing data and advanced analytics in order to fill the gaps they know exist in clinical workflows. They would be able to access a simple and easy-access funnel whereby solutions to existing problems are analyzed, co-created, iterated, and developed to pilot stage, supported with a budget allocated to innovation creation within a healthcare organization. Enabling this ecosystem of support would require healthcare organizations to dedicate teams, budgets, and events to encourage creativity and risk-taking amongst their staff, as well as agile and user-friendly pathways toward innovation collaborations, allowing commercial entities to partner fluidly with healthcare providers. We can see a few examples of these supportive ecosystems emerging across Europe.

[11] https://cdhi.ucsf.edu/

[12] https://digital-transformation.hee.nhs.uk/building-a-digital-workforce/digital-literacy/digital-capabilities-frameworks#digital1

One[13] of these collaborative innovation hubs is based in Wales, where the Tritech Institute supports the development of innovative solutions to unmet healthcare needs, prioritizing those proposed by healthcare professionals. The team of multidisciplinary experts give essential funding, tech and development guidance to clinicians, from early concept through to pilot and ready for market.

A second example[14] is the Berlin Institute of Health which supports employees from research and clinical departments at the University Hospital Charité to develop their pitches into market ready solutions. A similar hospital-based digital health unit can be found at the Sant Pau hospital campus in Barcelona.[15] A validation center for digital health solutions within the center has the aim of supporting and facilitating digital innovation in the hospital, allowing clinicians to bring their ideas to life, and matching clinical experts with the technology expertise they need to co-create successful solutions.

- **The ideal scenario: adaptive and continuous digital health upskilling**
  As technology evolves, so does the continuous and individualized training given to each healthcare professional, with insightful assessment and identification of each clinical group or individual needs, they are able to access rapid updating through advanced training platforms, using virtual reality (VR) to enhance learning and practice. By investing in the most advanced training with incorporated behavior change theory (i.e., learning enhanced by virtual reality interactive models), organizations can empower their staff to participate in the digital maturity advancement of their workplace and environment (Jallad and Işık 2022).

One example combining clinician innovation and advanced VR supported training is found in Denmark where the nurses at the Department of Gastroenterology at Odense University Hospital (OUH) are using VR Trainer[16] to acquire additional professional competences, especially in specialist nursing procedures that require learning new systems and operating with advanced technology.

The outlined scenarios illustrate the importance of fostering a digitally empowered workforce within the healthcare sector. However, it is essential to note that these scenarios focus primarily on the interaction between healthcare professionals and technology within the hospital setting, highlighting the need for future exploration of digital transformation beyond the hospital walls to encompass broader healthcare ecosystems.

[13] https://tritech.nhs.wales/case-studies/

[14] https://www.bihealth.org/en/translation/innovation-enabler/charite-bih-innovation/digital-labs/digital-health-accelerator

[15] https://www.santpau.cat/en/web/public/unitat-funcional-de-salut-digital1

[16] https://cimt.dk/gb/vr-trainer/

4.3 Leveraging Workforce and Citizen-Driven Digital Transformation

Wilandika et al. (2023) identified the significant role of clinicians in supporting health literacy of citizens and patients, due to their involvement in education, care navigation and support in utilizing digital solutions. Research also demonstrates that the trusted status of healthcare professionals can mitigate the barriers of mistrust and lack of understanding of digital technologies by citizens, leading to increased adoption and engagement with the digital solution. An Organization for the Review of Care and Health Apps (ORCHA 2021) found that patients were likely to perceive up to 100 times more satisfaction with a digital health application when they were recommended by a healthcare professional, than without. Acknowledging the benefit that digital health applications create for persons' health, especially in addressing the burden created by noncommunicable diseases (NCD), as mentioned earlier in the article, the United Kingdom's National Institute for Health and Care Excellence (NICE) published in 2023 its draft guidance[17] on digital programs to help the National Health Service (NHS) deliver specialist weight management services and support access to weight loss medications. More generally, these programs:

1. Offer support from a multidisciplinary team of healthcare professionals, including psychological support and monitoring
2. Provide the ability to prescribe weight management medication such as semaglutide (Wegovy, Novo Nordisk) or liraglutide (Saxenda, Novo Nordisk)
3. Collect and share data about weight medication use with an NHS team to support weight-management medication prescribing

NICE expects these technologies to reduce the demand of in-person specialist weight management services, thereby freeing up scarce resources, reducing waiting times, and improving access.

Germany set a world precedent in regulating the reimbursement of various digital healthcare technologies with the adoption of Digital Healthcare Act (DVG) in December 2019.[18] Healthcare professionals in Germany can routinely prescribe DiGA listed in the DiGA directory of digital health applications that have successfully completed the assessment by the German regulator, the Federal Institute for Drugs and Medical Devices (BfArM). According to Wangler and Jansky (2023), healthcare professionals deem that DiGA are reliable (67%) and secure (61%) applications. 22% are confident that they can give patients competent advice on DiGA, 14% have already prescribed DiGA, and 13% plan to do so. 83% of doctors with DiGA experience rate the prescribed applications as useful. 47 DiGAs are

[17] https://www.nice.org.uk/news/article/digital-services-to-enable-easier-access-to-weight-management-support

[18] https://diga.bfarm.de/de

currently available to 70 million statutory health insured patients and the list is growing.

In a nutshell, the evolving regulatory environment supporting the inclusion of digital health technologies in mainstream care is increasing the role of clinicians in promoting digital literacy for patients, recognizing the importance of understanding and effectively utilizing digital tools and resources for managing healthcare. Clinicians educate and prescribe patients the use of digital health tools, such as patient portals, digital health applications, or wearable devices to support health progress monitoring but also medication tracking and adherence. Conversely, patients also play a role in promoting digital literacy among clinicians by providing feedback to clinicians on their experiences with digital health technologies, including usability, effectiveness, and areas for improvement. Patients may also frequently share their knowledge and experiences with digital health technologies, including tips, tricks, and insights gained from using specific apps or devices, raising awareness on the benefits of digital health tools among clinicians and the need for integration and inclusion of digital solutions into routine care delivery. This mutual exchange of information enhances the understanding of digital tools and the need for improved access to digital health technologies within healthcare settings, ultimately enhancing patient engagement and improving health outcomes.

4.4 Recommendations

Creating an environment that recognizes and supports innovation and the desire to find solutions to enhance efficiency and capacity remains essential. Other elements to support workforce-driven advancement of digital maturity include:

- Established career pathways and support for advanced professional development for health professionals and allied health professionals (AHPs)
- Encourage and reward innovative digital solutions, i.e. establish a positive environment for innovation amongst health workers
- Ensure user-friendly and co-designed digital technology, participative creation and continuous feedback
- Ensure an agile and iterative IT culture, allowing smooth adaptation to feedback and implementation of improvements, ensuring user satisfaction and efficiency of workflows
- Supportive leadership and governance relating to digital solutions
- Establish digital centers of excellence and identify digital champions to drive motivation
- Effective patient engagement—user-friendly patient portals, co-design and continuous feedback loop between the health workforce and citizens
- Empowering citizens through digital inclusion and access support—enabling improved digital literacy by the use of digital navigators, education programs, and accessible support points (such as digital hubs in hospitals and in towns where citizens can find in-person support). One such initiative is the EU funded

Horizon program IDEAHL[19] aimed at activating empowered citizens in their own health literacy (especially targeting vulnerable populations such as elderly, homeless, and immigrants)

5 Conclusions and Perspectives

The journey toward digital health maturity at scale relies on empowering healthcare professionals to utilize technology adeptly in delivering care. Given the significant resource constraints faced by healthcare organizations, strengthening healthcare professionals' capacity, confidence, and trust in the transformative potential of digital tools would be paramount in optimizing resource utilization, enhancing workforce performance and promoting evidence-informed decision-making. This article explored ways to leverage data and artificial intelligence, especially through the prism of enhancing the effectiveness of healthcare delivery.

As we move toward the future, the composition of the health workforce would evolve to include additional stakeholders important to the digital transformation process, both within and outside the hospital. The authors of this article have offered three distinct user groups, including (1) Digital Health Workforce, (2). Health Workforce, and (3) Empowered Citizens—each group with specific roles and responsibilities in the digital ecosystem, contributing to improved healthcare delivery and outcomes.

The ongoing digital transformation presents both challenges and opportunities for healthcare organizations. Embracing digital health technologies requires not only technological investment but also a cultural shift within the workforce. Organizations need to prioritize training and upskilling programs to ensure that healthcare professionals are equipped with the necessary digital competencies. Once the healthcare workforce has been equipped with the suite of digital tools and knowledge, the next step in the successful integration of digital technologies into healthcare systems is the unleash of a workforce-driven advancement of digital maturity. This approach requires identifying key requirements for workforce development and empowerment. It also involves fostering collaboration between healthcare professionals, citizens, and digital experts to drive meaningful change.

Finally, citizen engagement and empowerment are critical components of a successful digital health ecosystem. Healthcare systems should strive to educate and involve citizens in their own care processes, leveraging digital tools to enhance communication, access to information, and self-management capabilities. Looking ahead, the convergence of digital health and workforce development will continue to shape the future of healthcare delivery. It is essential for stakeholders to collaborate and innovate to address emerging challenges and seize opportunities for improving patient care, operational efficiency, and overall health outcomes in the digital era.

[19] https://ideahl.eu/

References

Affinito L, Fontanella A, Montano N et al (2020) How physicians can empower patients with digital tools. J Public Health (Berl) 30:897–909. https://doi.org/10.1007/s10389-020-01370-4

Biomedical Alliance in Europe (2023) Reaction to draft ENVILIBE Report on EHDS from 10.02.2023. https://www.biomedeurope.org/images/news/2023/BioMed_Alliance_comments_draft_ENVI-LIBE_report_EHDS.pdf. Accessed 21 December 2023

Boniol M, Kunjumen T, Nair TS et al (2022) The global health workforce stock and distribution in 2020 and 2030: a threat to equity and 'universal' health coverage? BMJ Glob Health 7:e009316. https://doi.org/10.1136/bmjgh-2022-009316

Chaltikyan G, Fernandes AF (2022) WHO Europe's "Structured training program to support digitalization of the health information system"

Clarke A, Lindquist EA, Roy J (2017) Understanding governance in the digital era: an agenda for public administration research in Canada. The Institute of Public Administration of Canada 60(4):457–475. https://doi.org/10.1111/capa.12246

DigitalHealthEurope (2022) Boosting active cooperation between patients and healthcare professionals [White paper]. https://digitalhealtheurope.eu/results-and-publications/white-paper-on-boosting-active-cooperation-between-patients-and-healthcare-professionals/. Accessed 24 Nov 2023

eHealth Network (2023) 24th eHealth network meeting, minutes of meeting 1 June 2023 and 2 June 2023. https://health.ec.europa.eu/system/files/2023-09/ehealth_20230601_sr_en.pdf. Accessed 16 November 2023

European Commission (2022) Questions and answers - EU Health: European Health Data Space (EHDS). https://ec.europa.eu/commission/presscorner/detail/en/qanda_22_2712. Accessed 23 December 2023

Federal Ministry of Health (2023) Germany's digitalisation strategy for health and care. https://www.bundesgesundheitsministerium.de/fileadmin/Dateien/3_Downloads/D/Digitalisierungsstrategie/Germany_s_Digitalisation_Strategy_for_Health_and_Care.pdf. Accessed 29 December 2023

G20 & G7 Health Development Partnership (2023) The roadmap to sustainable finance in health, a toolkit for G20 & G7 countries and developing a taxonomy for health. https://g20healthpartnership.com/wp-content/uploads/2023/06/G20-REPORT-DIgital-V2.pdf. Accessed 27 December 2023

HCA Healthcare (2023) HCA healthcare collaborates with Google cloud to bring generative AI to hospitals. https://investor.hcahealthcare.com/news/news-details/2023/HCA-Healthcare-Collaborates-With-Google-Cloud-to-Bring-Generative-AI-to-Hospitals/default.aspx. Accessed 3 January 2024

Healthcare Information and Management Systems Society (2020) Health information and technology job descriptions. https://www.himss.org/sites/hde/files/media/file/2020/01/30/health-it-job-descriptions.pdf. Accessed 24 Nov 2023

Healthcare Information and Management Systems Society (2023a) Digital health indicator. https://www.himss.org/what-we-do-solutions/digital-health-transformation/digital-health-indicator. Accessed 24 Nov 2023

Healthcare Information and Management Systems Society (2023b) What are maturity models? https://www.himss.org/what-we-do-solutions/digital-health-transformation/maturity-models. Accessed 24 Nov 2023

Hood CM, Gennuso KP, Swain GR et al (2016) County health rankings: relationships between determinant factors and health outcomes. Am J Prev Med 50(2):129–135. https://doi.org/10.1016/j.amepre.2015.08.024

Hübner U, Thye J, Shaw T et al (2019) Towards the TIGER international framework for recommendations of core competencies in health informatics 2.0: extending the scope and the roles. Stud Health Technol Inform 264:1218–1222. https://doi.org/10.3233/SHTI190420

International Labour Office Geneva (2012) International standard classification of occupations ISCO-08 volume 1. https://www.ilo.org/sites/default/files/wcmsp5/groups/public/@dgreports/@dcomm/@publ/documents/publication/wcms_172572.pdf. Accessed 24 Nov 2023

Jallad ST, Işık B (2022) The effectiveness of virtual reality simulation as learning strategy in the acquisition of medical skills in nursing education: a systematic review. Ir J Med Sci 191:1407–1426. https://doi.org/10.1007/s11845-021-02695-z

Kingdom of Saudi Arabia, Vision 2030 (2021) Health sector transformation program. https://www.vision2030.gov.sa/media/x3jbf45y/2021-2025-health-sector-transformation-program-delivery-plan-en.pdf. Accessed 29 December 2023

Matthews B (2021) Digital literacy in UK health education: what can be learnt from international research? Contemp Educ Technol 13(4):ep317. https://doi.org/10.30935/cedtech/11072

National Digital Inclusion Alliance (n.d.) The digital navigator model. https://www.digitalinclusion.org/digital-navigator-model/. Accessed 24 Nov 2023

National Institute for Health and Care Excellence (2023) Digital services to enable easier access to weight management support. https://www.nice.org.uk/news/article/digital-services-to-enable-easier-access-to-weight-management-support. Accessed 30 December 2023

Norman CD, Skinner HA (2006) eHealth literacy: essential skills for consumer health in a networked world. J Med Internet Res 8(2):e9. https://doi.org/10.2196/jmir.8.2.e9

O'Reilly T (2011) Government as a platform. Innov Technol Gov Glob 6(1):13–40. https://doi.org/10.1162/INOV_a_00056

ORCHA (2021) ICS research report 2021. Digital health in the UK: national attitudes and behaviour research. https://orchahealth.com/wp-content/uploads/2021/07/2107_ICS_Research_Report_2021_National_v2.pdf. Accessed 2 July 2024

Socha-Dietrich K (2021) Empowering strategies to make the most of the digital revolution. OECD Health Working Papers 129. OECD Publishing, Paris. doi:https://doi.org/10.1787/37ff0eaa-en. https://www.oecd.org/health/health-systems/Empowering-Health-Workforce-Digital-Revolution.pdf. Accessed 24 Nov 2023

The Standing Committee of European Doctors - CPME (2022) Position on the European health data space. https://www.cpme.eu/api/documents/adopted/2022/11/cpme.2022-065.FINAL.CPME.position.EHDS.pdf. Accessed 21 December 2023

Tuurnas S, Stenvall J, Virtanen P et al (2019) Towards collaborative development culture in local government organisations. Int J Public Sect Manag 32(6):582–599. https://doi.org/10.1108/IJPSM-05-2018-0119

Wangler J, Jansky M (2023) Digitale Gesundheitsanwendungen (DiGA) in der Primärversorgung – Erfahrungen und Beobachtungen von Hausärzt*innen hinsichtlich der Anwendung von DiGA. Präv Gesundheitsf 18: 483–491. doi:https://doi.org/10.1007/s11553-022-00988-4. Accessed 12 March 2024

Whitehead M, Dahlgren G (1991) What can be done about inequalities in health? Lancet 338(8774):1059–1063. https://doi.org/10.1016/0140-6736(91)91911-d

WHO (2020) State of the world's nursing 2020: investing in education, jobs and leadership. Geneva: World Health Organization. https://www.who.int/publications/i/item/9789240003279. Accessed 26 Dec 2023

Wilandika A, Pandin MGR, Yusuf A (2023) The roles of nurses in supporting health literacy: a scoping review. Front Public Health 11. https://doi.org/10.3389/fpubh.2023.1022803

World Health Organization & International Telecommunication Union (2012) National eHealth strategy toolkit. International Telecommunication Union. https://iris.who.int/handle/10665/75211. Accessed 2 July 2024

Digital Maturity and AI: Ethical Considerations for Hospital Management

Eduard Buzila and Francisco Pedrosa Gil

Abstract

This chapter examines the implications of the use of artificial intelligence (AI) in inpatient psychiatric care and highlights the health, ethical, and legal risks involved. The first part of the chapter focuses on the challenges posed by the digital transformation in medicine. It is argued that the current economic and technical framework conditions and the resulting social changes are fuelling the increased use of AI in psychiatry (Zucchetti et al., Ital J Psychiatry 10(1):17–24, 2024, https://doi.org/10.36180/2421-4469-2024-5). This development reflects the general societal trend in which people increasingly consume digital services and move in digital worlds, leading to a decrease in interpersonal contact. The authors discuss how large as well as emerging technology companies are using "nudging" and "data mining" to drive the merging of the real and digital worlds into a "metaverse," which ultimately leads to a normalization of data disclosure and transforms people into a "homo vitreus digitalis"—a digitally transparent person.

The second part of the chapter looks at the risks from the perspective of hospital management. Practical recommendations for action are presented, based on the experience of a senior hospital doctor. In addition to management aspects, ethical and philosophical considerations are also included in order to develop strategies for dealing with the risks arising from the use of AI in psychiatry, particularly for staff.

The chapter concludes by stating that while the use of AI in medicine has the potential to improve mental health care, it requires careful consideration of the

E. Buzila (✉) · F. P. Gil
Otto-von-Guericke-University, Magdeburg, Germany

Zentrum für Psychische Gesundheit, Markkleeberg, Germany
e-mail: eduard.buzila@ovgu.de; info@zfpg-markkleeberg.com

A. Scheuer, J. Studzinski (eds.), *Digital Maturity in Hospitals*,
https://doi.org/10.1007/978-3-031-80704-6_19

risks. The authors emphasize the need for a multidisciplinary approach to harness the benefits of AI while protecting patient privacy and autonomy.

Keywords

Artificial intelligence · Psychiatry · Psychotherapy · Ethical risks · AI · Digital maturity

Abbreviations

AI	Artificial Intelligence
CBT	Cognitive Behavioral Therapy
CDO	Chief Digital Officer
EU	European Union
GDPR	General Data Protection Regulation
GP	General Practitioner
HCAI	Human-Centered Artificial Intelligence
ML	Machine Learning
NHS	National Health Service
OECD	Organisation for Economic Co-operation and Development
TAM	Technology Acceptance Model

1 Health Risks in Inpatient Psychiatric Treatment

The healthcare systems of most countries represented in the OECD have been under enormous cost pressure for many years. For the German healthcare system, and in particular for municipal hospitals, an exponential rise in costs has been demonstrable since 1996 at the latest (Federal Statistical Office 2022a). At the same time, their income is on average either just below or just above expenditure, i.e. profits and losses are not stable but fluctuate from year to year (Augurzky 2020, p. 188) which makes it difficult to build up a financial buffer through profits and medium- to long-term financial planning. Private hospital operators are also feeling this cost pressure, as of the four largest (Helios, Asklepios, Sana, and Rhön), only one, Helios, has been able to continuously increase its profits since 2007, while the others have generated constant profits on average (Statista 2022). The largest cost factor for hospitals in Germany, with a share of around 70% of total costs in 2021, is personnel costs totalling 77.4 billion euros (Federal Statistical Office 2022b). Added to this are the costs incurred due to the COVID-19 pandemic, which further increased the already existing cost pressure and even led to a wave of insolvencies among German hospitals (Medical Journal 2023). This situation is further exacerbated by the shortage of specialists, especially in psychiatry (Rumpf and Hoch 2024) where the Germany-wide minimum staffing requirements are only met by just under two thirds of psychiatric hospitals. In addition, there is a steadily increasing workload for existing staff.

As a result, medical directors and hospital managers are in favor of raising digital maturity, advocating for the use of digital technologies in general, and AI in particular (Alves et al. 2024). On the one hand, this is intended to relieve the burden on existing staff and, at the same time, reduce personnel costs (Grandi and Bruni 2024; Pham et al. 2022; Rony et al. 2024). While the use of AI in somatic medicine can have positive effects (Bhagat and Kanyal 2024; Gerdes 2024) in the psychiatric and psychotherapeutic field, this initially appears to be associated with considerable health risks for patients (Dube et al. 2024; Rogan et al. 2024). In addition to patients, the use of digitality and AI in psychiatry can also affect nursing (Nashwan et al. 2023) and medical staff (Zhang et al. 2023) and also pose psychological, ethical, and legal risks for them. If AI is used in psychotherapeutic treatment, it can have an impact on various relationship constructs in inpatient psychiatric treatment, depending on the exact area of application. If it is used in therapy as a supportive form of treatment, it initially has a direct effect on the doctor–patient relationship, as the AI is interposed, creating a doctor–AI–patient relationship (Allen et al. 2024). As soon as the patient interacts with an AI that is recommended or even prescribed by the doctor, such as elona therapy depression (Kalde et al. 2023), the question arises as to whether the doctor must monitor the patient's use of the AI at all times or whether he only learns about the patient–AI relationship from feedback from the patient and/or the AI, so that he can only react ex post to psychological changes in the patient. If the AI is not used as a supportive treatment tool, but as an almost complete replacement for treatment, the question of liability becomes even more important (Lord et al. 2016). If AI is not used as a supportive treatment tool but rather as a near-complete replacement for the treatment, the questions of liability (Lord et al. 2016), the physician's duty of care (Fiske et al. 2019), and the efficiency of the treatment become even more critical. This is because the physician is now entirely excluded from the patient–AI relationship, and the success of the treatment depends solely on the interaction between the patient and the AI. It is conceivable that the doctor in such a setting would only have a weekly or monthly "follow-up discussion" with the patient to check whether the AI has been able to generate treatment successes (Khawaja and Bélisle-Pipon 2023). It is also questionable for which psychiatric disorders the use of AI is actually indicated. Particularly in personality disorders characterized by a strong reduction in drive, such as mood, anxiety, and obsessive-compulsive disorders, the underlying disorder can be exacerbated by the AI if the patient can retreat to the comfort of their own home to interact with an AI, causing them to neglect the therapeutic sessions (Khawaja and Bélisle-Pipon 2023). In such a case, the therapist would have to motivate the patient even more than before to attend sessions, leave the home, and maintain interpersonal contact. Empirical studies demonstrate that AI complicates decision-making, as AI users tend to rely on the AI to make decisions for them. This reliance also amplifies the human tendency toward laziness (Fromm, 1999), as AI users relinquish their autonomy and transfer it to the AI, which then makes decisions on their behalf (Ahmad et al. 2023). Being unable to make decisions and having reduced drive

(Goldenberg et al. 2002) are two components that are at the core of many personality disorders. The use of AI in this area would therefore be counterproductive for the success of treatment, as digitality and AI exacerbate mental illness. In other words, technology would be used to treat a disorder that has previously exacerbated the disorder. This raises the question of how a therapist should use AI in therapy if the AI itself is part of the problem. In any case, this shows the uncertainty that every therapist is confronted with.

Various types of AI for the early detection of mental illness in social media are currently being investigated upstream of the treatment of an individual patient (Ellouze and Hadrich Belguith 2024). For example, machine learning (ML) is being used to analyze text posts on Facebook and X to identify signs of depression in users based on their choice of words and word frequency (Martins et al. 2021; Pande et al. 2023; Tao et al. 2024a, b). This creates a "predictable human being" (Eickhoff and Heinrichs 2021). However, the benefits of such an approach are questionable. Martins et al. (2021) used ML to identify possible depression from the posts of Twitter users with an accuracy of 98%. The following critical questions can be derived from this result:

- Do social media providers have a duty of care to identify all users who may be suffering from depression and refer them to appropriate services (Yıldırım 2023)?
- Will GPs be informed in the future that their patients may be suffering from depression because they post content on social media that could indicate this?
- How would the behavior of users on social platforms change if they knew that their posts could lead to them being suspected of being classified as depressive by the AI system and therefore identified as such by the social media provider?
- Would social media providers have to ask their users for explicit consent for their posts to be checked for possible depression?
- How would users feel and behave if they had to see a therapist for treatment because they were categorized as depressed on social media? Would they be open to therapy? How would this affect the therapist and the therapy if the patient was forced into therapy?
- Should social media providers block users' accounts if they have been categorized as depressed by the system and refuse therapy?
- Will we need a health certificate stating that we are free of mental illness to participate in social media in the future?

In general, the question must be asked whether every mental illness needs to be diagnosed and treated just because it might be technically possible. Nobody is completely free of mental illnesses and disorders. People can be more or less aware of this. They can often regulate themselves and do not need psychotherapeutic treatment. However, suppose they are "externally diagnosed" in their daily interactions, which, as described, mainly take place in digital worlds. In that case, this can lead to severe insecurity and a disruption of self-regulation, so that a therapeutic intervention becomes necessary that would not have been necessary before. Such an "external diagnosis" is also problematic for the therapist, as they have to work with a diagnosis that is given to them by an AI. If the therapist arrives at a different diagnosis, this could destabilize the doctor–patient relationship if the patient continues to adhere to the AI-generated diagnosis, but the doctor considers another illness to be more possible (Khawaja and Bélisle-Pipon 2023).

The doctor's duty of care toward the patient also gives rise to questions to which the medical profession and hospital management must provide answers:

- What happens to data that patients share with an AI application as part of a therapy? Where is it stored?
- Is every doctor, medical practice, hospital, or hospital ward responsible for setting up, maintaining, and securing a server to store, secure, and manage patient data? Or will there be a state or nationwide server on which all data is stored centrally?
- How and by whom may such data be analyzed? Are health insurance companies allowed to access it? If so, can they engage in "data mining," for example to adjust insurance premiums to patients' medical conditions (Kaushik et al. 2022)?

Such "psychometrics" or "psychographics" "is a data-driven procedure for creating a personality profile" (Han 2021, p. 32) that could be used to ensure that those who are mentally "healthy," i.e., never require psychiatric treatment, pay lower health insurance premiums than those who require such treatment more frequently and for longer periods of time. Based on such a "psychography" created by an AI, for example by analyzing posts in social media, those affected could, conversely, have the AI suggest treatment concepts even before they have consulted a doctor (Monteith et al. 2024). Diagnosis and treatment would then be carried out by the AI, which could save treatment costs. However, the problem here would be that patients with a predefined, AI-generated diagnosis and treatment, who would later have to see a therapist, would bring this diagnosis and treatment into the therapy setting, which could jeopardize the entire therapy if patients trust the competence of the AI more than that of the doctor (Khawaja and Bélisle-Pipon 2023). It is therefore clear that there is considerable potential for savings in automated diagnosis and therapy using AI. However, there are major risks for psychotherapeutic practice. Firstly, before the first contact with the therapist, there is a risk that patients will be suggested AI-generated diagnoses and treatment that could be wrong. Secondly, there are also many dangers in the therapy itself. For example, the doctor–patient relationship can be disrupted by the AI and the treatment of the patient can be jeopardized by the AI, as an AI does not have the ability to mentalize and probably never will (Abbate 2023; Fonagy et al. 2002), i.e. to recognize, understand, and interpret the thoughts, feelings, intentions, and motivations of other people. An ability that is essential in psychotherapeutic treatment (Bateman et al. 2023).

A report in the Belgian newspaper *La Libre* from 28 March 2023 shows just how fatal this inability can be. A man in Belgium committed suicide after talking to a chatbot called Eliza about climate change for 6 weeks and being encouraged to commit suicide in order to save the planet (Affsprung 2023). The AI not only perpetuated his worries about the state of the Earth, but even intensified them. In this vulnerable situation, an anthropomorphization of the AI probably took place

(Alabed et al. 2024) by the man, so that he was no longer able to contextualize the software's responses. He was apparently no longer able to distinguish between reality and virtuality. Lacking the ability to mentalize, the AI was unable to respond to the man's unstable mental and emotional state in order to stabilize him. If a patient suffering from depression were to commit suicide as a result of using the elona therapy depression app prescribed by a psychiatrist (Kalde et al. 2023), under German law, the question would need to be raised as to what extent the psychiatrist and/or the hospital could or should be held legally and/or morally responsible. As long as these questions have not been conclusively and clearly clarified, even the factual availability of such technology, i.e. the existence of digitality, can lead to uncertainty on the part of the doctor because he is aware of such technology but does not know whether he is allowed to use it and how he should use it. Patients can also be unsettled if they learn about such technology by proactively searching for it on the Internet, but are not prescribed such an app by their therapist (Khawaja and Bélisle-Pipon 2023). A debate between therapists and patients about the pros and cons of such an app could complicate and strain the doctor–patient relationship and ultimately delay or even completely destroy the success of therapy.

OpenAI, the company that created the chatbot ChatGPT, therefore deliberately switched off the option for ChatGPT to give psychotherapeutic advise shortly after making ChatGPT 3.5 available to the general public. The company was therefore aware that the chatbot could drive a person to suicide by giving the wrong advice. People with a distorted perception of reality, such as schizophrenics, represent a particular risk group, as they may find it difficult to contextualize and relativize the advice of a chatbot because the AI-generated texts can hardly be objectively distinguished from human-generated texts.

With the establishment of "digital psychiatry" (Dave et al. 2021) digital-ethical considerations should therefore be made, in particular regarding the responsibility of the behavior of doctors and nursing staff in the non-analog or digital space, in order to ensure the protection of patients from AI and the success of treatment. In addition, it is necessary to reflect on the framework conditions, especially in dealing with the digital environment and the actual reality, in order to ensure successful work. Such a "digital psychiatry" must create consistent standards, orientation aids, values and ethics-oriented guidelines that enable the so-called value-based ethical digital competence for all stakeholders. The question must also be asked as to how patients are informed about the use of AI by doctors, as doctors would need to have in-depth knowledge of the technical and digital effects of AI in order to be able to fully inform patients. Almost no doctor has such knowledge, so intensive training and further education would be required, which would also be associated with high costs, in order to enable doctors to make professionally competent assessments and decisions.

In addition, there are the general dangers of the digital appropriation of people through the use of AI and the "voluntary" disclosure of data and information, which can lead to a loss of reality (Bauer 2023) and even to totalitarian surveillance of all people (Han 2021) as more and more data could be generated, processed, analyzed, and used as an economic and/or political tool of power by different actors. Experts in medicine and philosophy have therefore already raised their critical voices in the literature and have been warning of these consequences for many years, particularly in relation to people's thinking and their social environment.

It can be stated that the digital transformation, caused by the economically induced consumption of digital products in the digital world, is leading to process modifications and re-evaluations in psychiatry. Systematic attempts are being made to persuade psychiatrists and psychotherapists in hospitals and in everyday practice that there is no time for reassuring conversations and empathy, and that empathizing with patients in particular is a sign of time inefficiency and waste. The so-called guidelines and algorithms are supposed to lead to better treatment outcomes and enable more efficient treatment regimes. But this raises the question of whether the industrial production ideal (mass production) is actually conducive to individually personalized medicine. This leads to an increasing apparent scientific externalization and objectification, which in fact leads to assembly line production, with the disintegration of holistic medicine and the delegitimization of the non-measurable and mentalization. These developments in turn lead to increased burnout rates (Bykov et al. 2022) and other crises of finding meaning in the medical professions and medical professionals.

In this area of tension between medicine, ethics, economics, law, and health policy, the topic of digital ethics (Dreier & Andina, 2022) must also be considered. There is a high risk for the patient if the freedom of medical decision-making is made dependent on instructions generated by algorithms (as part of an industrial complex). The doctor–patient relationship is denaturalized and reduced to an "account transaction." Documentation and procedural rules then take center stage rather than the human being. Even for Aristotle, the "practice of action was an end in itself" (e.g., meditation or art, poetic labor, to produce something calculated). However, medical action and existence is not limited to (technique with poiesis), but is dependent on phronesis (according to Aristotle, both can be translated as wisdom, reason, and thinking), which in turn influences practice. In the present day, the balance is increasingly swinging in the direction of technology, with all the consequences for medical practice, but also for the other healing professions in medicine. In medical ethics, economic understanding is increasingly being prioritized over self-knowledge, reflection, and dialogue. The economization of medicine is leading to increasing emotional stress situations, rationing, and conflicts of loyalty and conscience among employees (Strech et al. 2008).

The following stressors are becoming increasingly noticeable here: Employees feel overloaded by technology: high speed, frequent interruptions, expectation of fast response times of digital communication, the so-called techno-overload (Molino et al. 2020); anxiety states also occur, then "techno-insecurity" (Bail et al. 2024) in the sense that technology itself creates insecurity, as feelings of uncertainty and ambiguity often arise in the digital space, with a simultaneous feeling of constant surveillance. Constant updates and downloads lead to technical failures and errors, so that unreliability (techno-unreliability) occurs (Meyer et al. 2024).

To summarize, all these new technical possibilities, including digital surveillance technologies, lead to constant control and monitoring of the workplace, with the so-called monitored workers feeling helpless and at the mercy of others. Ultimately, this destroys the human ability to mentalize, so that psychiatrists and psychotherapists are no longer able to respond to the needs and concerns of their patients because they themselves are subjected to such pressure that they become numb.

For these reasons, it is extremely important to develop digital ethics, as the existing ethical standards no longer do justice to these new challenges. It is no longer sufficient, for example, to simply refer to a non-official Charter of Digital Fundamental Rights of the European Union (van der Sloot 2021). The seemingly secure guard rails, such as Article 2 "Everyone has the right to freedom of information and communication," lead to new questions and challenges in everyday implementation, especially in clinical practice. The goal here should be to develop adequate digital media literacy to prevent a loss of reality, as postulated by Bauer (2023).

2 Psychotherapeutic-Ethical Considerations and Therapy Strategies for Dealing with Digital Health Risks

In the first part of this specialist article, we looked at the health risks of digital transformation and the attacks on the genuinely human, namely thinking. In the following, psychotherapeutic-ethical considerations are made to show how therapists and nursing staff in hospitals can deal with the existing and potential health risks associated with digitality and the use of AI. Ultimately, following Friedrich Nietzsche and Dieter Henrich, this is about the question of the "future of the human" (Henrich 2016, p. 15) in general and in medicine in particular, which must be answered from the perspective of hospital management. It is therefore about the humane treatment of hospital staff, which also has an indirect effect on the treatment of patients. The human being must therefore be placed at the center of the discussion and must not be lost sight of by hospital managers and medical directors.

In the following, brief psychotherapeutic-ethical considerations will be made before concrete recommendations for action are developed on this basis from the practical perspective of hospital management.

An important goal in medicine and especially in psychiatry is to understand a person's biography, how they deal with themselves and their environment, because their life story is central to their medical history. However, this is increasingly being lost through digital processes because human–human interactions are diminishing due to digitalization and the use of AI. For a psychotherapist and psychoanalyst with many years of supervisory experience in many hospitals, this is a crucial aspect, as this work has allowed the second author of this article to experience for himself that digital procedures and transformation processes lead to the destruction and annihilation of rituals and medical concepts. From a depth-psychological and psychoanalytical perspective, a space for self-reflection is of great importance for finding one's identity. In view of the challenge of privacy in the digital age, authenticity and identity development (including autonomy) are under great pressure. In philosophical, ethical, and psychological terms, respect is part of the "human competence to be able to reflect on and master oneself and one's life, to develop oneself and one's society" (Grimm et al. 2019, p. 220).

In other words, the aim of the digital transformation of psychiatry should be to maintain mental autonomy with ethical standards. However, in order to have self-awareness and an understanding of ethics, attentional (consciousness) autonomy is also necessary. This presupposes inner truthfulness as a condition for autonomy. In other words, our first plea in the digital transformation for employee management means, for example, recommending the following:

Using relaxation techniques (e.g., autogenic training) to accompany the work processes from the outset in order to create islands of retreat and mental autonomy. Psychotherapeutic techniques, starting with autogenic training and other meditation techniques, are scientifically and clinically proven techniques. Otherwise, the digital transformation threatens the so-called self-reification of humans (Fuchs 2022) including mania: medial speed of light, so that decisions have to be made in a matter of seconds, without regard for immoderation and without regard for our own times and fleetingness in the sense of a manifest flight of ideas. Our bodies (including medical ones) thus become objects of electronics, monitored and recorded by smartwatches and apps, the so-called wearables (Abd-alrazaq et al. 2023). However, humanity must once again be revitalized ethically and philosophically. Spitzer (2002) noted back in 2002 that the constant collective level of excitement is increasingly leading to indignation management. Boredom as a necessary basis for attention is being abolished by sensory clicks; there is a threat of the decomposition of the symbolic, with a constant optical-acoustic barrage (Playstation, displays, touchscreens, etc.).

Here, psychotherapy with its concepts of mentalization versus digitalization can lead to a correction and the re-establishment of physicality. Continuous medial firing leads to a disturbance of the unconscious, also from a neurobiological point of view. That which was wrested from us in evolution as psyche and soul, also in the sense of sublimation, is increasingly dissolved and transformed into sensory-bodily-digital processes. Symbolization and separation are no longer possible, leaving individuals ultimately at the mercy of presence machines and presence destroyers. The experience of lack and boredom is crucial for the development of symbolic thinking. In this context, the idea that "the implementation of mental autonomy involves combining regular attention practice with systematic training in criticism and rationality"[1] is particularly significant. This approach is highly recommended when combined with psychotherapeutically trained supervision and intervision, including the practice of meditation techniques.

Psychotherapeutic-ethical considerations therefore place people, not only patients, but also nursing staff and doctors, at the center of medicine and psychiatry. When implementing AI in hospital processes and psychotherapeutic treatment, it is important not to lose sight of people and the human factors that contribute to the success of a hospital and the success of psychotherapeutic treatment.

[1] As noted by Professor Metzinger at the DGPPN Congress 2021.

In principle, four levels should be considered when implementing AI: (1) patients, (2) nursing and allied health staff, (3) medical staff, and (4) hospital administration.

According to the Technology Acceptance Model (TAM) (Davis et al. 1989; Venkatesh and Davis 1996), for a technology to be successful, and this is particularly true for AI, it must gain widespread acceptance and approval from users and those indirectly affected by it. Hospital management, which should always monitor and analyze the implementation, should also take into account the insights of "stakeholder theory" (Freeman 2010). According to this theory, "a stakeholder is defined as any entity (e.g., individual or group) who can influence or is influenced by mission-driven organizations (i.e., set out to accomplish objectives)" (Mahajan et al. 2023, p. 2). This means that all stakeholders should accept the technology, in this case AI, because only then can it be used successfully and efficiently. In order to increase this acceptance, trust in AI must also be built up (Asan et al. 2020). Those who use it and are indirectly affected by it must therefore be able to trust the technology and its results. It can therefore be said that hospital management must first identify the stakeholders (Yang et al. 2022) who will be directly and indirectly affected by AI before implementation, and then ensure that they will accept and trust the technology (Lambert et al. 2023). Acceptance and trust can only arise if the stakeholders understand how the AI works, if the models are transparent to them (Wang et al. 2023). All of this requires hospital management to provide training and continuing education courses for both nursing staff (Abuzaid et al. 2022) as well as for the medical profession (Chen et al. 2022). At the same time, it is important to ensure that they are willing and open to acquiring new knowledge and using new technologies (Lera et al. 2020). Without these steps, which must be taken before implementation, the use of AI cannot be successful. It is therefore initially up to hospital management to establish acceptance and trust in AI by communicating openly about the benefits and limitations of AI so that initial fears on the part of nursing staff and doctors can be dispelled (Chen et al. 2022). Hospital management must then define the framework within which AI should or may be used.

Studies show that patients also have concerns about the use of AI in hospitals and tend to be negative about it (Temple et al. 2023). A negative attitude of patients toward AI is very counterproductive for the success of therapy, especially in psychotherapy. Therefore, extensive patient education is needed to build acceptance and trust in AI. This in turn shows that both nursing staff and psychotherapists need to have a sound knowledge of the use and limitations of AI so that they can fully inform patients, also with regard to possible risks and dangers.

The topic of AI must therefore be increasingly included in the training of nursing staff as well as in medical studies (Grunhut et al. 2022) as further training can provide certain insights into the use of AI, but cannot replace thorough training. Incorporating this technology into training and the profession through continuing education is likely to lead to better handling and a better understanding of AI, but will increase the level of difficulty of the training and the profession, which will ultimately be reflected in higher salaries. The implementation of AI therefore inevitably leads to higher costs for the hospital and also for the state if professional training and medical studies are to be extended to include the aspect of AI, as professionally trained staff will have to be employed at universities and vocational schools, and they will also expect a corresponding salary. The costs to both hospitals and the state will then have to be financed through higher health insurance contributions. However, this will further aggravate the already strained financial situation, as it is not foreseeable that contributors will be willing to pay higher health insurance contributions, which are already at a relatively high level. At this point, it becomes clear that the general public also needs to be made more aware of the benefits of AI in medicine, psychiatry, and psychotherapy. Since treatment appointments in psychotherapy are currently scarce and there are long waiting times, the use of AI in psychotherapy could shorten this waiting time if patients are provided with an AI to bridge this time and start therapy with it.

In conclusion, it can be said that the use of AI in medicine in general and in psychiatry in particular is still associated with very high risks, and that the potential of this technology is already putting pressure on hospitals, as its availability tempts them to use it. The investment in AI is not limited to the purchase of the software, but also requires an investment in human capital, as the software can only be used optimally if it is understood and trusted. The use of AI in psychiatry therefore only makes sense if it is ensured that stakeholders are not overburdened during training, medical studies and subsequent practice.

Hospital managers should therefore have a thorough understanding of human–algorithm interaction before implementation. To do this, they need to draw on the models and theories already discussed, but also on the insights of the so-called human-centered AI (HCAI), as this approach "will increase the chance that technology will empower people rather than replace them" (Shneiderman 2022, p. 7). This brings the discussion back to its starting point: people—stakeholders, patients, nursing and allied health staff, doctors, and hospital management—must be at the center of all considerations.

3 Conclusion

Current research underlines the potential of artificial intelligence (AI) in medicine, including psychiatry and psychotherapy, but emphasizes the need to identify and minimize risks. This article emphasizes that the added value of AI can only be realized if corresponding risks are addressed and solutions developed. A key element is comprehensive education and training in the field of AI, both in vocational and academic training and in the workplace, in order to close knowledge gaps and avoid a skills mismatch. The public needs to be informed about the advantages and disadvantages of AI in order to promote acceptance and trust, which in turn increases the willingness to use and invest in these technologies.

Hospital management needs to carefully plan the implementation of AI, taking into account the latest research findings on acceptance and trust as well as approaches to "human-centered AI." This requires specialized staff, such as a Chief Digital Officer. In addition, ethics committees should be set up to develop guidelines and clarify ethical issues—especially where there may not yet be any guidelines from professional associations to build on. Transparent communication about the use and limitations of AI is vital to alleviate the concerns and fears of patients, carers, and doctors, ensuring that the journey toward digital maturity is beneficial, responsible, and ethical.

References

Abbate F (2023) Natural and artificial intelligence: a comparative analysis of cognitive aspects. Mind Mach 33(4):791–815. https://doi.org/10.1007/s11023-023-09646-w

Abd-alrazaq A, AlSaad R, Aziz S et al (2023) Wearable artificial intelligence for anxiety and depression: scoping review. J Med Internet Res 25:e42672. https://doi.org/10.2196/42672

Abuzaid MM, Elshami W, Fadden SM (2022) Integration of artificial intelligence into nursing practice. Heal Technol 12(6):1109–1115. https://doi.org/10.1007/s12553-022-00697-0

Affsprung D (2023) The ELIZA defect: constructing the right users for generative AI. Proceedings of the 2023 AAAI/ACM conference on AI, ethics, and society, pp 945–946. doi:https://doi.org/10.1145/3600211.3604744

Ahmad SF, Han H, Alam MM et al (2023) Impact of artificial intelligence on human loss in decision making, laziness and safety in education. Humanit Soc Sci Commun 10:311. https://doi.org/10.1057/s41599-023-01787-8

Alabed A, Javornik A, Gregory-Smith D, Casey R (2024) More than just a chat: a taxonomy of consumers' relationships with conversational AI agents and their well-being implications. Eur J Mark 58(2):373–409. https://doi.org/10.1108/EJM-01-2023-0037

Allen MR, Webb S, Mandvi A et al (2024) Navigating the doctor-patient-AI relationship-a mixed-methods study of physician attitudes toward artificial intelligence in primary care. BMC Prim Care 25(1):42. https://doi.org/10.1186/s12875-024-02282-y

Alves M, Seringa J, Silvestre T, Magalhães T (2024) Use of Artificial Intelligence tools in supporting decision-making in hospital management [Preprint]. https://doi.org/10.21203/rs.3.rs-4491119/v1

Asan O, Bayrak AE, Choudhury A (2020) Artificial intelligence and human trust in healthcare: focus on clinicians. J Med Internet Res 22(6):e15154. https://doi.org/10.2196/15154

Augurzky B (2020) Gewinne im Krankenhaus. In: Klauber J, Geraedts M, Friedrich J, Wasem J, Beivers A (eds) Krankenhaus-report 2020. Springer, Berlin, pp 185–204. https://doi.org/10.1007/978-3-662-60487-8_10

Bail C, Marquardt B, Harth V, Mache S (2024) Technostress experience in inpatient medical care in German and Swiss hospitals: Current state of research. Zentralblatt für Arbeitsmedizin, Arbeitsschutz und Ergonomie. https://doi.org/10.1007/s40664-024-00542-3

Bateman A, Fonagy P, Campbell C et al (2023) Cambridge guide to mentalisation-based treatment (MBT), 1st edn. Cambridge University Press. https://doi.org/10.1017/9781108896054

Bauer J (2023) Realitätsverlust: Wie KI und virtuelle Welten von uns Besitz ergreifen und die Menschlichkeit bedrohen (Original edition). Heyne

Bhagat SV, Kanyal D (2024) Navigating the future: the transformative impact of artificial intelligence on hospital management - a comprehensive review. Cureus 16(2):e54518. https://doi.org/10.7759/cureus.54518

Bykov KV, Zrazhevskaya IA, Topka EO et al (2022) Prevalence of burnout among psychiatrists: a systematic review and meta-analysis. J Affect Disord 308:47–64. https://doi.org/10.1016/j.jad.2022.04.005

Chen M, Zhang B, Cai Z et al (2022) Acceptance of clinical artificial intelligence among physicians and medical students: a systematic review with cross-sectional survey. Front Med 9:990604. https://doi.org/10.3389/fmed.2022.990604

Dave S, Abraham S, Ramkisson R et al (2021) Digital psychiatry and COVID-19: the Big Bang effect for the NHS? BJPsych Bull 45(5):259–263. https://doi.org/10.1192/bjb.2020.114

Davis FD, Bagozzi RP, Warshaw PR (1989) User acceptance of computer technology: a comparison of two theoretical models. Manag Sci 35(8):982–1003. https://doi.org/10.1287/mnsc.35.8.982

Dreier T, Andina T (eds) (2022) Digital ethics: the issue of images. Hart, Nomos

Dube AR, Ambrose AJH, Velez G et al (2024) Real concerns, artificial intelligence: Reality testing for psychiatrists. Int Rev Psychiatry:1–6. https://doi.org/10.1080/09540261.2024.2363374

Eickhoff SB, Heinrichs B (2021) The predictable person: opportunities and risks of AI-based prediction of cognitive abilities, personality traits and mental illness. Nervenarzt 92(11):1140–1148. https://doi.org/10.1007/s00115-021-01197-8

Ellouze M, Hadrich Belguith L (2024) Artificial Intelligence application for the analysis of personality traits and disorders in social media: a Survey. ACM Trans Asian Low-Resour Lang Inf Process 3674971. https://doi.org/10.1145/3674971

Federal Statistical Office (2022a) Adjusted costs of German hospitals in the years 1996 to 2021 (in billion euros) [Graph]. Statista. https://de.statista.com/statistik/daten/studie/3182/umfrage/kosten-deutscher-krankenhaeuser-seit-1996/. Accessed on 1 Aug 2024

Federal Statistical Office (2022b) Total personnel costs of German hospitals in the years 2000 to 2021 (in billion euros) [Graph]. Statista. https://de.statista.com/statistik/daten/studie/169304/umfrage/personalkosten-deutscher-krankenhaeuser-insgesamt-seit-2000/. Accessed on 1 Aug 2024

Fiske A, Henningsen P, Buyx A (2019) Your robot therapist will see you now: ethical implications of embodied artificial intelligence in psychiatry, psychology, and psychotherapy. J Med Internet Res 21(5):e13216. https://doi.org/10.2196/13216

Fonagy P, Gergely G, Jurist EL (eds) (2002) Affect regulation, mentalization and the development of the self, 1st edn. Routledge. https://doi.org/10.4324/9780429471643

Freeman RE (2010) Strategic management: a stakeholder approach. Cambridge University Press. https://doi.org/10.1017/CBO9781139192675

Fromm E (1999) Is man lazy by nature? In: Funk R (ed) Complete edition in twelve volumes: Vol. XII:161–192

Fuchs T (2022) Defence of the human being: basic questions of an embodied anthropology (Original edition, 4th ed.). Suhrkamp

Gerdes A (2024) The role of explainability in AI-supported medical decision-making. Discov Artif Intell 4:29. https://doi.org/10.1007/s44163-024-00119-2

Goldenberg G, Pössl J, Ziegler W (eds) (2002) Neuropsychology in everyday life. Georg Thieme Verlag. https://doi.org/10.1055/b-002-5216

Grandi LC, Bruni S (2024) Will the artificial intelligence touch substitute for the human touch? NeuroSci 5(3):254–264. https://doi.org/10.3390/neurosci5030020

Grimm P, Keber TO, Zöllner O (eds) (2019) Digitale Ethik. Leben in vernetzten Welten, Reclam

Grunhut J, Marques O, Wyatt ATM (2022) Needs, challenges, and applications of artificial intelligence in medical education curriculum. JMIR Med Educ 8(2):e35587. https://doi.org/10.2196/35587

Han B-C (2021) Infocracy: digitalisation and the crisis of democracy. Matthes & Seitz

Henrich D (2016) Thinking and selfhood: lectures on subjectivity. Suhrkamp

Kalde J, Atik E, Stricker J et al (2023) Enhancing the effectiveness of CBT for patients with unipolar depression by integrating digital interventions into treatment: a pilot randomised controlled trial. Psychother Res 34(8):1131–1146. https://doi.org/10.1080/10503307.2023.2277866

Kaushik K, Bhardwaj A, Dwivedi AD et al (2022) Machine learning-based regression framework to predict health insurance premiums. Int J Environ Res Public Health 19(13):7898. https://doi.org/10.3390/ijerph19137898

Khawaja Z, Bélisle-Pipon J-C (2023) Your robot therapist is not your therapist: understanding the role of AI-powered mental health chatbots. Front Digit Health 5:1278186. https://doi.org/10.3389/fdgth.2023.1278186

Lambert SI, Madi M, Sopka S et al (2023) An integrative review on the acceptance of artificial intelligence among healthcare professionals in hospitals. Npj Digit Med 6(1):111. https://doi.org/10.1038/s41746-023-00852-5
Lera M, Taxtsoglou K, Frantzana A et al (2020) Nurses' attitudes toward lifelong learning via new technologies. Asian/Pac Isl Nurs J 5(2):89–102. https://doi.org/10.31372/20200502.1088
Lord S, Moore SK, Ramsey A et al (2016) Implementation of a substance use recovery support mobile phone app in community settings: qualitative study of clinician and staff perspectives of facilitators and barriers. JMIR Ment Health 3(2):e24. https://doi.org/10.2196/mental.4927
Mahajan R, Lim WM, Sareen M et al (2023) Stakeholder theory. J Bus Res 166:114104. https://doi.org/10.1016/j.jbusres.2023.114104
Martins R, Almeida J, Henriques P et al (2021) Identifying depression clues using emotions and AI: proceedings of the 13th international conference on agents and artificial intelligence, pp 1137–1143. doi:https://doi.org/10.5220/0010332811371143
Medical Journal (2023) Wave of insolvencies among hospitals. https://www.aerzteblatt.de/nachrichten/146549/Insolvenzwelle-bei-den-Krankenhaeusern. Accessed on 1 Aug 2024
Meyer S-C, Tisch A et al (2024) Exploring the relationship between techno-unreliability at work and burnout. J Occup Environ Med 66(3):185–191. https://doi.org/10.1097/JOM.0000000000003008
Molino M, Ingusci E, Signore F et al (2020) Wellbeing costs of technology use during Covid-19 remote working: an investigation using the Italian translation of the technostress creators scale. Sustain For 12(15):5911. https://doi.org/10.3390/su12155911
Monteith S, Glenn T, Geddes JR et al (2024) Implications of online self-diagnosis in psychiatry. Pharmacopsychiatry 57(02):45–52. https://doi.org/10.1055/a-2268-5441
Nashwan AJ, Gharib S, Alhadidi M et al (2023) Harnessing artificial intelligence: strategies for mental health nurses in optimising psychiatric patient care. Issues Ment Health Nurs 44(10):1020–1034. https://doi.org/10.1080/01612840.2023.2263579
Pande SD, Hasane Ahammad SK, Gurav MN et al (2023) Depression detection based on social networking sites using data mining. Multimed Tools Appl 83:25951–25967. https://doi.org/10.1007/s11042-023-16564-7
Pham KT, Nabizadeh A, Selek S (2022) Artificial intelligence and chatbots in psychiatry. Psychiatry Q 93(1):249–253. https://doi.org/10.1007/s11126-022-09973-8
Rogan J, Bucci S, Firth J (2024) Health care professionals' views on the use of passive sensing, AI, and machine learning in mental health care: systematic review with meta-synthesis. JMIR Ment Health 11:e49577. https://doi.org/10.2196/49577
Rony MK, Parvin Mst R, Wahiduzzaman M et al (2024) "I wonder if my years of training and expertise will be devalued by machines": concerns about the replacement of medical professionals by artificial intelligence. Sage Open Nurs 10:23779608241245220. https://doi.org/10.1177/23779608241245220
Rumpf H-J, Hoch E (2024) "Der Fachkräftemangel trifft uns da, wo es weh tut" - Zur Notwendigkeit von Nachwuchsförderung in Suchtforschung und Suchttherapie: "The Lack of Qualified Personell Hits Us Where It Hurts" - The Need to Promote Young Professionals in Addiction Research and Addiction Therapy. Addiction 70(2):63–65. https://doi.org/10.1024/0939-5911/a000857
Shneiderman B (2022) Human-centred AI. Oxford University Press
Spitzer M (2002) Learning: brain research and the school of life. Spektrum Akademischer Verlag
Statista (2022) Profit of the largest private hospital operators in Germany in the years 2007 to 2022 (in millions of euros). Chart. https://de.statista.com/statistik/daten/studie/322501/umfrage/gewinn-der-groessten-privaten-klinikbetreiber-in-deutschland/. Accessed on 1 Aug 2024
Strech D, Börchers K, Freyer D et al (2008) Doctors' behaviour when resources are scarce: Results of a qualitative interview study. Ethics Med 20(2):94–109. https://doi.org/10.1007/s00481-008-0550-x

Tao Y, Yang M, Li H et al (2024a) DepMSTAT: multimodal spatio-temporal attentional transformer for depression detection. IEEE Trans Knowl Data Eng 36(7):2956–2966. https://doi.org/10.1109/TKDE.2024.3350071

Tao Y, Yang M, Wu Y et al (2024b) Depressive semantic awareness from vlog facial and vocal streams via spatio-temporal transformer. Digit Commun Netw 10(3):577–585. https://doi.org/10.1016/j.dcan.2023.03.007

Temple S, Rowbottom C, Simpson J (2023) Patient views on the implementation of artificial intelligence in radiotherapy. Radiography 29(Suppl 1):S112–S116. https://doi.org/10.1016/j.radi.2023.03.006

van der Sloot B (2021) The right to be let alone by oneself: narrative and identity in a data-driven environment. Law, Innovation and Technol 13(1):223–255. https://doi.org/10.1080/17579961.2021.1898315

Venkatesh V, Davis FD (1996) A model of the antecedents of perceived ease of use: development and test. Decis Sci 27(3):451–481. https://doi.org/10.1111/j.1540-5915.1996.tb01822.x

Wang Y-C, Chen T-CT, Chiu M-C (2023) An improved explainable artificial intelligence tool in healthcare for hospital recommendation. Healthcare Analytics 3:100147. https://doi.org/10.1016/j.health.2023.100147

Yang L, Ene IC, Arabi Belaghi R et al (2022) Stakeholders' perspectives on the future of artificial intelligence in radiology: a scoping review. Eur Radiol 32(3):1477–1495. https://doi.org/10.1007/s00330-021-08214-z

Yıldırım S (2023) The challenge of self-diagnosis on mental health through social media: a qualitative study. In: Battineni G, Mittal M, Chintalapudi N (eds) Computational methods in psychiatry. Springer Nature, Singapore, pp 197–213. https://doi.org/10.1007/978-981-99-6637-0_10

Zhang M, Scandiffio J, Younus S et al (2023) The adoption of AI in mental health care-perspectives from mental health professionals: qualitative descriptive study. JMIR Form Res 7:e47847. https://doi.org/10.2196/47847

Correction to: Digital Maturity Data: Extracting Insights for Health System Management

Justus Vogel, Johannes Hollenbach, Alexander Haering, David Ehlig, and Alexander Geissler

Correction to:
Chapter 4 in: A. Scheuer, J. Studzinski (eds.), Digital Maturity in Hospitals,
https://doi.org/10.1007/978-3-031-80704-6_4

The original version of the chapter was inadvertently published with an incorrect affiliation for the following authors: Justus Vogel and Alexander Haering. The chapter has been updated with correct affiliation of these authors as follows:

J. Vogel
Health Economics, Policy and Management, School of Medicine, University of St. Gallen, St. Gallen, Switzerland
email: justus.vogel@unisg.ch

A. Haering
RWI – Leibniz-Institut für Wirtschaftsforschung e.V, Essen, Germany
email: alexander.haering@rwi-essen.de

The updated version of this chapter can be found at
https://doi.org/10.1007/978-3-031-80704-6_4

A. Scheuer, J. Studzinski (eds.), *Digital Maturity in Hospitals*,
https://doi.org/10.1007/978-3-031-80704-6_20

The manufacturer's authorised representative in the EU is Springer Nature Customer Service Centre GmbH, Europaplatz 3, 69115 Heidelberg, Germany. If you have any concerns regarding our products, please contact ProductSafety@springernature.com

Printed and bound by CPI Group (UK) Ltd, Croydon, CR0 4YY
30/09/2026
02225568-0014